Quantum Antennas

Quantum Antennas

Harish Parthasarathy

Professor

Electronics & Communication Engineering

Netaji Subhas Institute of Technology (NSIT)

New Delhi, Delhi-110078

Manakin
PRESS

First published 2021
by CRC Press
2 Park Square, Milton Park, Abingdon, Oxon, OX14 4RN

and by CRC Press
6000 Broken Sound Parkway NW, Suite 300, Boca Raton, FL 33487-2742

© 2021, Manakin Press Pvt. Ltd.

CRC Press is an imprint of Informa UK Limited

The right of Harish Parthasarathy to be identified as author of this work has been asserted by him in accordance with sections 77 and 78 of the Copyright, Designs and Patents Act 1988.

Print edition not for sale in South Asia (India, Sri Lanka, Nepal, Bangladesh, Pakistan or Bhutan).

British Library Cataloguing-in-Publication Data
A catalogue record for this book is available from the British Library

Library of Congress Cataloging-in-Publication Data
A catalog record has been requested

ISBN: 978-0-367-75703-8 (hbk)
ISBN: 978-1-003-16362-6 (ebk)

Manakin
PRESS

Table of Contents

Preface

The quantum antenna consists of just electrons and positrons which satisfy the Dirac second quantized field equations. The current density of this field is obtained as a quadratic function of the Dirac field operators:

$$J^\mu(x) = -e\psi(x)^* \alpha^\mu \psi(s), \alpha^\mu = \gamma^0 \gamma^\mu$$

This current density produces a quantum electromagnetic field described by the retarded potentials

$$A^\mu((t,r) = \int \frac{J^\mu(t - |r - r'|, r')}{4\pi|r - r'|} d^3 r'$$

Thus, $A^\mu(t,r)$ is a Bosonic field represented as a quadratic functional of Fermionic field operators. Apart from this quantum em field produced by the quantum antenna, there is in space a free photon field described by a linear superposition of creation and annihilation operators of the photon field. Since the second quantized Dirac field $\psi(x)$ can be represented as a quadratic functional of the electron-positron creation and annihilation operator fields, it follows that the total electromagnetic field in space can be represented as a linear functional of photon creation and annihilation operators plus a quadratic functional of electron-positron creation and annihilation operators. This means that by using standard commutation rules for the photon creation and annihilation operators and anticommutation rules for electron-positron creation and annihilation operators, we can calculate the mean and mean square fluctuations of the quantum electromagnetic field produced by the antenna along with the free photon electromagnetic field in any given state, say for example, in a state in which there is a specific number of electrons and positrons with prescribed four momenta and spins and photons with specific four momenta and helicities. Or we may assume that the state of the photon component is a coherent state comprising af an infinite number of photons.

Chapter 1

Basic quantum electrodynamics required for the analysis of quantum antennas

1.1   Introduction

In this lecture, we discuss various quantum mechanical models for the current field produced in an antenna by the motion of electrons, positrons as well as non-Abelian gauge particles interacting with the gravitational field described the general relativistic metric tensor. Quantization of the electromagnetic field, second quantization of the Dirac field, computation of the current density of the Dirac field and its effect on quantum fluctuations in the radiated electromagnetic field is discussed. Other related problems like image processing algorithms on the antenna pattern are suggested. The Feynman path integral approach to computing the photon propagator in a gravitational field is proposed. This would enable us to study the effect of the gravitational field on quantum electromagnetic field fluctuations.

1

1.2   The problems to be discussed

1.2.1   Quantization of electromagnetic field and Dirac field in the presence of a background gravitational field

1.2.2   Lagrangian and Hamiltonian densities for photons interacting with electrons and positrons

1.2.3   Quantization in the Coulomb and Lorentz gauges using operator theory and using the Feynman path integral

1.2.4   Electrons, positrons and photons–scattering matrix using the interaction picture

1.2.5   photon creation and annihilation operator fields and electron-positron creation and annihilation operator fields in the momentum-spin/helicity domain

1.2.6   canonical commutation relations for the photon field and canonical anticommutation relations for the electron-positron field

1.2.7   Current in the Dirac field–the second quantization picture

1.2.8   photons in the second quantized electromagnetic field

1.2.9   Interaction Lagrangian and Hamiltonians between the photon field and electron-positron field

1.2.10   Second quantization using the Boson Fock space and Fermion Fock space

1.2.11   Coherent states of the photon field

1.2.12   Hudson-Parthasarathy quantum stochastic calculus for dynamically modeling photon bath noise

1.2.13   Dirac second quantized Hamiltonian in terms of creation and annihilation operator fields of electrons and positrons

1.2.14   Quantum electromagnetic four potential produced by the Dirac second quantized current field

Statistical moments of the electromagnetic potentials and fields produced by free photons and the radiation fields produced by electron and positron current density.

Approximate solution using time dependent perturbation theory, computation of the Dirac current after perturbation. Computation of the Dirac current moments in specific states of the electron-positron-photon fields.

The effect of photon noise on the gravitational field—Einstein field equations for gravitation perturbed by the quantum stochastic photon noise using the Hudson-Parthasarathy quantum stochastic calculus.

The gauge groups $U(1) \times SU(2)$ and $U(1) \times SU(2) \times SU(3)$.

Influence of gravitational waves on the photon field and on the electron-positron field.

1.3   EM field Lagrangian density

$$L_{EM} = K F_{\mu\nu} F^{\mu\nu}, F_{\mu\nu} = A_{\nu,\mu} - A_{\mu,\nu}, K = -1/16\pi$$

1.4   Electric and magnetic fields in special relativity

$$F_{0r} = E_r, F_{12} = -B_3, F_{23} = -B_1, F_{31} = -B_2$$

$$L_{EM} = (1/8\pi)(E^2 - B^2)$$

(CGS units with $c = 1$ are used so that $\epsilon_0 = 1/\mu_0 = 1/4\pi$). We prefer to use another system of units which gives $\epsilon_0 = 1/\mu_0 = 1, c = 1$). Then,

$$L_{EM} = (1/2)(E^2 - B^2) = (-1/4)F_{\mu\nu}F^{\mu\nu} = (-1/4)(2F_{0r}F^{0r} + F_{rs}F^{rs})$$

1.5   Canonical position and momentum fields in electrodynamics

$$\Pi_r = \frac{\partial L_{EM}}{\partial A^r_{,0}} = -F_{0r}$$

Canonical position fields are $A^r, r = 1, 2, 3$ and corresponding canonical momentu fields are $\Pi^r, r = 1, 2, 3$. In the Coulomb gauge, $A^r_{,r} = divA = 0$ and $\nabla^2 A^0 = -J^0$. So A^0 becomes a matter field. CCR's are

$$[A^r(t, r), \Pi_s(t, r')] = i\delta^r_s \delta^3(r - r')$$

Incorrect because of the constraints $A^r_{,r} = 0$. Total Lagrangian density for em field interacting with charged matter is

$$L_{EMM} = L_{EM} - J^\mu A_\mu = (-1/4)F_{\mu\nu}F^{\mu\nu} - J^\mu A_\mu$$

Another constraint is obtained from the equations of motion

$$\partial_r \frac{\partial L_{EMM}}{\partial A^0_{,r}} = \frac{\partial L_{EMM}}{\partial A^0}$$

1.6   The matter fields in electrodynamics

Note that J^μ, A^0 are matter fields while $A^r, \Pi_r, r = 1, 2, 3$ are em field fields. This equation of motion gives

$$F_{0r,r} - J^0 = 0$$

or equivalently,

$$div\Pi + J^0$$

ie

$$\Pi^r_{,r} + J^0 = 0$$

This is a constraint equation since it does not involve time derivatives.

1.7   The Dirac bracket in electrodynamics

$$\chi_1 = A^r_{,r}, \chi_2 = \Pi_{r,r} + J^0$$

constraints are $\chi_1 = 0, \chi_2 = 0$. To compute the Dirac bracket between A^r and Π_s, we use

$$[\chi_1(t, r), \chi_2(t, r')] = [A^r_{,r}(t, r), \Pi_{s,s}(t, r')] =$$

$$-i\delta^r_s \partial_r \partial_s \delta^3(r - r') = -i\nabla^2 \delta^3(r - r')$$

Its Fourier transform is ik^2. Inverse of the matrix

$$\begin{pmatrix} 0 & ik^2 \\ -ik^2 & 0 \end{pmatrix}$$

is

$$\begin{pmatrix} 0 & i/k^2 \\ -i/k^2 & 0 \end{pmatrix}$$

Also,

$$[A^r(t,r), \chi_1(t,r')] = 0, [A^r(t,r), \chi_2(t,r')] = -i\delta_s^r \partial_s \delta^3(r-r')$$

$$= -i\partial_r \delta^3(r-r')$$

$$[\chi_1(t,r), \Pi_s(t,r')] = [A^r_{,r}(t,r), \Pi_s(t,r')] = i\delta_s^r \partial_r \delta^3(r-r')$$

$$= i\partial_s \delta^3(r-r')$$

$$[\chi_2(t,r), \Pi_s(t,r')] = [\Pi_{s,s}(t,r), \Pi_s(t,r')] = 0,$$

The Fourier transform of the Dirac bracket of $A^r(t,r)$ and $\Pi_s(t,r')$ w.r.t $r - r'$ is therefore

$$[A^r(t,.), \Pi_s(t,.)](k) =$$

$$i\delta_s^r - [0, k^r] \begin{pmatrix} 0 & i/k^2 \\ -i/k^2 & 0 \end{pmatrix} \begin{pmatrix} -k^s \\ 0 \end{pmatrix}$$

$$= i\delta_s^r - ik^r k^s / k^2$$

Taking the inverse Fourier transform therefore gives the Dirac bracket as

$$[A^r(t,r), \Pi_s(t,r')] = i\delta_s^r \delta^3(r-r') - i\frac{\partial^2}{\partial x^r \partial x^s}(1/4\pi|r-r'|)$$

1.8   Hamiltonian of the em field

$$\mathcal{H} = \Pi_r A^r_{,0} - L_{EMM} =$$

$$-F_{0r}A^r_{,0} + (1/4)F_{\mu\nu}F^{\mu\nu} + J^\mu A_\mu$$

$$= -F_{0r}(A^r_{,0} + A^0_{,r}) + (1/4)(2F_{0r}F^{0r} + F_{rs}F^{rs}) + J^\mu A_\mu$$

$$+F_{0r}A^0_{,r}$$

Now,

$$F_{0r}A^0_{,r} = -\Pi_r A^0_{,r} = -(\Pi_r A^0)_{,r} + \Pi_{r,r}A^0$$

which after neglect of a total divergence (whose spatial integral will therefore vanish) we get on using the constraints $\Pi_{r,r} = -J^0$ that

$$\mathcal{H} = (1/2)(\Pi^2 + (\nabla \times A)^2) - (J^r A^r)$$

The Hamiltonian density of the matter field alone is $J^0 A^0$ and hence the total Hamiltonian of the field interacting with charged matter is

$$\mathcal{H}_{EMM} = (1/2)(\Pi^2 + (\nabla \times A)^2) + J^\mu A_\mu$$

In the general Lorentz gauge, the term $J^\mu A_\mu$ is taken as our field-charged matter interaction Hamiltonian. We write

$$\Pi_\perp = \Pi - \nabla A^0$$

Then,

$$= div\Pi_\perp = div\Pi - \nabla^2 A^0 = -J^0 + J^0 = 0$$

and further,

$$\int \Pi^2 d^3r = \int (\Pi_\perp^2 + (\nabla A^0)^2 + 2(\Pi_\perp, \nabla A^0)) d^3r$$

$$= \int (\Pi_\perp^2 - A^0 \nabla^2 A^0) d^3r = \int (\Pi_\perp^2 + J^0 A^0) d^3r$$

where we have used the fact that

$$\int (\Pi_\perp, \nabla A^0) d^3r = - \int A^0 div\Pi_\perp d^3r = 0$$

Thus, an alternative more useful expression for the Hamiltonian density after neglect of perfect spatial divergences (which do not contribute to the Hamiltonian) is given by

$$\mathcal{H}_{EMM} = (1/2)(\Pi_\perp^2 + (\nabla \times A)^2) + J^0 A^0 - J^r A^r$$

1.9 Interaction Hamiltonian between the current field and the electromagnetic field

and this justifies taking $H_I(t) = (\int J^\mu A_\mu d^3r$ as our Hamiltonian of interaction of the photon field with the electron-positron field. Note that the field commutation relations are not affected if Π is replaced by $\Pi_\perp$ since the difference between Π and $\Pi_\perp$ is a matter field in the Coulomb gauge and matter fields commute with the position and momentum fields of the em field.

Exercises: The free photon field $A_f^r(x)$ satisfies the wave equation

$$\Box A_f^r = 0$$

whose solution can be expressed as a superposition of plane waves:

$$A_f^r(t,r) = \int ((2|K|)^{-1} e^r(K,\sigma) c(K,\sigma) exp(-ik.x) + (2|K|)^{-1} \bar{e}^r(K,\sigma) c(K,\sigma)^* exp(ik.x)) d^3K$$

where

$$k = (|K|, K), k.x = k_\mu x^\mu = |K|t - K.r$$

1.10   The Boson commutation relations for the creation and annihilation operator fields for the EM field in momentum-spin domain

A simple calculation shows that the em field Hamiltonian is

$$H_{EM} = (1/2) \int ((A_{,t})^2 + (\nabla \times A)^2) d^3r$$

$$= \int |K| c(K,\sigma)^* c(K,\sigma) d^3K$$

This looks like the energy of a continuous system of quantum Harmonic oscillators, two of them associated with each spatial frequency K. Further justification follows from the assumption that if the operators $c(K,\sigma), c(K,\sigma)^*$ are assumed to satisfy the CCR

$$[c(K,\sigma), c(K',\sigma')^*] = \delta^3(K - K')\delta_{\sigma,\sigma'},$$

$$[c(K,\sigma), c(K',\sigma')] = 0, [c(K,\sigma)^*, c(K',\sigma')^*] = 0$$

then their dynamical behaviour described by Heisenberg's matrix mechanics is

$$dc_t(K,\sigma)/dt = i[H_{EM}, c(K,\sigma)] = -i|K| c_t(K,\sigma)$$

gives that

$$c_t(K,\sigma) = c(K,\sigma)exp(-i|K|t), c_t(K,\sigma)^* = c(K,\sigma)^* exp(i|K|t)$$

and hence we get the correct time dependence of $A_f^r(t,r)$. It can also be verified that these CCR's give the correct CCR's for $A_f^r(t,r)$ and $\Pi_s(t,r)$. That we leave as an exercise.

1.11   Electrodynamics in the Coulomb gauge

Remark: We have

$$\int \Pi_\perp^2 d^3r = \int (\Pi - \nabla A^0)^2 d^3r$$

$$= \int (F_{0r} + A_{0,r})^2 d^3r = \int (A_{r,0})^2 = \int (A_{,0})^2 d^3r$$

which is why we can express the em field Hamiltonian

$$(1/2) \int (\Pi_\perp^2 + (\nabla \times A)^2) d^3r$$

as

$$(1/2) \int ((A_{,0})^2 + (\nabla \times A)^2) d^3r$$

in the Coulomb gauge. Note that the Coulomb gauge condition $A^r_{f,r} = 0$ implies that the field polarization vector $e^r(K, \sigma)$ satisfies $K^r e^r = 0$ and hence there are just two independent directions of photon polarization for each wave vector K. These are specified by $\sigma = 1, 2$ and may be chosen so that $(e^r(K, 1)), (e^r(K, 2)), (K^r)$ are mutually orthogonal 3-vectors.

1.12 The Dirac second quantized field

The free Dirac equation for the electron is

$$(i\gamma^\mu \partial_\mu - m)\psi(x) = 0$$

where γ^μ are the Dirac matrices satisfying the anticommutation relations

$$\{\gamma^\mu, \gamma^\nu\} = 2\eta^{\mu\nu}$$

and $\psi(x) \in \mathbb{C}^4$. This equation on premultiplying by $(i\gamma^\nu \partial_\nu + m)$ gives the Klein-Gordon equation of special relativity:

$$(\Box + m^2)\psi(x) = 0$$

in agreement with the Einstein energy-momentum relation

$$(E^2 - P^2 - m^2)\psi = 0, E = i\partial_t, P = -i\nabla$$

The solution for free Dirac waves is

$$\psi(x) = \int (u(P, \sigma)a(P, \sigma)exp(-ip.x) + v(P, \sigma)b(P, \sigma)^* exp(ip.x))d^3P$$

where

$$p = (E, P), E = E(P) = \sqrt{P^2 + m^2}$$

and $u(P, \sigma), v(P, \sigma)$ satisfy in view of the Dirac equation

$$(\gamma^\mu p_\mu - m)u(P, \sigma) = 0,$$

$$(\gamma^\mu p_\mu + m)v(P, \sigma) = 0$$

The second implies

$$(\gamma^\mu p_\mu + m)v(-P, \sigma) = 0$$

Thus $u(P, \sigma), \sigma = 1, 2$ are orthogonal eigenvectors of the matrix $\gamma^\mu p_\mu$ with eigenvalues m and $v(-P, \sigma), \sigma = 1, 2$ are orthogonal eigenvectors of the matrix $\gamma^\mu p_\mu$ with eigenvalues $-m$. Thus, $a(P, \sigma)$ annihilates an electron of momentum P energy $E(P)$, spin σ mass m while $b(-P, \sigma)^*$ creates a positron of momentum P energy $E(P)$ and spin σ and mass m.

The Dirac equation in an electromagnetic field is given by the standard quantum mechanical rule of replacing p_μ by $p_\mu + eA_\mu$ and is thus given by

$$[\gamma^\mu(i\partial_\mu + eA_\mu(x)) - m]\psi(x) = 0$$

It is easily verified from this equation that the current

$$J^\mu = -e\psi(x)^* \alpha^\mu \psi(x), \alpha^\mu = \gamma^0 \gamma^\mu$$

is conserved ie

$$J^\mu_{,\mu} = 0$$

The Dirac equation can be derived from the Lagrangian density

$$L_D = \psi(x)^* \gamma^0 (\gamma^\mu(i\partial_\mu + eA_\mu) - m)\psi(x)$$

Exercise: Verify that the space-time integral of L_D is real.

Exercise: Verify by applying the Legendre transformation that the Hamiltonian of the Dirac equation interacting with the electromagnetic field is

$$H_{DEM} = (\alpha, P + eA) + \beta m - eA^0$$

where

$$(\alpha, P + eA) = \alpha^r(P^r + eA^r) = \gamma^0 \gamma^r(P^r + eA^r), \beta = \gamma^0$$

Exercise: Verify that if we define the Pauli spin matrices by

$$\sigma^0 = I_2, \sigma^1 = \begin{pmatrix} 0 & 1 \\ 1 & 0 \end{pmatrix},$$

$$\sigma^2 = \begin{pmatrix} 0 & -i \\ i & 0 \end{pmatrix}, \sigma^3 = \begin{pmatrix} 1 & 0 \\ 0 & -1 \end{pmatrix}$$

and

$$\sigma_\mu = \eta_{\mu\nu}\sigma^\nu$$

so that

$$\sigma_0 = \sigma^0, \sigma_r = -\sigma^r, r = 1, 2, 3$$

and finally the Dirac matrices by

$$\gamma^\mu = \begin{pmatrix} 0 & \sigma^\mu \\ \sigma_\mu & 0 \end{pmatrix}$$

then the desired anticommutation relations are satisfied by the γ^μ and moreover $\alpha^0 = I$ and α^r are Hermitian. This guarantees that the Dirac Hamiltonian is a Hermitian operator as it should be.

1.13 The Dirac equation in an EM field, approximate solution using Perturbation theory

The Dirac current perturbation in the presence of a weak em field. We assume A^μ to be of the first order of smallness. Then Write

$$\psi(x) = \psi_0(x) + \delta\psi(x)$$

Then the zeroth order part ψ_0 satisfies the free Dirac equation and is therefore expressible as a linear functional of $a(P, \sigma)$ and $b(P, \sigma)^*$. The first order part $\delta\psi[(x)$ satisfies

$$i\delta\psi_{,t}(x) = [(\alpha, -i\nabla) + \beta m]\delta\psi(x) + e[(\alpha, A) - A^0]\psi_0(x)$$

This can be solved in the four momentum domain as

$$\hat{\delta\psi}(p) = (ip^0 - (\alpha, P) - \beta m)^{-1}\mathcal{F}[(e(\alpha, A) - A^0)\psi_0](p)$$

1.14 Electromagnetically perturbed Dirac current

The above formula can be used to evaluate the first order shift in the Dirac current caused by the interaction of the quantum antenna with the photon field:

$$\delta J^\mu = -2eRe(\delta\psi(x)^*\alpha^\mu\psi_0(x))$$

where by Re, we mean "Hermitian part".

Chapter 2

Effects of the gravitational field on a quantum antenna and some basic non-Abelian gauge theory

2.1 The effect of a gravitational field on photon paths

On a single photon, we know from the GTR that its path bends in a gravitational field. Its paths are defined by null geodesics:

$$g_{\mu\nu}(x)(dx^\mu/d\lambda)(dx^\nu/d\lambda) = 0,$$

after solving the geodesic equations

$$d^2x^\mu/d\tau^2 + \Gamma^\mu_{\alpha\beta}(dx^\alpha/d\tau)(dx^\beta/d\tau) = 0$$

and then taking the limit $d\tau \to 0$. Suppose the metric does not depend on t. Then, one of the Euler-Lagrange equations is

$$d/d\lambda(g_{0\nu}(r)dx^\nu/d\tau) = 0$$

or equivalently,

$$g_{00}(r)dt/d\tau + g_{0m}(r)dx^m/d\tau = K$$

where K is a constant $\to \infty$ as $d\tau \to 0$. This equation can be used to eliminate $d/d\tau$ from the other equations. For example, if there is a spatial coordinate say x^s on which the metric does not depend. Then, we get another first integral

$$g_{s0}dt/d\tau + g_{sm}dx^m/d\tau = K'$$

where K' is another infinite constant. Hence, taking the ratio gives

$$\frac{g_{00} + g_{0m} dx^m/dt}{g_{s0} + g_{sm} dx^m/dt} = K/K' = \beta$$

where now β is a finite constant, although it is the ratio of two infinite constants. This fact can be used to determine the path of a light ray in a gravitational field.

2.2   Interaction of gravitation with the photon field

The interaction between the gravitational field and the electromagnetic field is contained in the Maxwell Lagrangian:

$$L_{EM} = (-1/4) F_{\mu\nu} F^{\mu\nu} = (-1/4) g^{\mu\alpha} g^{\nu\beta} F_{\mu\nu} F_{\alpha\beta} \sqrt{-g}$$

$$= L_{EM0} + L_{EMG}$$

For weak gravitational fields (as compared to the photon strength) we write

$$g_{\mu\nu} = \eta_{\mu\nu} + h_{\mu\nu}(x)$$

and then

$$g^{\mu\nu} = \eta_{\mu\nu} - h^{\mu\nu}$$

where

$$h^{\mu\nu} = \eta_{\mu\alpha} \eta_{\nu\beta} h_{\alpha\beta}$$

Specifically,

$$h^{00} = h_{00}, h^{0r} = h^{r0} = -h_{0r} = h_{r0}, h^{rs} = h_{rs}$$

Then, the interaction Lagrangian is

$$L_{EMG} = F_{\mu\nu}(x) F_{\alpha\beta}(x) C_{\mu\nu\alpha\beta}(x)$$

where $C_{\mu\nu\alpha\beta}(x)$ is the component in $(-1/4) g^{\mu\nu} g^{\alpha\beta} \sqrt{-g}$ that is linear in $h_{\mu\nu}(x)$. This is computed as

$$(-1/4) g^{\mu\nu} g^{\alpha\beta} \sqrt{-g} \approx$$

$$(-1/4)(\eta_{\mu\nu} - h^{\mu\nu})(\eta_{\alpha\beta} - h^{\alpha\beta})(1 - h/2)$$

$$\approx (-1/4) \eta_{\mu\nu} \eta_{\alpha\beta} - (1/4)(\eta_{\mu\nu} h^{\alpha\beta} + \eta_{\alpha\beta} h^{\mu\nu} + \eta_{\mu\nu} \eta_{\alpha\beta} h/2)$$

giving

$$C_{\mu\nu\alpha\beta}(x) = (-1/4)(\eta_{\mu\nu} h^{\alpha\beta} + \eta_{\alpha\beta} h^{\mu\nu} + \eta_{\mu\nu} \eta_{\alpha\beta} h/2)$$

where

$$h = \eta_{\mu\nu} h_{\mu\nu} = h^\mu_\mu$$

We can more generally calculate L_{EMG} upto quadratic orders in $h_{\mu\nu}$ writing the result as

$$C_{\mu\nu\alpha\beta}(x) = C_1(\mu\nu\alpha\beta\rho\sigma)h_{\rho\sigma}(x) + C_2(\mu\nu\alpha\beta\rho\sigma\delta\zeta)h_{\rho\sigma}(x)h_{\delta\zeta}(x)$$

where now $C_1(\mu\nu\alpha\beta\rho\sigma)$ and $C_2(\mu\nu\alpha\beta\rho\sigma\delta\zeta)$ are numerical constants. The effect of this interaction L_{EMG} on the classical wave propagation equation of photons is easily computed using the Euler Lagrange equations:

$$\partial_\mu \frac{\partial L_{EM0}}{\partial A^\nu_{,\mu}} = -\partial_\mu \frac{\partial L_{EMG}}{\partial A^\nu_{,\mu}}$$

or equivalently using the standard General relativistic Maxwell equations

$$(F^{\mu\nu}\sqrt{-g})_{,\nu} = 0$$

which approximates to

$$(\eta_{\mu\alpha} - h^{\mu\alpha})(\eta_{\nu\beta} - h^{\nu\beta})(1 - h/2)F_{\alpha\beta})_{,\nu} = 0$$

or equivalently, upto quadratic orders in the gravitational field

$$(\eta_{\mu\alpha}\eta_{\nu\beta}F_{\alpha\beta})_{,\nu} =$$

$$= [(D_1(\mu\nu\rho\sigma\alpha\beta)h_{\rho\sigma} + D_2(\mu\nu\rho\sigma\delta\zeta\alpha\beta)h_{\rho\sigma}h_{\delta\zeta})F_{\alpha\beta}]_{,\nu}$$

This equation can be solved using first and second order perturbation theory with the GTR Lorentz gauge condition

$$(A^\mu\sqrt{-g})_{,\mu} = 0$$

which approximates upto second order terms in the gravitational potentials to

$$\eta_{\mu\nu}A_{\nu,\mu} = [(E_1(\mu\alpha\beta)h_{\alpha\beta} + E_2(\mu\alpha\beta\rho\sigma)h_{\alpha\beta}h_{\rho\sigma})A_\nu]_{,\mu}$$

The method of perturbatively solving this is to expand

$$A_\mu = A^{(0)}_\mu + A^{(1)}_\mu + A^{(2)}_\mu + O(|h|^3)$$

where $A^{(0)}_\mu$ satisfies the usual special relativistic wave equation, $A^{(1)}_\mu$ is the correction to the em potentials that is a linear functional of $h_{\mu\nu}$ and $A^{(2)}_\mu$ is the correction that is a quadratic functional of $h_{\mu\nu}$.

2.3   Quantum description of the effect of the gravitational field on the photon propagator

The corrected photon propagator due to a background gravitational field is given by

$$< 0|T\{A_\mu(x)A_\nu(x')\}|0 >=$$

$$\int exp(i(S_{EM0} + S_{EMG}))A_\mu(x)A_\nu(x')\Pi_{\mu,z}dA_\mu(z)$$

$$\approx \int exp(iS_{EM0})(A_\mu(x)A_\nu(x') + iS_{EMG}A_\mu(x)A_\nu(x') - S^2_{EMG}A_\mu(x)A_\nu(x')/2)$$

$$\times \Pi_{\mu,z}\delta((A^\nu(z)\sqrt{-g(z)}),_\nu)dA_\mu(z)$$

where the δ-function accounts for the Lorentz gauge. This path integral can be evaluated using the standard theory of Gaussian integrals. The δ function part can be replaced by

$$\Pi_{\mu,z}\delta((A^\nu(z)\sqrt{-g(z)}),_\nu) = C \int exp(i \int \eta,_\mu(z)(A^\mu(z)\sqrt{-g(z)})d^4z)\Pi_z d\eta(z)$$

2.4 Electrons and positrons in a mixture of the gravitational field and an EM field Quantum antennas in a background gravitational field

The interaction of electrons, positrons and photons in a background gravitational field $g_{\mu\nu}(x)$ is described by the Lagrangian density

$$L_{EM} + L_{EMG} + L_D + L_{DG} + L_{DEM} =$$

$$(-1/4)F_{\mu\nu}F^{\mu\nu}\sqrt{-g} + Re(\psi^*(x)\gamma^0(\gamma^a V^\mu_a(x)(i\partial_\mu + eA_\mu(x) + i\Gamma_\mu(x)) - m)\psi(x))$$

where

$$\Gamma_\mu = (-1/2)V^{a\nu}(x)V^b_{\nu:\mu}(x)J^{ab}$$

where V^a_μ is a tetrad for the gravitational field, ie, a locally inertial frame:

$$g_{\mu\nu}(x) = \eta_{ab}V^a_\mu(x)V^b_\nu(x)$$

and

$$J^{ab} = (1/4)[\gamma^a, \gamma^b]$$

is the Dirac spinor Lie algebra representation of the Lorentz transformation basis elements

$$(\sigma^{ab})_{\mu\nu} = \delta^a_\mu\delta^b_\nu - \delta^a_\nu\delta^b_\mu$$

Writing the Dirac spinor representation of the Lorentz group as $\Lambda \to D(\Lambda)$, we have

$$J^{ab} = dD(\sigma^{ab})$$

and like σ^{ab}, they satisfy the standard commutation relations of the Lorentz algebra:

$$[J^{ab}, J^{cd}] = \eta^{ac}J^{bd} + \eta^{bd}J^{ac} - \eta^{ad}J^{bc} - \eta^{bc}J^{ad}$$

The gravitational connection satisfies the gauge covariance property:

$$D(\Lambda(x))V_a^\mu(x)(\partial_\mu + \Gamma_\mu(x)))D(\Lambda(x))^{-1} =$$

$$\Lambda(x)_a^b V_b^\mu(x)(\partial_\mu + \Gamma'_\mu(x))$$

for any local Lorentz transformation $\lambda(x)$. Here, $\Gamma'_\mu(x)$ is the transformed gravitational connection:

$$\Gamma'_\mu(x) = (-1/2)J^{ab}V_a^{\nu'}(x)V'_{b\nu:\mu}(x)$$

where

$$V_a^{\mu'}(x) = \Lambda_b^a(x)V_b^\mu(x)$$

This is verified by taking the local Lorentz transformation $\Lambda(x)$ to be infinitesimal:

$$\Lambda_b^a(x) = \delta_b^a + \omega_b^a(x)$$

where

$$\omega_{ab}(x) + \omega_{ba}(x) = 0, \omega_{ab}(x) = \eta_{ac}\omega_b^c(x)$$

The resulting Lagrangian above yields the Dirac equation for electrons and positrons in a gravitational field and electromagnetic field as

$$(\gamma^a V_a^\mu(x)(i\partial_\mu + eA_\mu(x) + i\Gamma_\mu(x)) - m)\psi(x) = 0$$

This equation is both diffeomorphic and Lorentz covariant. The Dirac current due to the gravitational correction can be evaluated using this equation and hence the effect of the gravitational field on the electromagnetic field radiated by the quantum antenna comprising electrons and positrons can be determined.

2.5 Dirac equation in a gravitational field and a quantum white noise photon field described in the Hudson-Parthasarathy formalism

The Hudson-Parthasarathy theory: The Dirac equation in a gravitational field and interacting with an electromagnetic field that is modeled as quantum noise as well as the Maxwell field comprising a classical component and a quantum noise component can jointly be expressed by the following quantum stochastic differential equation:

$$\gamma^a V_a^0(t,r)id\psi_t(r)) + \gamma^a V_a^m(t,r)i\partial_m\psi_t(r)dt$$

$$+\gamma^a V_a^\mu(t,r)L_{\mu\alpha}^\beta(t,r)\psi_t(r)d\Lambda_\alpha^\beta(t)$$

$$+\gamma^a V_a^\mu(t,r)i\Gamma_\mu(t,r)\psi_t(r)dt - m\psi_t(r)dt = 0$$

where $L^{\beta}_{\mu\alpha}(t,r)$ are system operators, ie built out of multiplication operators and spatial differential operators. It is more natural to consider them as only multiplication operators since then $e^{-1}L^{\beta}_{\mu\alpha}(t,r)d\Lambda^{\beta}_{\alpha}(t)/dt$ can formally be interpreted as the four vector potential $A_{\mu}(t,r)$ corresponding to the noisy photon bath. Here $\Lambda^{\alpha}_{\beta}, \alpha, \beta \geq 0$ are the standard Hudson-Parthasarathy noise operators satisfying the quantum Ito formula

$$d\Lambda^{\alpha}_{\beta}(t).d\Lambda^{\mu}_{\nu}(t) = \epsilon^{\alpha}_{\nu}d\Lambda^{\mu}_{\beta}(t)$$

with $\epsilon^{\alpha}_{\beta}$ being zero if either $\alpha = 0$ or $\beta = 0$ and δ^{α}_{β} otherwise. The operators $L^{\beta}_{\mu\alpha}(t,r)$ are forced to satsify certain constraints dictated by the fact that this equation describes a unitary evolution. These constraints can easily be deduced using the quantum Ito formula.

2.6 Dirac-Yang-Mills current density for non-Abelian gauge theories

When instead of electrons and positrons, we consider antennas having other kinds of elementary particles described by non-Abelian gauge theories of the Yang-Mills type, for example the weak and strong forces, then the dynamics is described by a Lagrangian density

$$L_{YM} = Tr(F_{\mu\nu}F^{\mu\nu}) + L_M(\psi, \nabla_{\mu}\psi)$$

where the connection is given by

$$\nabla_{\mu} = \partial_{\mu} + A_{\mu}(x)$$

with

$$A_{\mu}(x) \in \mathfrak{g}$$

$\mathfrak{g}$ being the Lie algebra of the gauge group $G \subset U(N)$ and $F_{\mu\nu}(x) \in \mathfrak{g}$, the field tensor being defined as the curvature of the connection:

$$F_{\mu\nu} = [\nabla_{\mu}, \nabla_{\nu}] = A_{\nu,\mu} - A_{\mu,\nu} + [A_{\mu}, A_{\nu}]$$

The matter field equations are

$$[i\gamma^{\mu}\nabla_{\mu} - m]\psi(x) = 0$$

which is to be interpreted as

$$[i\gamma^{\mu}\partial_{\mu} \otimes I_N + i\gamma^{\mu} \otimes A_{\mu}(x) - m]\psi(x) = 0$$

The Dirac current has to be generalized to this non-Abelian setting and the Yang-Mills field equations have to be solved using perturbation methods since unlike the Maxwell equations, the contain nonlinear terms too. It should be noted that under a local gauge transformation $g(x) \in G$, the matter field transforms as $\psi(x) \to g(x)\psi(x)$ while, the gauge field $A_\mu(x)$ has to transform to $A'_\mu(x)$ so that the field equations remain invariant, ie,

$$g(x)\nabla_\mu g(x)^{-1} = \nabla'_\mu$$

This gives

$$A'_\mu(x) = g(x)A_\mu(x)g(x)^{-1} + g(x)\partial_\mu(g(x)^{-1})$$

and generalizes the Lorentz gauge transformation of the electromagnetic field.

2.7 Dirac brackets

. Suppose the Lagrangian $L = L(q, Q, p, P)$ where (q, Q) are the position variables and (p, P) are the momentum variables with $q = (q_1, ..., q_n), p = (p_1, ..., p_n)$ and $Q = (Q_1, ..., Q_r), P = (P_1, ..., P_r)$. The constraints are $Q = Q(q, p), P = P(q, p)$ ie, Q, P are functions of q, p. In view of these constraints, we cannot assume that $\{Q, u\} = \partial u/\partial P$ and $\{u, P\} = \partial u/\partial Q$. So we must modify our definition of the Poisson bracket to take these constraints into account. If we write $Q = Q(q, p)$ and $P = P(q, p)$, and compute our Poisson brackets based on (q, p) only we find that

$$(u, v) =$$

$$(u_{,q}^T + u_{,Q}^T Q_{,q} + u_{,P}^T P_{,q})(v_{,p} + Q_{,p}^T v_{,Q} + P_{,p}^T v_{,P})$$

$$-(u_{,p}^T + u_{,Q}^T Q_{,p} + u_{,P}^T P_{,p})(v_{,q} + Q_{,q}^T v_{,Q} + P_{,q}^T v_{,P})$$

$$= u_{,q}^T v_{,p} - u_{,p}^T v_{,q} +$$

$$+ u_{,q}^T Q_{,p}^T v_{,Q} + u_{,q}^T P_{,p}^T v_{,P}$$

$$- u_{,p}^T Q_{,q}^T v_{,Q} - u_{,p}^T P_{,q}^T v_{,P}$$

$$+ v_{,p}^T Q_{,q}^T u_{,Q} + v_{,p}^T P_{,q}^T u_{,P}$$

$$- v_{,q}^T Q_{,p}^T u_{,Q} - v_{,q}^T P_{,p}^T u_{,P}$$

$$+ u_{,Q}^T (Q_{,q} Q_{,p}^T - Q_{,p} Q_{,q}^T) v_{,Q}$$

$$+ u_{,Q}^T (Q_{,q} P_{,p}^T - Q_{,p} P_{,q}^T) v_{,P}$$

$$+ u_{,P}^T (P_{,q} Q_{,p}^T - P_{,p} Q_{,q}^T) v_{,Q}$$

$$+ u_{,P}^T (P_{,q} P_{,p}^T - P_{,p} P_{,q}^T) v_{,P}$$

Now, define the matrix

$$D = D(q, p) = \begin{pmatrix} (Q, Q^T) & (Q, P^T) \\ (P, Q^T) & (P, P^T) \end{pmatrix}$$

Then, we can write

$$(u, v) = (u, v)_{(q,p)} + [u_{,Q}^T, u_{,P}^T].D. \begin{pmatrix} v_{,Q} \\ v_{,P} \end{pmatrix}$$

$$+(u, Q^T)v_{,Q} + (u, P^T)v_{,P} - (v, Q^T)u_{,Q} - (v, P^T)u_{,P} =$$

$$(u, v)_{(q,p)} + [u_{,Q}^T, u_{,P}^T].D. \begin{pmatrix} v_{,Q} \\ v_{,P} \end{pmatrix}$$

$$+[u_{,Q}^T, u_{,P}^T] \begin{pmatrix} (Q, v) \\ (P, v) \end{pmatrix}$$

$$+[(u, Q^T), (u, P^T)] \begin{pmatrix} v_{,Q} \\ v_{,P} \end{pmatrix}$$

In this formula, we are assuming that the observable u, v depend explicitly on (q, p, Q, P). If they depend only on q, p as is natural to expect since Q, P are both functions of (q, p) by virtue of the constraints, then it is natural to define

$$\begin{pmatrix} u_{,Q} \\ u_{,P} \end{pmatrix} =$$

$$D^{-1} \begin{pmatrix} (u, Q) \\ (u, P) \end{pmatrix}$$

for this is the same as saying that

$$(u, Q) = (Q, Q^T)u_{,Q} + (Q, P^T)u_{,P}$$

and likewise for v. Note that this equation is the same as

$$(u, Q)^T = (u, Q^T) = u_{,Q}^T(Q, Q^T) + u_{,P}^T(P, Q^T)$$

With this definition, we get from the above,

$$(u, v)_{(q,p)} = (u, v) - [(u, Q)^T, (u, P)^T]D^{-1} \begin{pmatrix} (v, Q) \\ (v, P) \end{pmatrix}$$

This is called the Dirac bracket and works very well in problems involving constraints.

Other related problems to be discussed: The spectral action principle in the unification of gravity with the electroweak and strong forces, ie, the fields of Leptons, quarks, photons gauge bosons and gluons starting from the heat equation.

Image processing for antenna patterns: The aim is to estimate the antenna parameters like the shape of the antenna and the distribution of current on its surface from measurements of the far field electromagnetic radiation pattern.

Scattering theory in quantum mechanics and its relation to antenna theory. Both of them involve solving Helmholtz equation with source.

2.8 Harish-Chandra's discrete series representations of $SL(2, \mathbb{R})$ and its application to pattern recognition under Lorentz transformations in the plane

.

Conclusions: In this lecture, we have proposed the analysis of quantum fluctuations in the electromagnetic field pattern produced by an antenna comprising electrons and positrons described by the second quantized Dirac field. We have also proposed the study of the Dirac field of electrons and positrons interacting with a gravitational field and the Maxwell electromagnetic field using the spinor connection of the gravitational field.

2.9 Estimating the shape of the antenna surface from the scattered EM field when an incident EM field induces a surface current density on the antenna that is determined by Pocklington's integral equation obtained by setting the tangential component of the total incident plus scattered electric field on the surface to zero

Assume that the antenna surface is parametrized by $(u, v) \rightarrow R(u, v)$. Assume that an incident electromagnetic field $E_i(\omega, r), H_i(\omega, r)$ falls on this surface. The induced surface current density is $J_s(u, v)$. The electric field (scattered) produced by this surface current density is given by

$$E_s(\omega, r) = \int G_m(\omega, r, u, v) J_{sm}(u, v) dS(u, v)$$

where G_m is a 3×1 complex vector valued function and the summation is over $m = 1, 2$. Here $J_{sm}, m = 1, 2$ are the components of the surface current density relative to the surface tangent basis $R_{,u}, R_{,v}$. This means that

$$J_s(u, v) = J_{s1}(u, v) R_{,u}(u, v) + J_{s2}(u, v) R_{,v}(u, v)$$

The Kernels $G_m, m = 1, 2$ are determined as follows: First the scattered magnetic vector potential is

$$A_s(\omega, r) = (\mu/4\pi) \int J_s(u, v) exp(-jk|r - R(u, v)|) dS(u, v)/|r - R(u, v)|$$

The scattered electric field is then

$$E_s(\omega, r) = \nabla \times (\nabla \times A_s)/j\omega\epsilon$$

$$= (\nabla(\nabla.A_s) + k^2 A_s)/j\omega\epsilon$$

of if the medium is inhomogeneous and anisotropic, then $\epsilon(\omega, r)$ is a 3×3 matrix and

$$E_s(\omega, r) = (j\omega\epsilon(\omega, r))^{-1} \nabla \times (\nabla \times A_s(\omega, r))$$

Thus, we get

$$G_1(\omega, r, u, v) = (j\omega\epsilon)^{-1}(\mu/4\pi)\nabla \times (\nabla(exp(-jk|r-R(u,v)|)/|r-R(u,v)|) \times R_{,u}(u,v)),$$

$$= (j\omega\epsilon)^{-1}(\mu/4\pi)[(R_{,u}(u,v), \nabla) + k^2](exp(-jk|r-R(u,v)|)/|r-R(u,v)|) ----(1)$$

and

$$G_2(\omega, r, u, v) = (j\omega\epsilon)^{-1}(\mu/4\pi)\nabla \times (\nabla(exp(-jk|r$$
$$-R(u,v)|)/|r-R(u,v)|) \times R_{,v}(u,v)) ----(2)$$

The components $J_{sm}(u,v), m = 1, 2$ of the surface current density are now determined by setting the tangential components of the electric field on the antenna surface equal to zero, ie,

$$(E_s(\omega, R(u,v)) + E_i(\omega, R(u,v)), R_{,u}(u,v)) = 0,$$

$$(E_s(\omega, R(u,v)) + E_i(\omega, R(u,v)), R_{,v}(u,v)) = 0,$$

These equations can be expressed as integral equations:

$$\sum_{m=1,2} \int_S (G_m(\omega, r, u', v'), R_{,u}(u,v)) J_{sm}(u', v') dS(u', v')$$

$$= (E_i(\omega, R(u,v)), R_{,u}(u,v)) ---(3),$$

$$\sum_{m=1,2} \int_S (G_m(\omega, r, u', v'), R_{,v}(u,v)) J_{sm}(u', v') dS(u', v')$$

$$= (E_i(\omega, R(u,v)), R_{,v}(u,v)) ---(4),$$

From these equations, we can determine the surface induced current density $J_s(u,v)$ on the antenna surface and hence the scattered electromagnetic field $E_s(\omega, r), H_s(\omega, r)$. Now, if the surface shape is unknown, we assume it to be given in parametric form as

$$R(u,v) = \sum_{k=1}^p \theta_k \psi_k(u,v) = R(u,v,\theta)$$

where ψ_k are test/basis functions as in the method of moments and the $\theta'_k s$ are unknown parameters to be estimated. We next observe that the Green's functions $G_m, m = 1, 2$ defined by (1) and (2) can be expressed explicitly as functions of $\omega, r, R(u,v), R_{,u}(u,v)$ and $R_{,v}(u,v)$. We can thus express this relation as

$$G_m(\omega, r, u, v) = F_m(\omega, r, R(u,v), R_{,u}(u,v), R_{,v}(u,v))$$

$$= F_m(\omega, r, u, v, \theta), \theta = (\theta_1, ..., \theta_p)^T$$

Thus, the integral equations (3) and (4), when solved, yield the surface current densities $J_{sm}(u, v)$ as functions of (u, v, θ) and thus the scattered electric field

$$E_s(\omega, r) = \int F_m(\omega, r, u, v, \theta) J_{sm}(u, v, \theta) |R_{,u}(u, v, \theta) \times R_{,v}(u, v, \theta)| du dv$$

is known as a function of θ. θ may then be estimated by matching this scattered electric field to given measurements and hence the shape of the object can be obtained.

2.10 Surface current density operator induced on the surface of a quantum antenna when a quantum EM field is incident on it

Problem: If the incident field is a quantum electromagnetic field built out of a superposition of creation and annihilation operators, then the surface current density $J_s(u, v)$ will also be a superposition of creation and annihilation operators. This can be seen from the equations (3) and (4) which tell us that $J_s(u, v)$ is a linear functional of E_i which is in turn a linear functional of the creation and annihilation operators. We can thus write

$$J_s(\omega, u, v) = \sum_{k=1}^{N} (a_k \chi_k(\omega, u, v) + a_k^* \bar{\chi}_k(\omega, u, v))$$

where

$$[a_k, a_j^*] = \delta_{kj}, [a_k, a_m] = 0, [a_k^*, a_m^*] = 0$$

Now suppose that the incident field is in a coherent state

$$|\phi(u) >= exp(-|u|^2/2) \sum_{n_1,...,n_p \geq 0} a_1^{*n_1} ... a_p^{*n_p} |0 > /n_1! ... n_p!$$

where $u \in \mathbb{C}^p$. Then the average value of the surface current density in this state is given by

$$< \phi(u)|J_s(\omega, u, v)|\phi(u) >=$$

$$\sum_k (\chi_k(\omega, u, v) < \phi(u)|a_k|\phi(u) > + \bar{\chi}_k(\omega, u, v) < \phi(u)|a_k^*|\phi(u) >)$$

$$= \sum_k (\chi_k(\omega, u, v) u_k + \bar{\chi}_k(\omega, u, v) \bar{u}_k)$$

Problem: Compute the higher moments of the surface current density in this coherent state:

$$< \phi(u)|J_s(\omega_1, u_1, v_1) \otimes ... \otimes J_s(\omega_m, u_m, v_m) \otimes \bar{J}_s(\omega_1', u_1', v_1') \otimes ... \otimes \bar{J}_s(\omega_n', u_n', v_n')|\phi(u) >$$

In order to do this after applying the appropriate commutation relations, you will need to compute things such as

$$< \phi(u)|a_1^{*m_1}...a_p^{*m_p}a_1^{n_1}...a_p^{n_p}|\phi(u) >=$$

$$u_1^{n_1}...u_p^{n_p}\bar{u}_1^{m_1}...\bar{u}_p^{m_p}$$

Problem: Compute using the formula for the moments of the surface current density in a coherent state of the incident field, the moments of the scattered em potential and field in the far field zone:

$$E_s(\omega,r) = (\mu exp(-jkr)/4\pi r) \int J_s(\omega,u,v)exp(jk\hat{r}.R(u,v))dS(u,v)$$

Now consider the Dirac second quantized current density field

$$J^\mu(x) = -e\psi(x)^*\alpha^\mu\psi(x)$$

We write

$$\psi(x) = \int (u(P,\sigma)a(P,\sigma)exp(-ip.x) + v(P,\sigma)b(P,\sigma)^*exp(ip.x))d^3P$$

Express $J^\mu(\omega,r)$, ie the current density in the frequency domain using the product theorem for Fourier transforms:

$$J^\mu(\omega,r) = (-e/2\pi) \int_{\mathbb{R}} \hat{\psi}(\omega',r)^*\alpha^\mu\psi(\omega+\omega',r)d\omega'$$

Now express this Fourier transformed current density in terms of the creation and annihilation operators of the electrons and positrons:

$$\hat{\psi}(\omega,r) = \int [u(P,\sigma)a(P,\sigma)\delta(\omega+E(P))exp(iP.r)+v(P,\sigma)b(P,\sigma)^*\delta(\omega-E(P))exp(-iP.r)]d^3P$$

The final expression for $J^\mu(\omega,r)$ will be of the form

$$J^\mu(\omega,r) = \int [K_1^\mu(\omega,P,P',\sigma,\sigma',r)a(P,\sigma)^*a(P',\sigma')+$$

$$K_2^\mu(\omega,P,P',\sigma,\sigma',r)a(P',\sigma')^*b(P,\sigma)$$

$$+K_3^\mu(\omega,P,P',\sigma,\sigma',r)b(P,\sigma)a(P',\sigma')^*$$

$$+K_4^\mu(\omega,P,P',\sigma,\sigma',r)b(P,\sigma)b(P',\sigma')^*]d^3Pd^3P'$$

As mentioned in the introduction, apart from the electromagnetic field generated by this current field, there is a free photon electromagnetic field built out of a linear superposition of photon creation and annihilation operators $c(K,s)$. So the total electromagnetic four potential will be of the form

$$A^\mu(\omega,r) = (\mu/4\pi) \int J^\mu(\omega,r')exp(-j\omega|r - r'|/c)d^3r'/|r - r'|+$$

$$\int (c(P,s)F^\mu_\zeta \omega, P, s, r) + c(P,s)^* \bar{F}^\mu(\omega, P, s, r))d^3P$$

$$= \int [K^\mu_1(\omega, P, P', \sigma, \sigma', r')a(P, \sigma)^*a(P', \sigma')+$$

$$K^\mu_2(\omega, P, P', \sigma, \sigma', r')a(P', \sigma')^*b(P, \sigma)$$

$$+K^\mu_3(\omega, P, P', \sigma, \sigma', r')b(P, \sigma)a(P', \sigma')^*$$

$$+K^\mu_4(\omega, P, P', \sigma, \sigma', r')b(P, \sigma)b(P', \sigma')^*]G(\omega, r - r')d^3P d^3P' d^3r'$$

$$+ \int (c(P,s)F^\mu_\zeta \omega, P, s, r) + c(P,s)^* \bar{F}^\mu(\omega, P, s, r))d^3P$$

where

$$G(\omega, r) = (\mu.exp(-jK|r|)/4\pi|r|), K = \omega, c = 1$$

Equivalently, writing

$$L^\mu_r(\omega, P, P', \sigma, \sigma', r) = \int G(\omega, r - r')K^\mu_r(\omega, P, P', \sigma, \sigma', r')d^3r'$$

we have the following final expression for the total quantum electromagnetic four potential:

$$A^\mu(\omega, r) =$$

$$= \int [L^\mu_1(\omega, P, P', \sigma, \sigma', r)a(P, \sigma)^*a(P', \sigma')+$$

$$L^\mu_2(\omega, P, P', \sigma, \sigma', r)a(P', \sigma')^*b(P, \sigma)$$

$$+L^\mu_3(\omega, P, P', \sigma, \sigma', r)b(P, \sigma)a(P', \sigma')^*$$

$$+L^\mu_4(\omega, P, P', \sigma, \sigma', r)b(P, \sigma)b(P', \sigma')^*]d^3P d^3P'$$

$$+ \int (c(P,s)F^\mu_\zeta \omega, P, s, r) + c(P,s)^* \bar{F}^\mu(\omega, P, s, r))d^3P$$

We now introduce the following notations:

$$|p_{1e}, \sigma_{1e}, ..., p_{ne}, \sigma_{ne}, p_{1p}, \sigma_{1p}, ..., p_{mp}, \sigma_{mp} >$$

stands for the state in which there is are n electrons of four momenta p_{ke} with corresponding spins $\sigma_{ke}, k = 1, 2, ..., n$ and m positrons of four momenta p_{kp} with corresponding spins $\sigma_{kp}, k = 1, 2, ..., m$. Further we know that (See reference [1])

$$a(P, \sigma)|p_e, \sigma_e >= \delta(\sigma, \sigma_e)\delta^3(P - P_e)|0 >$$

$$a(P, \sigma_e)^*|0 >= |p_e, \sigma_e >$$

and likewise for positrons. More generally, we have

$$a(P_e, \sigma_e)|p_{1e}, \sigma_{1e}, ..., p_{ne}, \sigma_{ne}, p_{1p}, \sigma_{1p}, ..., p_{mp}, \sigma_{mp} >$$

$$= \sum_{k=1}^{n}(-1)^{k-1}\delta(\sigma_e, \sigma_{ke})\delta^3(P_e, P_{ke})|p_{1e}, \sigma_{1e}, ..., \hat{p}_{ke}, \hat{\sigma}_{ke},$$

$$..., p_{ne}, \sigma_{ne}, p_{1p}, \sigma_{1p}, ..., p_{mp}, \sigma_{mp} >$$

where a hat above p_{ke}, σ_{ke} indicates that these variables do not occur and

$$a(P_e, \sigma_e)^*|p_{1e}, \sigma_{1e}, ..., p_{ne}, \sigma_{ne}, p_{1p}, \sigma_{1p}, ..., p_{mp}, \sigma_{mp} >$$

$$= |p_e, \sigma_e, p_{1e}, \sigma_{1e}, ..., \hat{p}_{ke}, \hat{\sigma}_{ke}, ..., p_{ne}, \sigma_{ne}, p_{1p}, \sigma_{1p}, ..., p_{mp}, \sigma_{mp} >$$

and likewise for positrons. We also have the anticommutation rules:

$$\{a(P, \sigma), a(P', \sigma')^*\} = \delta^3(P - P')\delta(\sigma, \sigma')$$

$$\{b(P, \sigma), b(P', \sigma')^*\} = \delta^3(P - P')\delta(\sigma, \sigma')$$

and all the other anticommutation relations are zero. These anticommutation rules and the action of creation and annihilaition operators can in principle be realized in a mathematically rigorous way using the Fermion Fock space (Reference [4]).

2.11   Summary of the second quantized Dirac field

Deriving the anticommutation rules for the Dirac field operators. The free Dirac equation is

$$(i\gamma^\mu \partial_\mu - m)\psi(x) = 0$$

or equivalently in terms of the anticommuting Hermitian matrices

$$\beta = \gamma^0, \alpha^r = \gamma^0\gamma^r, 1 \leq r \leq 3$$

we have

$$i\partial_0\psi = ((\alpha, -i\nabla) + \beta m)\psi$$

The solution for $\psi(x)$ is a representation as a superposition of plane waves:

$$\psi(x) = \int (u(P, \sigma)a(P, \sigma)exp(-ip.x) + v(P, \sigma)b(P, \sigma)^*exp(ip.x))d^3P$$

where

$$p^0 = E(P) = \sqrt{m^2 + P^2} > 0$$

and in the second quantized picture, $a(P, \sigma), b(P, \sigma)$ are operators in a Fermion Fock space and likewise their adjoints $a(P, \sigma)^*, b(P, \sigma)^*$. Note that

$$E(-P) = E(P)$$

For the above wave field to satisfy the Dirac equation, we must evidently have the purely algebraic relations

$$(\gamma^\mu p_\mu - m)u(P, \sigma) = 0, (\gamma^\mu p_\mu + m)v(P, \sigma) = 0, \sigma = 1, 2$$

or equivalently,

$$((\alpha, P) + \beta m)u(P, \sigma) = E(P)u(P, \sigma), \sigma = 1, 2$$

$$((\alpha, P) + \beta m)v(-P, \sigma) = -E(P)v(-P, \sigma), \sigma = 1, 2$$

Since the eigenvectors of a Hermitian matrix can be chosen to be an orthonormal basis for the underlying vector space and since the eigenvectors of a Hermitian matrix corresponding to distinct eigenvalues are orthogonal, we have that

$$u(P, \sigma)^* v(P, \sigma') = 0, \sigma, \sigma' = 1, 2$$

and we can ensure that

$$u(P, \sigma)^* u(P, \sigma') = 0, \sigma \neq \sigma' \; --- (a)$$

$$v(P, \sigma)^* v(P, \sigma') = 0, \sigma \neq \sigma' \; --- (b)$$

The second quantized Hamiltonian of the Dirac field is given by

$$H_D = \int \psi(x)^* ((\alpha, -i\nabla) + \beta m)\psi(x) d^3 r$$

where

$$x = (t, r) = (t, x^1, x^2, x^3)$$

We observe that $p.x = E(P)t - P.r$ and

$$((\alpha, -i\nabla) + \beta m)u(P, \sigma)exp(iP.r) = exp(iP.r)((\alpha, P) + \beta m)u(P, \sigma),$$

$$= E(P)u(P, \sigma).exp(iP.r),$$

$$((\alpha, -i\nabla) + \beta m)v(P, \sigma)exp(-iP.r) = exp(-iP.r)(-(\alpha, P) + \beta m)v(P, \sigma)$$

$$= -E(P)v(P, \sigma).exp(-iP.r)$$

and hence, the Hamiltonian at time $t = 0$ is given by

$$H_D = \int (u(P, \sigma)a(P, \sigma)exp(iP.r) + v(P, \sigma)b(P, \sigma)^* exp(-iP.r))^*$$

$$.E(P')(u(P', \sigma')a(P', \sigma')exp(iP'.r) - v(P', \sigma')b(P', \sigma')exp(-iP'.r))d^3 P d^3 P' d^3 r$$

$$= \int E(P')[u(P, \sigma)^* u(P', \sigma')a(P, \sigma)^* a(P', \sigma')exp(i(P' - P).r)+$$

$$v(P, \sigma)^* u(P', \sigma')b(P, \sigma)a(P', \sigma')exp(i(P' + P).r)$$

$$-u(P, \sigma)^* v(P', \sigma')a(P, \sigma)^* b(P', \sigma')^* exp(-i(P' + P).r)$$

$$-v(P, \sigma)^* v(P', \sigma')b(P, \sigma)b(P', \sigma')^* exp(i(P - P').r)]d^3 P d^3 P' d^3 r$$

$$= (2\pi)^3 \int E(P)[u(P, \sigma)^* u(P, \sigma)a(P, \sigma)^* a(P, \sigma) - v(P, \sigma)^* v(P, \sigma)b(P, \sigma)b(P, \sigma)^*]d^3 P$$

where we have used the identity

$$\int exp(i(P - P').r)d^3r = (2\pi)^3 \delta^3(P - P')$$

and the orthogonality relations (a) and (b). We now wish to decide the normalizations of u and v, ie, to evaluate $u(P,\sigma)^* u(P,\sigma)$ and $v(P,\sigma)^* v(P,\sigma)$. These are evaluated as follows. First note that the Lagrangian density of the free Dirac field is

$$L_D(\psi, \psi^*, \psi_{,\mu}, \psi^*_{,\mu}) =$$

$$\psi(x)^* \gamma^0 (i\gamma^\mu \partial_\mu - m)\psi(x)$$

It follows that the momentum conjugate to ψ is

$$\pi = \frac{\partial L_D}{\partial \psi_{,0}} = i\psi^*$$

Thus, the canonical anticommutation rules

$$\{\psi_l(t,r), \pi_m(t,r)\} = i\delta_{lm}\delta^3(r - r')$$

give us

$$\{\psi_l(t,r), \psi_m(t,r')^*\} = \delta_{lm}\delta^3(r - r') - - - (c)$$

So we must have

$$\int [u_l(P,\sigma)u_m(P',\sigma')^* \{a(P,\sigma), a(P',\sigma')^*\} exp(i(P.r - P'.r'))$$

$$+u_l(P,\sigma)v_m(P',\sigma')^* \{a(P,\sigma), b(P',\sigma')\} exp(i(P.r + P'.r'))$$
$$+v_l(P,\sigma)u_m(P',\sigma')^* \{b(P,\sigma)^*, a(P',\sigma')^*\} exp(-i(P.r + P'.r'))$$
$$+v_l(P,\sigma)v_m(P',\sigma')^* \{b(P,\sigma)^*, b(P',\sigma')\} exp(-i(P.r - P'.r'))]d^3P d^3P'$$

$$= \delta_{lm}\delta^3(r - r') - - - (d)$$

This forces us to introduce the anti-commutation relations

$$\{a(P,\sigma), a(P',\sigma')^*\} = \delta_{\sigma,\sigma'}\delta^3(P - P')$$

$$\{b(P,\sigma), b(P',\sigma')^*\} = \delta_{\sigma,\sigma'}\delta^3(P - P')$$

and all the other anti-commutators vanish, along with the normalizations

$$\sum_{\sigma=1,2} (u_l(P,\sigma)u_m(P,\sigma)^* + v_l(-P,\sigma)v_m(-P,\sigma)^*) = \delta_{lm}$$

We note that

$$\Pi(P) = \sum_{\sigma=1,2} u(P,\sigma)u(P,\sigma)^{*T}$$

is the orthogonal projection onto the space of eigenvectors of the momentum space free Dirac Hamiltonian corresponding to the energy eigenvalue $E(P)$ while

$$I - \Pi(P) = \sum_{\sigma=1,2} v(-P,\sigma)v(-P,\sigma)^{*T}$$

is the orthogonal projection onto the space of eigenvectors of the free Dirac Hamiltonian in the momentum space corresponding to the energy eigenvalue $-E(P)$.

2.12 Electron propagator computation

The electron propagator is defined as

$$S_{lm}(x|x') = <0|T(\psi_l(x)\psi_m(x')^*)|0>$$

where T denotes the time ordering operator. In other words, with θ denoting the Heavy-side step function, we have

$$S_{lm}(t,r|t',r') = \theta(t-t') <0|\psi_l(t,r)\psi_m(t',r')^*|0>$$

$$-\theta(t'-t) <0|\psi_m(t',r')\psi_l(t,r)|0>$$

We shall evaluate this propagator using two methods. First, the purely operator theoretic method combined with properties of Dirac eigenvectors and second the differential equation method combined with the operator theoretic method.

The first method: We note that

$$\psi_l(t,r)\psi_m(t',r')^* = \left(\int [u_l(P,\sigma)a(P,\sigma)exp(-ip.x)+v_l(P,\sigma)b(P,\sigma)^*exp(ip.x)]d^3P \right)$$

$$\times \int (u_m(P',\sigma')^*a(P',\sigma')^*exp(ip'.x')+v_m(P',\sigma')^*b(P',\sigma')exp(-ip'.x'))d^3Pd^3P'$$

so that

$$<0|\psi_l(t,r)\psi_m(t',r')^*|0>=$$

$$\int [u_l(P,\sigma)u_m(P',\sigma')^*exp(-ip.x+ip'.x') <0|a(P,\sigma)a(P',\sigma')^*|0> d^3Pd^3P'$$

$$= \int u_l(P,\sigma)u_m(P',\sigma')^*exp(-ip.x+p'.x')\delta^3(P-P')\delta_{\sigma,\sigma'}d^3Pd^3P'$$

$$= \int u_l(P,\sigma)u_m(P,\sigma)^*exp(-iE(P)(t-t')-iP.(r-r'))d^3P$$

$$= \int \Pi_{lm}(P)exp(-i(E(P)(t-t')+P.(r-r')))d^3P$$

Likewise,

$$< 0|\psi_m(t', r')^* \psi_l(t, r)|0 >$$

$$= \int v_m(P', \sigma')^* v_l(P, \sigma) < 0|b(P', \sigma')b(P, \sigma)^*|0 > exp(-ip'.x' + ip.x)d^3 P d^3 P'$$

$$= \int v_l(P, \sigma)v_m(P, \sigma)^* exp(i(E(P)(t - t') - P.(r - r')))d^3 P$$

$$= \int (\delta_{lm} - \Pi_{lm}(P))exp(i(E(P)(t - t') + P.(r - r')))d^3 P$$

Thus, our final expression for the electron propagator is

$$S_{lm}(x|x') = \theta(t - t') \int \Pi_{lm}(P)exp(-i(E(P)(t - t') + P.(r - r')))d^3 P$$

$$-\theta(t' - t) \int (\delta_{lm} - \Pi_{lm}(P))exp(i(E(P)(t - t') + P.(r - r')))d^3 P$$

We can thus change the integration variable P to $-P$ and express this propagator in matrix form as

$$S(x|x') = \theta(t-t') \int \Pi(-P)exp(-ip.(x-x'))d^3 P - \theta(t'-t) \int (I-\Pi(-P))exp(ip.(x-x'))d^3 P$$

where the integration is over the mass shell, ie $p^0 = E(P)$.

The second method: We write

$$S_{lm}(x|x') = S_{lm}(t, r|t', r') = \theta(t - t') < 0|\psi_l(t, r)\psi_m(t', r')^*|0 >$$

$$-\theta(t' - t) < 0|\psi_m(t', r')^* \psi_l(t, r)|0 >$$

Thus,

$$(i\gamma^\mu \partial_\mu - m)S(x|x') =$$

$$i\gamma^0 \partial_0 S(x|x') + (i\gamma^r \partial_r - m)S(x|x')$$

$$= i\gamma^0 \delta(t - t') < 0|\psi_l(t, r)\psi_m(t, r')^*|0 > +i\gamma^0 \delta(t - t') < 0|\psi_m(t, r')^* \psi_l(t, r)|0 >$$

$$+\theta(t - t') < 0|((i\gamma^\mu \partial_\mu - m)\psi(t, r))_l \psi_m(t', r')^*|0 >$$

$$-\theta(t' - t) < 0|\psi_m(t', r')^* ((i\gamma^\mu \partial_\mu - m)\psi(t, r))_l|0 >$$

$$= i\gamma^0 \delta(t't') < 0|\{\psi_l(t, r), \psi_m(t, r')^*\}|0 >= -i\gamma^0 \delta(t - t')\delta_{lm}\delta^3(r - r')$$

$$= i\gamma^0 \delta^4(x - x')$$

where we have used the fact that $\psi(t, r)$ satisfies the free Dirac equation. Taking the four dimensional space-time Fourier transform followed by Fourier inversion then gives us

$$S(x|x') = (2\pi)^{-4} \int i\gamma^0(i\gamma^\mu p_\mu - m - i\epsilon)^{-1} exp(ip.(x - x'))d^4 p$$

The reason for including a term $i\epsilon$ with $\epsilon \to 0+$ is for ensuring regularity of the propagator as $t - t' \to \pm\infty$. We shall elaborate on this using contour integrals.

2.13 Quantum mechanical tunneling of a Dirac particle through the critical radius of the Schwarzchild blackhole

.

It is well known [Steven Weinberg, Gravitation and Cosmology-Principles and applications of the general theory of relativity] that Dirac's relativistic equation for a particle of mass m in curved space-time with metric $g_{\mu\nu}(x)$ is given by

$$[\gamma^a V_a^\mu(x)(i\partial_\mu + i\Gamma_\mu(x)) - m]\psi(x) = 0$$

where $V_a^\mu(x)$ is a tetrad for the metric $g_{\mu\nu}$, ie,

$$\eta^{ab} V_a^\mu(x) v_b^\nu(x) = g^{\mu\nu}(x)$$

and $\Gamma_\mu(x)$ is the spinor connection of the gravitational field. It is calculated as

$$\Gamma_\mu(x) = (-1/2) J^{ab} V_a^\nu(x) V_{b\nu:\mu}(x)$$

where

$$J^{ab} = (1/4)[\gamma^a, \gamma^b]$$

are the standard Lie algebra generators of the Dirac spinor representation of the Lorentz group. It is easily shown that this Dirac equation is invariant under local Lorentz transformations.

Now let $g_{\mu\nu}$ be the metric of a Schwarzchild blackhole having mass M expressed in the cartesian system, ie, in the coordinate system x, y, z, t where

$$x = r.cos(\phi)sin(\theta), y = r.sin(\phi)sin(\theta), z = r.cos(\theta)$$

We determine a tetrad $V_a^\mu(x)$ corresponding to this metric and the corresponding spinor connection $\Gamma_\mu(x)$. Solve the time dependent Dirac equation with this metric and assuming that at time $t = 0$, the Dirac wave function $\psi(t, r)$ is zero for $|r| > 2GM/c^2$, find the probability density of the particle at time $t > 0$ at points r with $|r| > 2GM/c^2$. This probability density is simply $\psi(t, r)^* \psi(t, r)$. More generally, determine the Dirac four current density

$$J^\mu(t, r) = \psi(t, r)^* \gamma^0 \gamma^\mu \psi(t, r)$$

for $|r| > 2GM/c^2$. This problem illustrates the fact that although classical general relativity mechanics prevents a particle in escaping through the event horizon of the blackhole in finite coordinate time, quantum general relativity yields a small tunnelling probability for the particle through the event horizon.

2.14 Supersymmetry–supersymmetric current in an antenna comprising superpartners of elementary particles

Abstract: Supersymmetry provides a nice mathematical basis for the unification of various field theories. Many kinds of fields like the scalar Klein-Gordon field with Higgs potential, the]Dirac Fermionic field, the Maxwell electromagnetic field, the non-Abelian matter and gauge field and the gravitational field appear in supersymmetry as components of a big superfield that is a polynomial in Majorana Fermionic variables. Apart from these fields, in supersymmetry, we are forced to introduce other component fields like the gaugino field and the gravitino field which are called the superpartners of the corresponding gauge field and the gravitational metric tensor field and also other auxiliary fields. The introduction of superpartners compels us to postulate that every particle has a superpartner. More precisely, a Boson has a superpartner that is a Fermion and vice versa. The importance of supersymmetry in the context of antennas comes from the fact that when one writes down the supersymmetric Lagrangian of a Chiral superfield, then this Lagrangian contains both the kinematic part of the Lagrangian of the Dirac field and the Lagrangian of the scalar KG field along with some other auxiliary fields. When one sets the variational derivative of this Lagrangian density w.r.t. the auxiliary fields to zero, then the resulting Lagrangian density contains the Dirac Lagrangian with a mass dependent on the scalar field through a superpotential. Thus, the current density of the Dirac field depends on the scalar KG field and the corresponding radiation produced by such a current will therefore also be dependent upon the scalar field. There is another aspect to supersymmetric effect in antennas. This is based on the notion of supercurrent. Noether's theorem states that when the Lagrangian density is invariant under a Lie group of transformations of the field, then there is an associated conserved current. The supersymmetric action integral is invariant under the transformations of the component fields defined by the Salam Strathdee supersymmetry generators and hence, we can associate a Noether current density to this. This Noether current density associated with transformations of the component fields under a supersymmetric transformation of the superfield will have a four divergence equal to the change in the Lagrangian density under an infinitesimal supersymmetric transformation and this guarantees that the action integral is invariant under supersymmetric transformations. This relation is satisfied provided that we assume that the component superfields satisfy the Euler-Lagrange equations of motion for the given supersymmetric Lagrangian density. This current density will not be conserved since it is only the action integral that is invariant under supersymmetric transformations and not the Lagrangian density. The four divergence of this Noether current density gives us the infinitesimal change in the Lagrangian density under supersymmetric transformation. On the other hand, the Lagrangian density itself being a function of the component fields, undergoes a transformation under supersymmetry and this infinitesimal change is a four divergence of another current. Therefore, the difference between the four divergences of the two currents, one the Noether current and two the current arising from the change in the Lagrangian under supersymmetry must be zero. Thus, we get a conservation law. This law is the conservation of the supercurrent and we would like to evaluate the radiation fields produced by such super currents. There is another aspect to supersymmetry when one considers gauge fields and their superpartners the gaugino fields. The electromagnetic field is a $U(1)$ gauge field. On the other hand, there exist non-Abelian gauge fields which arise in the weak and strong nuclear interactions. These non-Abelian gauge fields are messenger fields which communicate the nuclear forces just as the electromagnetic field is an Abelian gauge field which communicates the electromagnetic forces between charges. In supersymmetry we talk of invariance under supersymmetric gauge transformations. Just as the total Lagrangian density of the matter and gauge field is invariant under gauge transfromations in the Dirac theory and its generalized non-Abelia Yang-Mills theory, so also we must look for a Lagrangian density built solely out of the gauge fields and the gaugino fields that is invariant non only under supersymmetric gauge transformations but also under general supersymmetry transformations. Such a Lagrangian density does exist (Reference [1]) and it contains apart from the Lagrangian of the Maxwell field and the Lagrangian of the Yang-Mills non-Abelian gauge field, the Lagrangian of the gaugino and auxiliary fields. When we talk of radiation by matter in quantum antennas as mentioned earlier, we should apart from taking the matter current as the Dirac current of electrons and positrons and the corresponding radiation field as the Maxwell field, also talk about taking the matter current coming from other components of the superfield like the scalar KG field and the corresponding radiation fields being not only the Maxwell field, but the other gauge fields, gaugino fields and auxiliary fields. The relation between the matter fields and the gauge, gaugino and auxiliary fields will come from the total Lagrangian of the matter fields and gauge, gaugino and auxiliary fields. Such a Lagrangian can be derived from a superfield and it respects both supersymmetry invariance and supersymmetric gauge invariance (Reference III).

Majorana Fermion:

$$\theta = \begin{pmatrix} \zeta \\ -e\zeta^* \end{pmatrix}$$

where

$$e = i\sigma^2 = \begin{pmatrix} 0 & 1 \\ -1 & 0 \end{pmatrix}$$

Then, we have

$$\theta^* = \begin{pmatrix} \zeta^* \\ -e\zeta \end{pmatrix}$$

$$= \begin{pmatrix} 0 & e \\ -e & 0 \end{pmatrix} \theta$$

We assume that the components of the Majorana Fermion $\theta = (\theta^a, a = 0, 1, 2, 3)$ mutually anticommute:

$$\theta^a \theta^b + \theta^b \theta^a = 0$$

Define

$$\epsilon = \begin{pmatrix} e & 0 \\ 0 & e \end{pmatrix}$$

Note that

$$\sigma^{2T} = -\sigma^2, e^T = -e,$$

$*$ stands for complex conjugate, not for conjugate transpose. For conjugate transpose, we shall use the notation $a \rightarrow a^H$. Also define the Dirac gamma matrices:

$$\gamma^\mu = \begin{pmatrix} 0 & \sigma^\mu \\ \sigma_\mu & 0 \end{pmatrix}$$

where

$$\sigma^0 = \sigma_0 = I, \sigma^1 = -\sigma_1 = \begin{pmatrix} 0 & 1 \\ 1 & 0 \end{pmatrix},$$

$$\sigma^2 = -\sigma_2 = \begin{pmatrix} 0 & -i \\ i & 0 \end{pmatrix},$$

$$\sigma^3 = -\sigma_3 = \begin{pmatrix} 1 & 0 \\ 0 & -1 \end{pmatrix}$$

are the usual Pauli spin matrices. We have

$$\sigma^{\mu H} = \sigma^\mu,$$

$$\sigma^{0T} = \sigma^0, \sigma^{1T} = \sigma^1, \sigma^{2T} = -\sigma^2, \sigma^{3T} = \sigma^3$$

$$\sigma^2 \sigma^0 \sigma^2 = \sigma^0, \sigma^2 \sigma^1 \sigma^2 = i\sigma^2 \sigma^3 = -\sigma^1 = \sigma_1,$$

$$\sigma^2 \sigma^2 \sigma^2 = \sigma^2 = -\sigma_2, \sigma^2 \sigma^3 \sigma^2 = i\sigma^1 \sigma^2 = -\sigma^3 = \sigma_3$$

where we have used the standard relations

$$\sigma^1 \sigma^2 = -\sigma^2 \sigma^1 = i\sigma^3, \sigma^2 \sigma^3 = -\sigma^3 \sigma^2 = i\sigma^1,$$

$$\sigma^3 \sigma^1 = -\sigma^1 \sigma^3 = i\sigma^2$$

along with

$$\sigma^{\mu 2} = I, \mu = 0, 1, 2, 3$$

We also have the conjugation relations:

$$\sigma^2 \sigma^{1*} \sigma^2 = \sigma_1,$$

$$\sigma^2 \sigma^{2*} \sigma^2 = \sigma_2,$$

$$\sigma^2 \sigma^{3*} \sigma^2 = \sigma_3$$

Thus,

$$\sigma^2 \sigma^{\mu *} \sigma^2 = \sigma_\mu$$

and hence

$$e \sigma^{\mu *} e = -\sigma_\mu$$

where we use

$$\sigma^{0*} = \sigma^0, \sigma^{1*} = \sigma^1, \sigma^{2*} = -\sigma^2, \sigma^{3*} = \sigma^3$$

We also have the fact that $\epsilon, \gamma_5 \epsilon, \epsilon \gamma^\mu$ are six linearly independent 4×4 ansisymmetric matrices and hence form a basis for the vector space of all 4×4 antisymmetric matrices over the complex field. Here, we define

$$\gamma^5 = \gamma^0 \gamma^1 \gamma^2 \gamma^3 = \begin{pmatrix} I & 0 \\ 0 & -I \end{pmatrix}$$

Note that

$$\gamma^5 \epsilon = \begin{pmatrix} e & 0 \\ 0 & -e \end{pmatrix}$$

Exercise: Verify that the Dirac gamma matrices satisfy the Clifford algebra anticommutation relations with bilinear form $((\eta^{\mu\nu})) = diag[1, -, 1-, 1-,]$:

$$\gamma^\mu \gamma^\nu + \gamma^\nu \gamma^\mu = 2\eta^{\mu\nu}$$

Now let M be any symmetric 4×4 matrix (bosonic). Then

$$\theta^T M \theta = 0$$

since

$$\theta^T M \theta = \sum_{a,b} M_{ab} \theta^a \theta^b = \sum M_{ba} \theta^a \theta^b = \sum M_{ba} (-\theta^b \theta^a) = -\theta^T M \theta$$

Thus, any quadratic function of θ can be expressed as

$$\theta^T M \theta$$

where M is an antisymmetric matrix. It follows that since $\gamma^5 \epsilon, \epsilon, \epsilon \gamma^\mu, \mu = 0, 1, 2, 3$ form a basis for the six-dimensional complex vector space of 4×4 antisymmetric matrices, that any quadratic form in θ can be expressed as a linear combination of the six quadratic forms $\theta^T \epsilon \theta, \theta^T \gamma^5 \epsilon \theta, \theta^T \epsilon \gamma^\mu \theta, \mu = 0, 1, 2, 3$. In particular, we can write

$$\theta \theta^T = c_1 . \epsilon \theta^T \epsilon \theta + c_2 . \gamma^5_\mu \epsilon \theta^T \gamma^5 \epsilon \theta + c_3 \gamma_\mu \epsilon \theta^T \epsilon \gamma^\mu \theta$$

Note that the rhs is an antisymmetric matrix since the Fermionic variables θ^a anticommute. To verify this identity and also calculate the scalar coefficients c_1, c_2, c_3, we multiply both sides by $\epsilon, \gamma^5 \epsilon$ and $\epsilon \gamma^\nu$ and take traces. Then we get using $\epsilon^2 = -I_4 = (\gamma^5 \epsilon)^2$

$$\theta^T \epsilon \theta = -4 c_1 \theta^T \epsilon \theta$$

$$\theta^T \gamma^5 \epsilon \theta = -4c_2 \theta^T \gamma^5 \epsilon \theta$$

so that

$$c_1 = c_2 = -1/4$$

and finally,

$$\theta^T \epsilon \gamma^\nu \theta = c_3 . Tr(\epsilon \gamma^\nu \gamma_\mu \epsilon) \theta^T \epsilon \gamma^\mu \theta$$

$$= -c_3 Tr(\gamma^\nu \gamma_\mu) \theta^T \epsilon \gamma^\mu \theta$$

$$= -4c_3 \delta^\nu_\mu \theta^T \epsilon \gamma^\mu \theta$$

$$= -4c_3 \theta^T \epsilon \gamma^\nu \theta$$

so that

$$c_3 = -1/4$$

Note that we are making use of the identities

$$\theta^T \epsilon \theta . \theta^T \gamma^5 \epsilon \theta = 0,$$

$$\theta^T \epsilon \theta . \theta^T \epsilon \gamma^\mu \theta = 0,$$

$$\theta^T \gamma^5 \epsilon \theta . \theta^T \epsilon \gamma^\mu \theta = 0$$

Remark: It is easy to verify that

$$\epsilon \gamma^{\mu T} \epsilon = -\gamma^\mu$$

and hence

$$\epsilon \gamma^\mu \epsilon = -\gamma^{\mu T}$$

Also,

$$\gamma^{0*} = \gamma^0, \gamma^{1*} = \gamma^1, \gamma^{2*} = -\gamma^2, \gamma^{3*} = \gamma^3,$$

$$\gamma^{0T} = \gamma^0, \gamma^{1T} = -\gamma^1, \gamma^{2T} = \gamma^2, \gamma^{3T} = -\gamma^3$$

and hence

$$\gamma^{0H} = \gamma^0, \gamma^{1H} = -\gamma^1, \gamma^{2H} = -\gamma^2, \gamma^{3H} = -\gamma^3$$

or equivalently,

$$\gamma^{\mu H} = \gamma_\mu$$

Supersymmetry generators: Note that

$$(\gamma^5 \epsilon)^2 = \epsilon^2 = -I_4, (\gamma^5 \epsilon)^T = -\gamma^5 \epsilon, \gamma^{5T} = \gamma^5, \epsilon^T = -\epsilon, \gamma^5 \epsilon = \epsilon \gamma^5$$

Now,

$$L = \gamma^5 \epsilon \partial/\partial\theta + \gamma^\mu \theta \partial/\partial x^\mu$$

$$\bar{L} = -\gamma^5 \epsilon L = \partial/\partial\theta - \gamma^5 \epsilon \gamma^\mu \theta \partial/\partial x^\mu$$

or equivalently,

$$\bar{L}^T = L^T \gamma^5 \epsilon$$

Define

$$\bar{\theta} = -\gamma^5 \epsilon \theta$$

or equivalently,

$$\bar{\theta}^T = \theta^T \gamma^5 \epsilon$$

Now, we've seen that

$$\epsilon \gamma^\mu \epsilon = -\gamma^{\mu T}$$

and hence,

$$\gamma^5 \epsilon \gamma^\mu \gamma^5 \epsilon = \gamma^{\mu T}$$

(since γ^5 anti-commutes with the $\gamma^{\mu'}s$ and its square is I. Note that $\gamma^{5T} = \gamma^5$. Thus,

$$\gamma^5 \epsilon \gamma^\mu = \gamma^{\mu T} \gamma^5 \epsilon$$

and so

$$\bar{L} = \partial/\partial\theta + \gamma^{\mu T} \gamma^5 \epsilon \theta \partial/\partial x^\mu$$
$$= -\gamma^5 \epsilon \partial/\partial\bar{\theta} - \gamma^{\mu T} \bar{\theta} \partial/\partial x^\mu$$

Anticommutation relations between the supersymmetry generators: We use the fundamental identities

$$\{\partial/\partial\theta^a, \theta^b\} = \delta_a^b,$$
$$\{\theta^a, \theta^b\} = 0, \{\partial/\partial\theta^a, \partial/\partial\theta^b\} = 0$$

Then,

$$\{L_a, \bar{L}_b\} = \{(\gamma^5 \epsilon)_{ac} \partial/\partial\theta^c, (\gamma^{\mu T} \gamma^5 \epsilon)_{bd} \theta^d \partial/\partial x^\mu\}$$
$$+ \{(\gamma^\mu)_{ac} \theta^c \partial/\partial x^\mu, \partial/\partial\theta^b\}$$
$$= (\gamma^5 \epsilon)_{ac} (\gamma^{\mu T} \gamma^5 \epsilon)_{bd} \delta_c^d \partial/\partial x^\mu + (\gamma^\mu)_{ac} \delta_b^c \partial/\partial x^\mu$$
$$= [(\gamma^5 \epsilon)_{ac} (\gamma^{\mu T} \gamma^5 \epsilon)_{bc} + (\gamma^\mu)_{ab}] \partial/\partial x^\mu$$
$$= ((\gamma^5 \epsilon (\gamma^{\mu T} \gamma^5 \epsilon)^T) + \gamma^\mu)_{ab} \partial/\partial x^\mu$$
$$= 2\gamma_{ab}^\mu \partial/\partial x^\mu$$

Also,

$$\{L_a, L_b\} = \{(\gamma^5 \epsilon)_{ac} \partial/\partial\theta^c, \gamma_{bd}^\mu \theta^d \partial/\partial x^\mu\}$$
$$+ \{\gamma_{ac}^\mu \theta^c \partial/\partial x^\mu, (\gamma^5 \epsilon)_{bd} \partial/\partial\theta^d\}$$
$$= [(\gamma^5 \epsilon)_{ac} \gamma_{bd}^\mu \delta_c^d + \gamma_{ac}^\mu (\gamma^5 \epsilon)_{bd} \delta_d^c] \partial/\partial x^{mu}$$
$$= [\gamma^5 \epsilon \gamma^{\mu T} + \gamma^\mu (\gamma^5 \epsilon)^T]_{ab} \partial/\partial x^\mu = 0$$

since

$$\epsilon \gamma^{\mu T} \epsilon = -\gamma^\mu$$

implies

$$\gamma^5 \epsilon \gamma^{\mu T} \gamma^5 \epsilon = -\gamma^5 \gamma^\mu \gamma^5 = \gamma^\mu$$

implies

$$\gamma^5 \epsilon \gamma^{\mu T} = -\gamma^\mu \gamma^5 \epsilon = \gamma^\mu (\gamma^5 \epsilon)^T$$

These equations immediately imply that

$$\{\bar{L}_a, \bar{L}_b\} = 0$$

Superfields: A superfield is a function $S(x, \theta)$ of $x = (x^\mu)$ and $\theta = (\theta^a)$. Expanding S in powers of θ with coefficients being functions of x alone, we find that all the coefficients of the fifth and higher powers of θ may be taken as zero since any fifth and higher degree monomial in the θ is zero. Thus, we may expand a general superfield as

$$S[x, \theta] = C(x) + \theta^T \epsilon \omega(x) + \theta^T \epsilon \theta M(x)$$

$$+\theta^T \gamma^5 \epsilon \theta N(x) + \theta^T \epsilon \gamma^\mu \theta V_\mu(x) + \theta^T \epsilon \theta \theta^T \gamma^5 \epsilon (\lambda(x) + a\gamma.\partial\omega(x))$$

$$+(\theta^T \epsilon \theta)^2 (D(x) + b\square C(x))$$

where $\lambda(x), \omega(x) \in \mathbb{C}^4$ are spinorial. Let α be a Majorana spinor variable. An infinitesimal supersymmetry transformation is defined as the operator

$$\alpha^T \gamma^5 \epsilon L = \bar{\alpha}^T L = \delta L$$

We wish to examine how the bosonic components $C, \omega, M, N, V_\mu, \lambda, D$ transform under a supersymmetry transformation. The constants a, b will be chosen appropriately. We have

$$\delta L C(x) = \alpha^T \gamma^\mu \theta C_{,\mu}(x)$$

$$L\theta^T \epsilon \omega(x) = -\gamma^5 \omega(x) + \gamma^\mu \theta \theta^T \epsilon \omega_{,\mu}(x)$$

$$= -\gamma^5 \omega(x) + (-1/4)\gamma^\mu (\theta^T \epsilon \theta \epsilon + \theta^T \gamma^5 \epsilon \theta \gamma^5 \epsilon$$

$$+\theta^T \epsilon \gamma^\nu \theta \gamma_\nu \epsilon) \epsilon \omega_{,\mu}(x)$$

$$L\theta^T \epsilon \theta M(x) = 2\gamma^5 \epsilon.\epsilon\theta M(x) + \gamma^\mu \theta \theta^T \epsilon \theta M_{,\mu}(x)$$

$$= -2\gamma^5 \theta M(x) + \gamma^\mu \theta \theta^T \epsilon \theta M_{,\mu}(x)$$

so that

$$\delta L\theta^T \epsilon \theta M(x) = \alpha^T \gamma^5 \epsilon L\theta^T \epsilon \theta M(x)$$

$$= -2\alpha^T \epsilon \theta M(x) + \alpha^T \gamma^5 \epsilon \gamma^\mu \theta \theta^T \epsilon \theta M_{,\mu}(x)$$

$$= 2\theta^T \epsilon \alpha M(x) - \theta^T \epsilon \theta \theta^T \gamma^{\mu T} \gamma^5 \epsilon \alpha M_{,\mu}(x)$$

$$= 2\theta^T \epsilon \alpha M(x) + \theta^T \epsilon \theta \theta^T \gamma^5 \epsilon \gamma^\mu \alpha M_{,\mu}(x)$$

Next,

$$L\theta^T \gamma^5 \epsilon \theta N(x) = 2\gamma^5 \epsilon \gamma^5 \epsilon \theta N(x)$$

$$+\gamma^\mu \theta \theta^T \gamma^5 \epsilon \theta N_{,\mu}(x)$$

$$= -2\theta N(x) + \theta^T \gamma^5 \epsilon \theta \gamma^\mu \theta N_{,\mu}(x)$$

and hence

$$\delta L\theta^T \gamma^5 \epsilon \theta N(x) =$$

$$-2\alpha^T\gamma^5\epsilon\theta N(x) + \theta^T\gamma^5\epsilon\theta\alpha^T\gamma^5\epsilon\gamma^\mu\theta N_{,\mu}(x)$$

$$= 2\theta^T\gamma^5\epsilon\alpha N(x) - \theta^T\gamma^5\epsilon\theta\alpha^T\gamma^{\mu T}\gamma^5\epsilon\theta N_{,\mu}(x)$$

$$= 2\theta^T\gamma^5\epsilon\alpha N(x) + \theta^T\gamma^5\epsilon\theta\theta^T\gamma^5\epsilon\gamma^\mu\alpha N_{,\mu}(x)$$

Next,

$$L\theta^T\epsilon\gamma^\mu\theta V_\mu(x) =$$

$$2\gamma^5\epsilon\epsilon\gamma^\mu\theta V_\mu(x) + \gamma^\nu\theta\theta^T\epsilon\gamma^\mu\theta V_{\mu,\nu}(x)$$

$$= -2\gamma^5\gamma^\mu\theta V_\mu(x) + \gamma^\nu\theta\theta^T\epsilon\gamma^\mu\theta V_{\mu,\nu}(x)$$

so

$$\delta L\theta^T\epsilon\gamma^\mu\theta V_\mu(x) =$$

$$-2\alpha^T\epsilon\gamma^\mu\theta V_\mu(x) + \theta^T\epsilon\gamma^\mu\theta\alpha^T\gamma^5\epsilon\gamma^\nu\theta V_{\mu,\nu}(x)$$

$$= 2\theta^T\epsilon\gamma^\mu\alpha V_\mu(x) - \theta^T\epsilon\gamma^\mu\theta\alpha^T\gamma^{\nu T}\gamma^5\epsilon\theta V_{\mu,\nu}(x)$$

$$= 2\theta^T\epsilon\gamma^\mu\alpha V_\mu(x) + \theta^T\epsilon\gamma^\mu\theta\theta^T\gamma^5\epsilon\gamma^\nu\alpha V_{\mu,\nu}(x)$$

To simplify these formulas further, we make use of the fact that if M is any antisymmetric skew symmetric 4×4 matrix, then $\theta^T M\theta\theta^T\beta$ can be expressed as $\theta^T\epsilon\theta\theta^T\delta$ since there are only four linearly independent monmomials of the third degree in θ. We shall now determine $\delta \in \mathbb{C}^4$ in terms of M and $\beta \in \mathbb{C}^4$. To do so, we first consider a matrix M of the form

$$M = \begin{pmatrix} 0 & M_{12} \\ -M_{12}^T & 0 \end{pmatrix}$$

Then

$$\theta^T M\theta\theta^T\beta =$$

$$2\theta_{0:1}^T M_{12}\theta_{2:3}(\theta_{0:1}^T\beta_{0:1} + \theta_{2:3}^T\beta_{2:3})$$

$$= 2\beta_{0:1}^T\theta_{0:1}\theta_{0:1}^T M_{12}\theta_{2:3}$$

$$-2\beta_{2:3}^T\theta_{2:3}\theta_{2:3}^T M_{12}^T\theta_{0:1}$$

$$= 2\theta^0\theta^1\beta_{0:1}^T eM_{12}\theta_{2:3}$$

$$-2\theta^2\theta^3\beta_{2:3}^T eM_{12}^T\theta_{0:1}$$

Further,

$$\theta^T\epsilon\theta\theta^T\delta =$$

$$(\theta_{0:1}^T e\theta_{0:1} + \theta_{2:3}^T e\theta_{2:3})(\theta_{0:1}^T\delta_{0:1} + \theta_{2:3}^T\delta_{2:3})$$

$$= \theta_{0:1}^T e\theta_{0:1}\theta_{2:3}^T\delta_{2:3} + \theta_{2:3}^T e\theta_{2:3}\theta_{0:1}^T\delta_{0:1}$$

$$= 2\theta^0\theta^1\theta_{2:3}^T\delta_{2:3} + 2\theta^2\theta^3\theta_{0:1}^T\delta_{0:1}$$

Comparing the two expressions gives us

$$\delta_{0:1} = M_{12}e\beta_{2:3}, \delta_{2:3} = -M_{12}^T e\beta_{0:1}$$

Thus,

$$\delta = \begin{pmatrix} \delta_{0:1} \\ \delta_{2:3} \end{pmatrix} =$$

$$= \begin{pmatrix} 0 & M_{12}e \\ -M_{12}^T e & 0 \end{pmatrix} \beta$$

$$= \begin{pmatrix} 0 & M_{12} \\ -M_{12}^T & 0 \end{pmatrix} \begin{pmatrix} e & 0 \\ 0 & e \end{pmatrix} \beta$$

$$= M\epsilon\beta$$

In particular,

$$\theta^T \epsilon\gamma^\mu \theta\theta^T \beta = \theta^T \epsilon\theta\theta^T \epsilon\gamma^\mu \epsilon\beta$$

$$= -\theta^T \epsilon\theta\theta^T \gamma^{\mu T} \beta$$

Now let M be a 4×4 antisymmetric matrix of the form

$$M = \begin{pmatrix} M_{11} & 0 \\ 0 & M_{22} \end{pmatrix}$$

where

$$M_{11}^T = -M_{11}, M_{22}^T = -M_{22}$$

Then,

$$\theta^T M\theta\theta^T \beta =$$

$$(\theta_{0:1}^T M_{11}\theta_{0:1} + \theta_{2:3}^T M_{22}\theta_{2:3})(\theta_{0:1}^T \beta_{0:1} + \theta_{2:3}^T \beta_{2:3})$$

$$= (2(M_{11})_{12}\theta^0\theta^1 + 2(M_{22})_{12}\theta^2\theta^3)(\theta_{0:1}^T \beta_{0:1} + \theta_{2:3}^T \beta_{2:3})$$

$$= 2(M_{11})_{12}\theta^0\theta^1\theta_{2:3}^T \beta_{2:3}$$

$$+2(M_{22})_{12}\theta^2\theta^3\theta_{0:1}^T \beta_{0:1}$$

Comparing this to

$$\theta^T \epsilon\theta\theta^T \delta = 2\theta^0\theta^1\theta_{2:3}^T \delta_{2:3} + 2\theta^2\theta^3\theta_{0:1}^T \delta_{0:1}$$

gives us

$$\delta_{0:1} = (M_{22})_{12}\beta_{0:1},$$

$$\delta_{2:3} = (M_{11})_{12}\beta_{2:3}$$

and hence

$$\delta = \begin{pmatrix} (M_{22})_{12}I_2 & 0 \\ 0 & (M_{11})_{12}I_2 \end{pmatrix} \beta$$

In particular if $M = \gamma^5\epsilon$, $M_{11} = e$, $M_{22} = -e$ and we get

$$\theta^T \gamma^5\epsilon\theta\theta^T \beta = \theta^T \epsilon\theta\theta^T \delta$$

where

$$\delta = \begin{pmatrix} -I_2 & 0 \\ 0 & I_2 \end{pmatrix} \beta = -\gamma^5\beta$$

In short, we summarize our discussion as follows:

$$\delta L\theta^T \epsilon\gamma^\mu\theta V_\mu(x) =$$

$$= 2\theta^T \epsilon\gamma^\mu \alpha V_\mu(x) + \theta^T \epsilon\gamma^\mu\theta\theta^T \gamma^5\epsilon\gamma^\nu \alpha V_{\mu,\nu}(x)$$

$$= 2\theta^T \epsilon\gamma^\mu \alpha V_\mu(x)$$

$$-\theta^T \epsilon\theta\theta^T \gamma^{\mu T}\gamma^5\epsilon\gamma^\nu \alpha V_{\mu,\nu}(x)$$

$$= 2\theta^T \epsilon\gamma^\mu \alpha V_\mu(x) - \theta^T \epsilon\theta\theta^T \gamma^5\epsilon\gamma^\mu\gamma^\nu \alpha V_{\mu,\nu}(x)$$

$$= 2\theta^T \epsilon\gamma^\mu \alpha V_\mu(x)$$

$$-\theta^T \epsilon\theta\theta^T \gamma^5\epsilon([\gamma^\mu,\gamma^\nu]f_{\nu\mu}(x) + \gamma.\partial(\gamma.V))\alpha$$

and

$$\delta L\theta^T \epsilon\theta\theta^T M(x) =$$

$$\delta L\theta^T \epsilon\theta M(x) =$$

$$= 2\theta^T \epsilon\alpha M(x) + \theta^T \epsilon\theta\theta^T \gamma^5\epsilon\gamma.\partial M(x)\alpha,$$

$$\delta L\theta^T \gamma^5\epsilon\theta N(x) =$$

$$= 2\theta^T \gamma^5\epsilon\alpha N(x)+$$

$$-\theta^T \epsilon\theta\theta^T \epsilon(\gamma.\partial N)\alpha$$

Further,

$$L\theta^T \epsilon\theta\theta^T (\lambda + a\gamma.\partial\omega) =$$

From these equations we get the following formulas for the change in the components of the superfield under an infinitesimal supersymmetry transformation:

$$\delta C(x) = -\alpha^T \gamma^5\epsilon\gamma^5\omega(x) = -\alpha^T \epsilon\omega(x)$$

$$\epsilon\delta\omega(x) = 2\epsilon\alpha M(x) + 2\gamma^5\epsilon\alpha N(x) + 2\epsilon\gamma.V(x)\alpha$$

or equivalently,

$$\delta\omega(x) = 2(M(x) + \gamma^5 N(x) + \gamma.V(x))\alpha$$

Likewise,

$$L\theta^T \epsilon\theta\theta^T \beta(x)$$

$$= -2\gamma^5\theta\theta^T \beta(x) + \theta^T \epsilon\theta\gamma^5\epsilon\beta$$

$$+\theta^T \epsilon\theta\gamma^\mu\theta\theta^T \beta_{,\mu}$$

$$= (1/2)(\gamma^5\theta^T \epsilon\theta\epsilon\beta(x) + \gamma^5\theta^T \gamma^5\epsilon\theta\gamma^5\epsilon\beta(x)$$

$$+\gamma^5\theta^T \epsilon\gamma^\nu\theta\gamma_\nu\epsilon\beta(x))$$

$$+\theta^T \epsilon\theta\gamma^5\epsilon\beta(x) - (1/4)(\theta^T \epsilon\theta)^2\gamma^\mu\epsilon\beta_{,\mu}(x)$$

It follows therefore that

$$\delta L(\theta^T \epsilon \theta \theta^T \gamma^5 \epsilon \beta(x)) =$$
$$\alpha^T \gamma^5 \epsilon L(\theta^T \epsilon \theta \theta^T \gamma^5 \epsilon \beta(x)) =$$
$$-(1/2)(\alpha^T \epsilon \theta^T \epsilon \theta \gamma^5 \beta(x) + \alpha^T \epsilon \theta^T \gamma^5 \epsilon \theta \beta(x)$$
$$+\alpha^T \epsilon \theta^T \epsilon \gamma^\nu \theta \gamma_\nu \gamma^5 \beta(x))$$
$$-\theta^T \epsilon \theta \alpha^T \gamma^5 \epsilon \beta(x) + (1/4)(\theta^T \epsilon \theta)^2 \alpha^T \gamma^5 \epsilon \gamma^\mu \gamma^5 \beta_{,\mu}(x) =$$
$$(1/2)(\theta^T \epsilon \theta (\gamma^5 \epsilon \alpha)^T \beta(x) + \theta^T \gamma^5 \epsilon \theta (\epsilon \alpha)^T \beta(x)$$
$$+\theta^T \epsilon \gamma^\mu \theta (\gamma^5 \epsilon \gamma_\mu \alpha)^T \beta(x))$$
$$+\theta^T \epsilon \theta (\gamma^5 \epsilon \alpha)^T \beta(x) + (1/4)(\theta^T \epsilon \theta)^2 (\epsilon \gamma^\mu \alpha)^T \beta_{,\mu}(x)$$

By setting $\beta = \gamma + a.\gamma.\partial \omega$, we therefore get

$$\delta M(x) = (3/2)(\gamma^5 \epsilon \alpha)^T \beta(x) + (1/4)(\gamma^5 \epsilon \alpha)^T \gamma.\partial \omega,$$

$$\delta N(x) = (1/2)(\epsilon \alpha)^T \beta(x) + (1/4)(\epsilon \alpha)^T \gamma.\partial \omega,$$

$$\delta V_\mu(x) = (1/2)(\gamma^5 \epsilon \gamma_\mu \alpha)^T \beta(x) + (1/4)\alpha^T \gamma^5 \epsilon \gamma^\nu \gamma_\mu \omega_{,\nu}$$
$$= (1/2)\alpha^T \gamma^5 \epsilon \gamma_\mu \beta(x) + (1/4)\alpha^T \gamma^5 \epsilon \gamma^\nu \gamma_\mu \omega_{,\nu}$$
$$= (1/2)\alpha^T \gamma^5 \epsilon \gamma_\mu (\lambda + a.\gamma.\partial \omega)+$$
$$(1/4)\alpha^T \gamma^5 \epsilon (\{\gamma^\nu, \gamma_\mu\} - \gamma_\mu \gamma^\nu)\omega_{,\nu} =$$
$$(1/2)\alpha^T \gamma^5 \epsilon \gamma_\mu \lambda+$$
$$+(a/2)\alpha^T \gamma^5 \epsilon \gamma_\mu \gamma.\partial \omega$$
$$+(1/4)\alpha^T \gamma^5 \epsilon (2\delta^\nu_\mu \omega_{,\nu} - \gamma_\mu \gamma.\partial \omega)$$
$$= (1/2)\alpha^T \gamma^5 \epsilon \gamma_\mu \lambda+$$
$$+(a/2)\alpha^T \gamma^5 \epsilon \gamma_\mu \gamma.\partial \omega$$
$$+(1/4)\alpha^T \gamma^5 \epsilon (2\omega_{,\mu} - \gamma_\mu \gamma.\partial \omega)$$

Taking $a = 1/2$ gives us

$$\delta V_\mu(x) = (1/2)\alpha^T \gamma^5 \epsilon \gamma_\mu \lambda + (1/2)\alpha^T \gamma^5 \epsilon \omega_{,\mu}$$

It follows that

$$\delta f_{\mu\nu} = \delta V_{\nu,\mu} - \delta V_{\mu,\nu} =$$
$$(1/2)\alpha^T \gamma^5 \epsilon (\gamma_\nu \lambda_{,\mu}(x) - \gamma_\mu \lambda_{,\nu}(x))$$

We now compute $\delta \lambda(x)$ and $\delta D(x)$. We have with $\beta(x) = \lambda(x) + a.\gamma.\partial \omega(x)$,

$$\theta^T \epsilon \theta \theta^T \gamma^5 \epsilon \delta \beta(x)$$
$$= \alpha^T \gamma^5 \epsilon \partial_\theta (\theta^T \epsilon \theta)^2 (b \Box C(x) + D(x))$$
$$-\theta^T \epsilon \theta \gamma^5 \epsilon \gamma.\partial M(x)\alpha,$$

$$-\theta^T \epsilon \theta \theta^T \epsilon (\gamma . \partial N)\alpha$$

$$-\theta^T \epsilon \theta \theta^T \gamma^5 \epsilon ([\gamma^\mu, \gamma^\nu] f_{\nu\mu}(x) + \gamma . \partial(\gamma . V))\alpha$$

$$= -2\gamma^5 (\theta^T \epsilon \theta)(b \Box C(x) + D(x))$$

$$-\theta^T \epsilon \theta \gamma^5 \epsilon \gamma . \partial M(x)\alpha,$$

$$-\theta^T \epsilon \theta \theta^T \epsilon (\gamma . \partial N)\alpha$$

$$-\theta^T \epsilon \theta \theta^T \gamma^5 \epsilon ([\gamma^\mu, \gamma^\nu] f_{\nu\mu}(x) + \gamma . \partial(\gamma . V))\alpha$$

We deduce that

$$\gamma^5 \epsilon \delta(\lambda(x) + a.\gamma . \partial \omega(x)) =$$

$$= -2\gamma^5 (b \Box C(x) + D(x))$$

$$-\gamma^5 \epsilon \gamma . \partial M(x)\alpha$$

$$-\epsilon(\gamma . \partial N)\alpha$$

$$-\gamma^5 \epsilon([\gamma^\mu, \gamma^\nu] f_{\nu\mu}(x) + \gamma . \partial(\gamma . V))\alpha$$

Chapter 3

Conducting fluids as quantum antennas

3.1 A short course in basic non-relativistic and relativistic fluid dynamics with antenna theory applications

3.1.1 The basic physical quantities of a fluid

Density, pressure, velocity field, momentum density, mass flux, momentum flux, energy density, entropy density, enthalpy density, temperature field.

Density $\rho(t, r), r = (x, y, z)$ is the mass per unit volume in a fluid at the point r at time t. If $M(t, r, \delta V)$ is the amount of fluid mass at time t within a volume δV that surrounds r, then

$$\rho(t, r) = lim_{|\delta V| \to 0} \frac{M(t, r, \delta V)}{|\delta V|}$$

3.1.2 Parameters of a fluid

Viscosity, conductivity, density if the fluid is incompressible, parameters on which the basic physical quantities can depend and which can be estimated from the equations of motion.

3.1.3 Ordinary time derivative and material derivative of a physical quantity associated to a fluid

q is a physical quantity whose density is $Q(t, r)$. The rate of change of Q at the fixed point r in the fluid as different fluid particles enter this point and leave this point is given by $\frac{\partial Q(t,r)}{\partial t}$. Its rate of change along the trajectory of a fixed fluid particle which is at r at time t and at $r + dr$ at time $t + dt$ is given by

$$DQ/Dt = lim_{\delta t \to 0} \frac{Q(t + dt, r + dr) - Q(t, r)}{dt}$$

$$= \frac{\partial Q(t, r)}{\partial t} + (dr/dt, \nabla_r)Q(t, r)$$

$$= \frac{\partial Q(t, r)}{\partial t} + (v(t, r), \nabla_r)Q(t, r)$$

This is called the material derivative of Q.

3.1.4 The mass conservation equation/equation of continuity

The general conservation equation for the density of any physical quantity taking into account the rate of its generation per unit volume.

3.1.5 The momentum equation starting from Newton's second law of motion

. Eulerian equations for a non-viscous fluid. Momentum equation taking into account the stress tensor of a fluid, special case of diagonal stress tensor-the pressure term.

3.1.6 The momentum equation derived from conservation of momentum flux

The momentum flux tensor is

$$T^{ij} = \rho v^i v^j$$

and the momentum density or equivalently the mass flux is

$$T^{0i} = T^{i0} = \rho v^i$$

Let σ^{ij} denote the stress tensor due to pressure and viscous forces. Then we have the obvious momentum conservation law/second law of motion,

$$T^{i0}_{,0} + T^{ij}_{,j} = \sigma^{ij}_{,j}$$

or in the dyad notation,

$$(\rho v^i)_{,0} + div\mathbf{T} = div\sigma$$

This is the differential form of the integral form

$$\frac{d}{dt} \int_V T^{i0} d^3r + \int_{\partial V} T^{ij} n_j dS = \int_{\partial V} \sigma^{ij} n_j dS$$

where (n_j) is the unit normal to the surface ∂V that bounds the volume V. For a fluid, we can express the stress tensor as

$$\sigma^{ij} = \eta(v^i_{,j} + v^j_{,i}) + \chi(divv)\delta^{ij} - p\delta^{ij}$$

where η, χ are coefficients of viscosity and p is the pressure. Then, the above momentum conservation law gives

$$(\rho v^i)_{,0} + (\rho v^i v^j)_{,j} = \eta\nabla^2 v^i + (\eta + \chi)(divv)_{,i} - p_{,i}$$

and on combining this with the equation of continuity

$$\rho_{,0} + (\rho v^i)_{,i} = 0$$

we get the Navier-Stokes equation

$$\rho v^i_{,0} + \rho v^j v^i_{,j} = \eta\nabla^2 v^i + (\eta + \chi)(divv)_{,i} - p_{,i}$$

or in vector form

$$\rho v_{,0} + \rho(v, \nabla)v = -\nabla p + \eta\nabla^2 v + (\eta + \chi)\nabla(divv)$$

3.1.7 Viscosity, the strain rate, shear and bulk viscosity, the Navier-Stokes equation for a viscous fluid, Expression of the Navier-Stokes equation in different orthogonal curvilinear coordinate systems

3.1.8 The fluid dynamical equations taking viscous and thermal effects into account, thermal conductivity, the energy equation for a fluid based on the first and second laws of thermodynamics

3.1.9 Vorticity, the vorticity equation, irrotational flows, the velocity potential, Bernoulli equation

3.1.10 Incompressible flows, the stream function, Bernoulli equation along a streamline for rotational flows

Here we have the following situation:

$$divv = 0,$$

which follows from the equation of continuity with $\rho = constt.$ and the Navier-Stokes equation

$$(v, \nabla)v + v_{,t} = -\nabla p/\rho + \nu\nabla^2 v$$

where
$$\nu = \eta/\rho$$

The incompressibility condition implies the existence of a stream function field $\psi(t, r) \in \mathbb{R}^3$ such that
$$v = \nabla \times \psi$$

Also, we may assume that $div\psi = 0$, for replacing ψ by $\psi + \nabla f$, we do not alter $v = \nabla \times \psi$ and we can choose f so that
$$0 = div(\psi + \nabla f) = div\psi + \nabla^2 f$$

ie
$$f(t, r) = (4\pi)^{-1} \int |r - r'|^{-1} div\psi(t, r') d^3 r'$$

The Navier-Stokes equation can be expressed as
$$\Omega \times v + \nabla v^2/2 + v_{,t} = -\nabla p/\rho + \nu \nabla^2 v - \nabla V$$

where
$$\Omega = \nabla \times v$$

is the vorticity and $V(t, r)$ is the potential of the external force, like for example, gravity. We note that this equation is still valid if the fluid is not incompressible. Further, in the incompressible case, it follows by taking the scalar product with a unit vector field $\hat{l}$ parallel to the velocity field that

$$(1/2)\frac{\partial v^2}{\partial l} + v_{,t}.\hat{l} + \rho^{-1}\frac{\partial p}{\partial l} + \nu\hat{l}.\nabla^2 v - \frac{\partial V}{\partial l} = 0$$

In the special case when apart from being incompressible, the fluid is in a steady state, ie, $v_{,t} = 0$ and is also non-viscous, ie, $\eta = 0$, it follows from the above that
$$\frac{\partial}{\partial l}(v^2/2 + p/\rho + V) = 0$$

and hence the quantity
$$p/\rho + v^2/2 + V$$

is a constant along each streamline. This is a slightly restricted form of Bernoulli's principle. If apart from being incompressible, the fluid is irrotational so that $\Omega = 0$ and hence $v = -\nabla\phi$ where ϕ is the velocity potential, then we get
$$\nabla(p/\rho + v^2/2 - \phi_{,t}V + \eta\nabla^2\phi) = 0$$

This is valid even if the fluid is viscous and not in steady state. But the incompressiblity condition gives $\nabla^2\phi = divv = 0$ and hence we get the following version of the Bernoulli principle for incompressible irrotational vlows that are not necessarily in steady state:
$$p/\rho + v^2/2 - \phi_{,t} + V = constt.$$

Actually, the constant on the rhs can depend on time but not on space and writing this constant as $c(t)$, we can redefine our velocity potential as $\phi(t, r) - \int_0^t c(s)ds$ without affecting the velocity field to get

$$p/\rho + v^2/2 - \phi_{,t} + V = 0$$

at all points in the fluid and at all times. This is the most general form of the Bernoulli equation.

3.1.11 Some examples of fluid flows

[a] Derivation of Stokes' law for the damping force on a sphere moving inside a viscous fluid.

Exercise: Assume that a sphere of radius R is placed in a viscous fluid having a constant velocity field $v_0\hat{z}$. After placing the sphere, the velocity field of the fluid changes to $v(r, \theta) = v_r(r, \theta)\hat{r} + v_\theta(r, \theta)\hat{\theta}$. The boundary conditions on this modified velocity field is that $lim_{r\to\infty} v(r, \theta) = v_0\hat{z}$ and $v_r = 0$ when $r = R$. Assuming the fluid field to be irrotational, we can derive it from a velocity potential $\phi(r, \theta)$ as

$$v = \nabla\phi = \phi_{,r}\hat{r} + r^{-1}\phi_{,\theta}\hat{\theta}$$

The incompressibility condition $\nabla.v = 0$ becomes

$$\nabla^2\phi = 0$$

and hence in view of the azimuthal symmetry of ϕ, it can be expanded using the Legendre polynomials as

$$\phi(r, \theta) = v_0 r cos(\theta) + \sum_{l\geq 0} c(l) r^{-l-1} P_l(cos(\theta)), r \geq R$$

Note that as $r \to \infty$, we must have $\nabla\phi \to v_0\hat{z}$ which means that $\phi \to v_0 z = v_0 r cos(\theta)$ which is why we have omitted the terms $r^l P_l(cos(\theta)), l \geq 2$ in the Legendre series expansion. Application of the boundary conditions gives that $\phi_{,r}(R, \theta) = 0$ (ie the normal component of the velocity field on the surface of the sphere vanishes)

$$(v, \nabla)v = -\nabla p/\rho + (\eta/\rho)\nabla^2 v$$

becomes

$$\Omega \times v + \nabla v^2/2 + \nabla p/\rho = (\eta/\rho)\nabla^2 v$$

where $\Omega = \nabla \times v = 0$ and $\nabla^2 v = \nabla\nabla^2\phi = 0$ so that this equation can be satisfies by satisfying

$$\nabla^2(v^2/2 + p/\rho) = 0$$

or equivalently,

$$p = -\rho v^2/2 + C$$

where C is a constant. This determines the pressure field once the velocity field is known. Now applying the above boundary condition $\phi_{,r}(R, \theta) = 0$ gives us

$$c(0) = 0, v_0 - 2c(1)R^{-3} = 0, c(l) = 0, l \geq 2$$

Thus,

$$c(1) = v_0 R^3 / 2$$

Thus, the complete solution for the velocity potential is given by

$$\phi = v_0 r cos(\theta) + v_0 R^3 cos(\theta)/2r^2$$

$$= v_0 cos(\theta)(r + R^3/2r^2)$$

Hence the non-vanishing components of the velocity field in the spherical polar coordinate system are given by

$$v_r(r, \theta) = \phi_{,r}(r, \theta) = v_0 cos(\theta)(1 - R^3/r^3)r \geq R$$

$$v_\theta(r, \theta) = r^{-1}\phi_{,\theta}(r, \theta) = -v_0 sin(\theta)(1 + R^3/2r^3), r \geq R$$

Now using this velocity field, we can compute the spherical polar components of the viscous stress tensor, ie, the force per unit area on the surface of the sphere at each point defined as

$$\sigma_{nm} = \sigma_{ij} n_i m_j$$

where n, m are unit vectors and the summation convention is adopted. In particular, after integrating over the surface area of the sphere, we are left with only the z component of the viscous force on the sphere, namely,

$$f_z = \int_S (\sigma_{zx} n_x + \sigma_{zy} n_y + \sigma_{zz} n_z) dS$$

where

$$\sigma_{ab} = \eta(v_{a,b} + v_{b,a})$$

in cartesian coordinates and the components of the stress tensor are evaluated at $r = R, \theta, \phi, dS = sin(\theta)d\theta d\phi$ and

$$n_x = cos(\phi)sin(\theta), n_y = sin(\phi)sin(\theta), n_z = cos(\theta)$$

Perform this computation and hence prove Stokes' formula

$$f_z = 6\pi\eta R v_0$$

   Remark: Suppose we impose the boundary condition that $v_\theta = 0$ when $r = R$ rather than $v_r = 0$ when $r = R$. This would imply that $\phi(R, \theta) = constt.$ and we would get a different solution, namely,

$$\phi(r, \theta) = v_0 cos(\theta)(r - R^2/r)$$

This boundary condition implies that the surface of the sphere is an equipotential surface. However, it is natural to believe that the normal component of the velocity of the fluid on the sphere surface is zero for otherwise, a locally turbulent flow would result.

[b] Steady flow from a conical jet. The velocity field in spherical polar coordinates has the form

$$v = v_r(r, \theta)\hat{r} + v_\theta(r, \theta)\hat{\theta}$$

and we have

$$(v, \nabla)v = v_r(v_{r,r}\hat{r} + v_{\theta,r}\hat{\theta})$$

$$+ r^{-1}v_\theta(v_{r,\theta}\hat{r} + v_{\theta,\theta}\hat{\theta})$$

$$+ r^{-1}v_\theta(v_r\hat{\theta} - v_\theta\hat{r})$$

where we use

$$\hat{\theta}_{,\theta} = -\hat{r}, \hat{r}_{,\theta} = \hat{\theta}$$

$$\hat{r}_{,r} = 0, \hat{\theta}_{,r} = 0$$

$$\nabla^2 v = v_{,rr} + 2v_{,r}/r + r^{-2}v_{,\theta\theta} + r^{-2}cos(\theta)v_{,\theta}$$

Now observe that

$$v_{,r} = v_{r,r}\hat{r} + v_{\theta,r}\hat{\theta},$$

$$v_{,rr} = v_{r,rr}\hat{r} + v_{\theta,rr}\hat{\theta},$$

$$v_{,\theta} = v_{r,\theta}\hat{r} + v_{\theta,\theta}\hat{\theta}$$

$$+ v_r\hat{\theta} - v_\theta\hat{r}$$

$$= (v_{r,\theta} - v_\theta)\hat{r} + (v_{\theta,\theta} + v_r)\hat{\theta}$$

$$v_{,\theta\theta} = (v_{r,\theta\theta} - 2v_{\theta,\theta} - v_r)\hat{r}$$

$$+ (v_{r,\theta} - v_\theta - v_{\theta,\theta\theta} + v_r)\hat{\theta}$$

Also, for the pressure $p(r, \theta)$, we have

$$\nabla p = p_{,r}\hat{r} + r^{-1}p_{,\theta}\hat{\theta}$$

Further,

$$divv = r^{-2}(r^2 v_r)_{,r} + (r.sin(\theta))^{-1}(sin(\theta)v_\theta)_{,\theta}$$

Exercise:Substitute these expressions into the Navier-Stokes and incompressibility equations

$$(v, \nabla)v = -\nabla p/\rho + \nu\nabla^2 v, divv = 0$$

to derive three partial differential equations in r, θ for the three functions v_r, v_θ, p. Derive numerical algorithms for solving these.

Remark:Using the incompressibility equation, we can replace the two functions v_r, v_θ by a single function $\psi(r, \theta)$ so that

$$r^2 sin(\theta) v_r = \psi_{,\theta}, \quad r.sin(\theta) v_\theta = -\psi_{,r}$$

Then we get two pde's for two functions ψ, p of (r, θ).

[c] Flow in a cylindrical cup after stirring. Here in terms of cylindrical coordinates (ρ, ϕ, z), by cylindrical symmetry, we introduce the velocity field

$$v = v_\rho(t, \rho, z)\hat\rho + v_\phi(t, \rho, z)\hat\phi + v_z(t, \rho, z)\hat z$$

Then note that

$$\nabla^2 v = (\rho^{-1}\partial_\rho + \partial_\rho^2 + \rho^{-2}\partial_\phi^2 + \partial_z^2)(v_\rho\hat\rho + v_\phi\hat\phi + v_z\hat z)$$

$$= \hat\rho(\rho^{-1}v_{\rho,\rho} + v_{\rho,\rho\rho} + v_{\rho,zz} - \rho^{-2}v_\rho)$$

$$+\hat\phi(\rho^{-1}v_{\phi,\rho} + v_{\phi,\rho\rho} + v_{\phi,zz} - \rho^{-2}v_\phi)$$

$$+\hat z(\rho^{-1}v_{z,\rho} + v_{z,\rho\rho} + v_{z,zz})$$

Also,

$$(v, \nabla)v = (v_\rho\partial_\rho + \rho^{-1}v_\phi\partial_\phi + v_z\partial_z)(v_\rho\hat\rho + v_\phi\hat\phi + v_z\hat z)$$

$$= \hat\rho(v_\rho v_{\rho,\rho} - \rho^{-1}v_\phi^2 + v_z v_{\rho,z})$$

$$+\hat\phi(v_\rho v_{\phi,\rho} + \rho^{-1}v_\phi v_\rho + v_z v_{\phi,z})$$

$$+\hat z(v_\rho v_{z,\rho} + v_z v_{z,z})$$

Now substitute these expressions into the Navier-Stokes equations

$$(v, \nabla)v + v_{,t} = -\nabla p/\rho_0 + \nu\nabla^2 v - g\hat z$$

where $p = p(t, \rho, z)$ so that

$$\nabla p = p_{,\rho}\hat\rho + p_{,z}\hat z$$

to obtain three pde's for the four functions v_ρ, v_ϕ, v_z, p. The fourth equation is provided by the incompressibility condition:

$$\rho^{-1}(\rho v_\rho)_{,\rho} + v_{z,z} = 0$$

This equation implies that the two functions v_ρ, v_z can be replaced by a single function $\psi(t, \rho, z)$ using

$$\rho v_\rho = \psi_{,z}, \quad \rho v_z = -\psi_{,\rho}$$

Doing so, we end up with three pde's for the three functions ψ, p, v_ϕ of (t, ρ, z).

3.1.12 Two dimensional incompressible, irrotational flows described using the velocity potential and the stream function, solution with different boundary conditions using analytic functions of a complex variable, Cauchy-Riemann equations and proof of the orthogonality of the streamlines and constant velocity potential lines, formulation of incompressible and irrotational flows as solutions to the 2-D Laplace equation, sources and sinks

Exercise: Show by integrating $r^{-1} = (\rho^2 + z^2)^{-1}$ w.r.t. z over $\mathbb{R}$ and by using the fact that

$$\nabla^2 r^{-1} = -4\pi\delta(x)\delta(y)\delta(z)$$

that

$$\nabla^2 log(\rho) = 2\pi\delta(x)\delta(y)$$

Now consider a two dimensional incompressible flow. incompressibility implies the existence of a stream function $\psi(t, x, y)$ such that

$$v = \nabla\psi \times \hat{z}$$

or equivalently,

$$v_x(t, x, y) = \psi_{,y}, v_y(t, x, y) = -\psi_{,x}$$

For the vorticity, we have

$$\Omega = \nabla \times v = -\nabla^2\psi\hat{z}$$

and hence, we get on taking the curl of the Navier-Stokes equation,

$$\nabla \times (\Omega \times v) + \Omega_{,t} = \nu\nabla^2\Omega + g$$

where

$$g(t, x, y) = \nabla \times f(t, x, y)/\rho$$

with f as the external force per unit volume (in the xy plane). We can express this equation as

$$\nabla^2\psi_{,t} + (\hat{z}, \nabla \times (\nabla^2\psi\hat{z} \times (\nabla\psi \times \hat{z}))) = \nu(\nabla^2)^2\psi - g$$

Explicitly, show that this evaluates to

$$\nabla^2\psi_{,t} + (\hat{z}, \nabla^2\nabla\psi \times \nabla\psi) - \nu(\nabla^2)^2\psi - g$$

or equivalently,

$$\nabla^2\psi_{,t} + \psi_{,y}\nabla^2\psi_{,x} - \psi_{,x}\nabla^2\psi_{,y}$$
$$-\nu(\nabla^2)^2\psi + g = 0$$

Deduce that if K is the linear operator acting on functions on $\mathbb{R}^2$ with kernel

$$K(x - x', y - y') = 4\pi log((x - x')^2 + (y - y')^2)$$

then

$$\psi_{,t}(t, x, y) + K.(\psi_{,y}\nabla^2\psi_{,x} - \psi_{,x}\nabla^2\psi_{,y})(t, x, y) - \nu\nabla^2\psi(t, x, y) + g(t, x, y) = 0$$

Now introduce a perturbation parameter δ into the nonlinear term in the above equation and show that by expanding the stream function in a perturbation series as

$$\psi(t, x, y) = \sum_{n \geq 0} \delta^n \psi_n(t, x, y)$$

that

$$\psi_{0,t} - \nu\nabla^2\psi_0 + g = 0,$$

$$\psi_{n,t} - \nu\nabla^2\psi_n + \sum_{k=0}^{n-1} K.(\psi_{k,y}\nabla^2\psi_{n-k,x} - \psi_{k,x}\nabla^2\psi_{n-k,y}) = 0, n \geq 1$$

3.2   Flow of a 2-D conducting fluid

Now consider a 2-D conducting fluid in an electric field having the form

$$E = E_x(t, x, y)\hat{x} + E_y(t, x, y)\hat{y}$$

and a magnetic field of the form

$$B = B_z(t, x, y)\hat{z}$$

Write down the planar equations of motion

$$(v, \nabla v) + v_{,t} = -\nabla p/\rho + (\sigma/\rho)(E + v \times B) + \nu\nabla^2 v$$

in terms of $\psi(t, x, y), E_x(t, x, y), E_y(t, x, y), B_z(t, x, y)$ along with the Maxwell equations

$$E_{x,x} + E_{y,y} = 0$$

$$E_{y,x} - E_{x,y} = -B_{z,t},$$

$$-B_{z,x} = E_{y,t} + \sigma(E_y - v_x B_z), B_{z,y} = E_{x,t} + \sigma(E_x + v_y B_z)$$

The first Maxwell equation implies the existence of a scalar field $\chi(t, x, y)$ such that

$$E_x = \chi_{,y}, E_y = -\chi_{,x}$$

Now show using the above that we have exactly three equations for the three functions ψ, χ, B_z. Note that p has been eliminated by taking the curl of the Navier-Stokes equation, ie, we use $\Omega = -\nabla^2\psi\hat{z}$ and

$$\nabla \times (\Omega \times v) + \Omega_{,t} = -\nu\nabla^2\Omega + \nabla \times (E + v \times B)$$

Note that

$$\nabla \times (E + v \times B) = \hat{z}(-B_{z,t} - (v_x B_z)_{,x} - (v_y B_z)_{,y})$$

$$= -\hat{z}(B_{z,t} + v_x B_{z,x} + v_y B_{z,y})$$

with

$$v_x = \psi_{,y}, v_y = -\psi_{,x}$$

Explain using the $K = (\nabla^2)^{-1}$ kernel, how we get dynamical equations for ψ, χ, B_z that are first order in time.

3.2.1 Boundary conditions for non-viscous and viscous conducting fluids

3.2.2 The Reynold number, onset of turbulence, dimensional analysis of the conducting fluid equations

3.3 Finite element method for solving the fluid dynamical equations

Consider for example, the flow of a fluid in one dimension. Its velocity field is $v(t, x) \in \mathbb{R}$ and its density field is $\rho(t, x) \in \mathbb{R}$. Let the equation of state be $p = p(\rho)$. Then, the equations of motion and, equation of continuity can be expressed as

$$\rho(v_{,t} + vv_{,x}) = -p'(\rho)\rho_{,x} + \eta v_{,xx} + f(t, x)$$

$$(\rho v)_{,x} + \rho_{,t} = 0,$$

where f is the external force per unit volume. We can assume that in the unperturbed state ρ and v are constants with $f = 0$ and write the perturbed values of these as $\rho_0 + \delta\rho(t, x)$ and $v_0 + \delta v(t, x)$ respectively. Then expanding upto second degree terms and discretizing the space and time indices results in second order Volterra difference equation for $\delta\rho$ and δv. Keeping this in mind, we study finite register effects in the simulation of vector valued two dimensional second degree Volterra difference equations of the form

$$Y[n, m] = \sum_{k,m=1}^{p} A_1[k, l]Y[n-k, m-l] + \sum_{k,l,r,s=1}^{p} A_2[k, l, r, s](Y[n-k, m-l] \otimes Y[n-r, m-s])$$

$$+ \sum_{k,l=0}^{p} B_1[k, l]X[n-k, m-l] + \sum_{k,l,r,s=0}^{p} B_2[k, l, r, s](X[n-k, m-l] \otimes X[n-r, m-s])$$

The effects of truncation and rounding in the implementation of this difference equation on a digital computer is to modify this difference equation to

$$Y[n, m] = \sum_{k,l=1}^{p} A_1[k, l]Y[n-k, m-l] + \sum_{k,l,r,s=1}^{p} A_2[k, l, r, s](Y[n-k, m-l] \otimes Y[n-r, m-s])$$

$$+ \sum_{k,l=0}^{p} B_1[k,l]X[n-k,m-l] + \sum_{k,l,r,s=0}^{p} B_2[k,l,r,s](X[n-k,m-l] \otimes X[n-r,m-s])$$

$$+ \sum_{k,l} E_{1,kl}[n,m] + \sum_{k,l,r,s} A_2[k,l,r,s]E_{2,klrs}[n,m] + \sum_{klrs} E_{3,klrs}[n,m] +$$

$$\sum_{k,l} E_{4,kl}[n,m] + \sum_{klrs} B_2[k,l,r,d]E_{5,klrs}[n,m] + \sum_{klsr} E_{6,klrs}[n,m]$$

where $E_{j,kl}[n,m]$, $E_{j,klrs}[n,m]$ are truncation and rounding error processes assumed to have independent components uniformly distributed over $[0, \Delta]$ or $[-\Delta/2, \Delta/2]$ according as truncation or rounding is used where $\Delta = 2^{-b}$ with b denoting the number of bits in each register. Writing

$$Y[n,m] = Y_x[n,m] + Y_e[n,m]$$

where Y_x is the signal part of the output that has a larger amplitude than the noise part $Y_e[n,m]$ of the output, we get on equating the respectively the signal part and first order noise parts,

$$Y_x[n,m] = \sum_{k,l=1}^{p} A_1[k,l]Y[n-k,m-l] + \sum_{k,l,r,s=1}^{p} A_2[k,l,r,s](Y[n-k,m-l] \otimes Y[n-r,m-s])$$

$$+ \sum_{k,l=0}^{p} B_1[k,l]X[n-k,m-l] + \sum_{k,l,r,s=0}^{p} B_2[k,l,r,s](X[n-k,m-l] \otimes X[n-r,m-s])$$

$$Y_e[n,m] =$$

$$\sum_{klrs} A_2[k,l,r,s](Y_x[n-k,m-l] \otimes Y_e[n-r,m-s] + Y_e[n-k,m-l] \otimes Y_x[n-r,m-s]) +$$

$$+ \sum_{k,l} E_{1,kl}[n,m] + \sum_{k,l,r,s} A_2[k,l,r,s]E_{2,klrs}[n,m] + \sum_{klrs} E_{3,klrs}[n,m] +$$

$$\sum_{k,l} E_{4,kl}[n,m] + \sum_{klrs} B_2[k,l,r,d]E_{5,klrs}[n,m] + \sum_{klsr} E_{6,klrs}[n,m]$$

The second equation can be used to derive a difference equation for the output noise autocorrelation function.

3.4   Elimination of pressure, incompressible fluid dynamics in terms of just a single stream function vector field with vanishing divergence

$$div\,v = 0$$

This is the incompressibility equation. Thsube Navier-Stokes equation is

$$v_{,t} + (v, \nabla)v = -\nabla p/\rho + \nu \nabla^2 v + f$$

where f is the external force per unit mass. This equation can be rearranged as

$$v_{,t} + \Omega \times v + (1/2)\nabla(v^2) = -\nabla p/\rho + \nu \nabla^2 v + f$$

where

$$\Omega = \nabla \times v$$

is the fluid vorticity. The incompressibility condition implies

$$v = \nabla \times \psi$$

where we may assume $div\psi = 0$ in view of the fact that ψ may be replaced by $\psi + \nabla \eta$ where η is any scalar field without affecting v. Hence, we get

$$\Omega = -\nabla^\psi$$

and our N.S. equation on taking the curl gives us a nonlinear pde for the stream function vector field ψ:

$$\Omega_{,t} + \nabla \times (\Omega \times v) = \nu \nabla^2 \omega + g, g = \nabla \times f$$

or

$$\nabla^2 \psi_{,t} + \nabla \times ((\nabla^2 \psi) \times (\nabla \times \psi)) = \nu(\nabla^2)^2 \psi - g$$

which can be inverted using the Green's function $G(r) = -1/4\pi|r|$ for ∇^2:

$$\nabla^2 \int G(r - r')\eta(r')d^3r' = \eta(r)$$

We get

$$\psi_{,t} + G.(\nabla \times ((\nabla^2 \psi) \times (\nabla \times \psi))) = -\nu \nabla^2 \psi - G.g$$

3.5   Fluids driven by random external force fields

If in the above equation f is a random field, then so is $g(t, r)$ and assuming g to be a zero mean Gaussian field with autocorrelation $R_g(t, r|t', r') = <g(t, r).g(t', r')>$, our aim is to calculate all the moments of the stream function field $< \psi(t_1, r_1) \otimes ...\psi(t_n, r_n) >$ from which all the statistical moments of the velocity field can be evaluated. This is achieved by solving for ψ using perturbation theory:

$$\psi_{,t} + \delta.G.\nabla \times ((\nabla^2 \psi) \times (\nabla \times \psi)) = \nu(\nabla^2)\psi - G.g$$

$$\psi = \psi_0 + \delta.\psi_1 + \delta^2.\psi_2 + \dots$$

so that equating coefficients of $\delta^n, n = 0, 1, \dots$ successively gives us

$$\psi_{0,t} - \nu\nabla^2\psi_0 + G.g = 0$$

$$\psi_{n+1,t} - \nu\nabla^2\psi_{n+1} = G.\nabla \times ((\nabla^2\psi_n) \times (\nabla \times \psi_n)), n = 0, 1, \dots$$

so that if $K(t, r)$ denotes the Green's function of the diffusion equation, ie,

$$K_{,t}(t, r) - \nu\nabla^2 K(t, r) = \delta^3(r)$$

then we have

$$\psi_0(t, r) = -K.G.g(t, r) = -\int K(t - t', r - r')G.g(t', r')d^3r'd^3r',$$

$$\psi_{n+1}(t, r) = \nu.K.G.\nabla \times ((\nabla^2\psi_n) \times (\nabla \times \psi_n))(t, r), n = 1, 2, \dots$$

The convergence of this perturbation series needs to be investigated.

3.6 Relativistic fluids, tensor equations

Let $T^{\mu\nu}$ denote the energy-momentum tensor of the fluid field without taking viscous and thermal effects into account and let $\Delta T^{\mu\nu}$ denote the contribution to the energy-momentum tensor due to viscous and thermal effects. Then the basic special relativistic fluid equations are obtained by energy-momentum conservation:

$$(T^{\mu\nu} + \Delta T^{\mu\nu})_{,\nu} = f^\mu$$

where f^μ is the externally applied four force. Let v^μ denote the four velocity $dx^\mu/d\tau$ with $d\tau = dt\sqrt{1 - u^2}$ where $u = (u^r)^3_{r=1}$ with $u^r = dx^r/dt = v^r d\tau/dt$. Then

$$T^{\mu\nu} = (\rho + p)v^\mu v^\nu - p\eta^{\mu\nu}$$

We leave it as an exercise to write down the above energy-momentum conservation equations and separate it out into the momentum conservation part and the mass conservation part. A nice account of how to calculate $\Delta T^{\mu\nu}$ using particle number conservation $(nv^\mu)_{,\mu} = 0$ and the first and second laws of thermodynamics

$$TdS = dU + pdV, dS \geq 0$$

in the form

$$Td\sigma = d(\rho/n) + pd(1/n)$$

where σ is the entropy per particle and n is the number of particles per unit volume is given by Steven Weinberg, "Gravitation and Cosmology, Principles and applications of the general theory of relativity", Wiley.

3.7 General relativistic fluids, special solutions

For radially moving matter fields, the energy-momentum tensor in the coordinate system (t, r, θ, ϕ) has the form

$$T^{00} = (\rho + p)v^{02} - pg^{00}, T^{11} = (\rho + p)v^{12} - pg^{11},$$

$$T^{22} = -pg^{22}, T^{33} = -pg^{33}, T^{01} = T^{10} = (\rho + p)v^0 v^1$$

where

$$g_{00}v^{02} + g_{11}v^{12} = 1$$

ie,

$$v^0 = g_{00}^{-1/2}(1 - g_{11}(v^1)^2)^{1/2}$$

We are assuming a radially symmetric metric, ie, a metric of the form

$$d\tau^2 = A(t, r)dt^2 - B(t, r)dr^2 - r^2(d\theta^2 + sin^2(\theta)d\phi^2)$$

The Einstein field equations are

$$R_{\mu\nu} - (1/2)Rg_{\mu\nu} = -8\pi GT_{\mu\nu}$$

ρ, p are assumed to be functions of (t, r) only. The only non-trivial Einstein field equations are corresponding to the indices $(0, 0), (0, 1), (1, 1), (2, 2)$. These are four equations for the four functions A, B, ρ, v^1. Here, v^1 is assumed to be a function of (t, r) only. Note that the $(3, 3)$ component of the Einstein field equation is the same as the $(2, 2)$ component since

$$R_{33} = R_{22}sin^2(\theta), T_{33} = T_{22}sin^2(\theta), T_{22} = -pg^{22} = p/r^2$$

Problem: When in addition, there is an EM field having radial symmetry, then what are the Einstein-Maxwell field equations ?

3.8 Galactic evolution using perturbed fluid dynamics, dispersive relations. The unperturbed metric is the Robertson-Walker metric corresponding to a homogeneous and isotropic universe

$$g_{00} = 1, g_{11} = -f(r)S^2(t), g_{22} = -S^2(t)r^2, f(r) = 1/(1 - kr^2)$$

Let $\delta g_{\mu\nu}$ denote the perturbed metric tensor due to inhomogeneities. We may perturb our coordinate system slightly to ensure that $\delta g_{0\mu} = 0$. Then, the number of non-trivial perturbed metric tensor components are six, ie, $\delta g_{rs}, 1 \leq r \leq s \leq 3$. These satisfy the perturbed Einstein field equations with the perturbed energy-momentum tensor:

$$\delta R_{\mu\nu} = -8\pi G(\delta T_{\mu\nu} - \delta \Delta T_{\mu\nu} - (1/2)\delta(T + \Delta T)g_{\mu\nu} - (1/2)(T + \Delta T)\delta g_{\mu\nu})$$

The number of these equations is ten. The functions to be solved for are $\delta g_{rs}, \delta v^r$ and $\delta\rho$, namely ten in number. Note that the pressure is determined from the density using the equation of state. If we include the electromagnetic field, by regarding quadratic components of the EM field as being of first order, then we must add to the rhs of the above perturbed Einstein field equation the energy-momentum tensor of the EM field and in addition, take into account the Maxwell equations

$$(F^{\mu\nu}\sqrt{-g})_{,\nu} = 0$$

where we treat A_μ as the fundamental EM four potentials which are of $(1/2)^{th}$ order of smallness with raising and lowering of indices including computation of the metric determinant g being done using the unperturbed Robertson-Walker metric. These give us extra four equations for the EM four potential. We note that the perturbed Einstein field equations imply the perturbed fluid equations

$$[\delta T^{\mu\nu} + \delta\Delta T^{\mu\nu} + S^{\mu\nu}]_{:\nu} = 0$$

where the covariant derivatives are evaluated using the unperturbed metric.

3.9   Magnetohydrodynamics–diffusion of the mag-netic field and vorticity

The nonrelavitistic MHD equations are

$$v_{,t} + (v,\nabla)v = -\nabla p/\rho + \nu\nabla^2 v + (\sigma/\rho)(J \times B)$$

$$J = \sigma(E + v \times B)$$

along with the incompressibility condition

$$div\, v = 0,$$

and the Maxwell equations

$$div\, E = \rho_q/\epsilon,\, div\, B = 0,\, curl\, E = -B_{,t},\, curl\, B = \mu J + \mu\epsilon E_{,t}$$

where

$$\rho_{q,t} = -div\, J$$

We can write

$$B = curl\, A,\, E = -\nabla\Phi - A_{,t},\, A(t,r) = (\mu/4\pi)\int J(t - |r - r'|/c, r')d^3r'/|r - r'|,$$

$$\Phi(t,r) = (1/4\pi\epsilon)\int \rho_q(t - |r - r'|/c, r')d^3r'/|r - r'|$$

Thus,

$$B = (\mu/4\pi) \int J(t-|r-r'|/c, r')(r'-r)d^3r'/|r-r'|^3$$
$$+ (\mu/4\pi) \int J_{,t}(t-|r-r'|/c, r')(r-r')/|r-r'|^2)d^3r'$$

If the EM field perturbation with respect to zero electric field and constant magnetic field are weak, then we may assume these to be of the first order of smallness say $\delta E, \delta B$ and the fluid velocity field as a small perturbation of a constant velocity field V. Thus,

$$v(t,r) = V + \delta v(t,r), B(t,r) = B_0 + \delta B(t,r), E(t,r) = \delta E(t,r)$$

Then, upto linear orders, we have taking into account density fluctuations,

$$\delta v_{,t} + (V, \nabla)\delta v = -\nabla \delta p/\rho + (p/\rho^2)\nabla\delta\rho + \nu\nabla^2\delta v +$$
$$(\sigma/rho)[(\delta E + V \times \delta B) + \delta v \times B_0] \times B_0 - (\sigma/\rho^2)(V \times B_0)\delta\rho$$
$$+ (\sigma/\rho)(V \times B_0) \times \delta B,$$
$$div\delta B = 0, curl\delta E = -\mu\delta B_{,t}, curl\delta B = \mu\delta J + \mu\epsilon\delta E_{,t}$$

where

$$\delta J = \sigma(\delta E + V \times \delta B + \delta V \times B_0)$$

3.10 Galactic equation using perturbed Newtonian fluids

According to Hubble's law, the unperturbed velocity field is of the form $V(t,r) = H(t)r$. Let $\delta v(t,r)$ denote its perturbation. Let $\delta\rho$ denote the density perturbation. Then, the Navier-Stokes and matter conservation equations are

$$\delta v_{,t}(t,r) + H(t)(r, \nabla)\delta v(t,r) + H(t)\delta v(t,r) = -p'(\rho)\nabla\delta\rho(t,r) + \nu\nabla^2\delta v(t,r)$$
$$\rho div\delta v(t,r) + H(t)r.\nabla\rho + \delta\rho_{,t}(t,r) = 0$$

3.11 Plotting the trajectories of fluid particles

If $v(t,r)$ is the velocity field, then the trajectories of a fluid particle are obtained by solving the ode's

$$dr(t)/dt = v(t, r(t))$$

and we can approximate this by a difference equation

$$r(t+\Delta) = r(t) + \Delta v(t, r(t)) + \Delta^2(\partial/\partial t + (v(t,r(t)), \nabla))v(t,r(t))/2! +$$
$$+ ... + (\Delta^{n+1}/(n+1)!)(\partial/\partial t + (v(t,r(t)), \nabla))^n v(t,r(t)) + O(\Delta^{n+2})$$

In this way, the trajectories of the fluid particles can be plotted.

3.12   Statistical theory of fluid turbulence, equations for the velocity field moments, the Kolmogorov-Obhukov spectrum

In terms of components, the incompressible steady state fluid equations are

$$v_{k,k}(r) = 0, \, v_k(r)v_{i,k}(r) = -p_{,i}(r)/\rho + \nu v_{i,kk}(r)$$

with the summation convention being adopted. The velocity field and the pressure field are assumed to depend only on the spatial location $r = (x, y, z)$ and not on time. We assume homogeneous turbulence, ie,

$$< v_i(r)v_j(r') >= B_{ij}(r - r')$$

$$< v_i(r)v_j(r')v_k(r'') >= C_{ijk}(r - r', r' - r'')$$

etc. Eliminating the pressure gives us

$$(v_k v_{i,k})_{,j} - (v_k v_{j,k})_{,i} = \nu(v_{i,jkk} - v_{j,ikk})$$

Alternately assuming

$$< v_i(r)p(r') >= A_i(r - r'), < v_i(r)v_j(r')p(r'') >= A_{ij}(r - r', r' - r'')$$

etc, we obtain from the incompressibility equation,

$$< v_{k,k}(r)v_j(r') >= 0$$

or

$$B_{kj,k}(r) = 0,$$

$$< v_{k,k}(r)v_i(r')v_j(r'') >= 0$$

or

$$C_{kij,k}(r, r') = 0$$

and from the Navier-Stokes equations,

$$< v_k(r)v_{i,k}(r) >= - < p_{,i}(r) > /\rho + \nu < v_{i,kk}(r) >$$

Assuming zero mean pressure and zero mean velocity, this gives

$$B_{ik,k}(0) = 0$$

Also,

$$< v_{i,k}(r)v_k(r)v_j(r') >= - < p_{,i}(r)v_j(r') > /\rho + \nu < v_{i,kk}(r)v_j(r') >$$

or

$$C_{ikj,k}(0, r - r') - A_{j,i}(r' - r) - \nu B_{ij,kk}(r - r') = 0$$

We can now obtain more concrete results by assuming that the velocity correlation $B_{ij}(r - r')$ can be expressed in the isotropic form as

$$B_{ij}(r - r') = <v_i(r)v_j(r')> = P(|r - r'|)n_i n_j + Q(|r - r'|)\delta_{ij}$$

where (n_i) is the unit vector along $r - r'$, ie,

$$n = (n_i) = (r - r')/|r - r'|$$

and further assume that

$$A_i(r - r') = <v_i(r)p(r')> = S(|r - r'|)n_i$$

and $C_{ijk}(r, r') = 0$.

Exercise: Derive the equations satisfied by the functions $P(|r|), Q(|r|), S(|r|)$.

3.13 Estimating the velocity field of a fluid subject to random forcing using discrete space velocity measurements based on discretization and the Extended Kalman filter

3.14 Quantum fluid dynamics. Quantization of the fluid velocity field by the introduction of an auxiliary Lagrange multiplier field

The fluid equations are

$$(v, \nabla)v + v_{,t} = -\nabla p/\rho + \nu\nabla^2 v + f \ - - - \ (1)$$

where f is the external force field. This is to be supplemented with the incompressibility condition

$$div\, v = 0 \ - - - \ (2)$$

Using Lagrange multiplier fields $\lambda(t, r) \in \mathbb{R}^3$ and $\mu(t, r) \in \mathbb{R}$, the Lagrangian density from which the fluid equations are derived is

$$L(\mu, \lambda, v, p) = \lambda^T((v, \nabla)v + v_{,t} + \nabla p/\rho - \nu\nabla^2 v - f) - \mu\, div\, v \ - - - \ (3)$$

or equivalently, using the Einstein summation convention after integrating by parts,

$$L = \lambda_i v_j v_{i,j} + \lambda_i v_{i,t} + \lambda_i p_{,i}/\rho + \nu\lambda_{i,j}v_{i,j} - \lambda_i f_i - \mu v_{j,j}$$

The state equations are the constraint equations:

$$\frac{\partial L}{\partial \lambda} = 0, \ \frac{\partial L}{\partial \mu} = 0$$

and these are (1) and (2). The costate equations

$$\partial_t \frac{\partial L}{\partial v_{i,t}} + \partial_j \frac{\partial L}{\partial v_{i,j}} = \frac{\partial L}{\partial v_i}$$

give

$$\lambda_{i,t} + (\lambda_i v_j)_{,j} + \nu \lambda_{i,jj} - \mu_{,i} - \lambda_j v_{j,i} = 0$$

This can be expressed in vector notation as

$$\lambda_{,t} + \lambda.divv + (v, \nabla)\lambda + \nu \nabla^2 \lambda - \nabla\mu - \nabla_v((\lambda, v)) = 0$$

where the differential operator ∇_v acts only on v. Finally, the costate equation

$$\partial_i \frac{\partial L}{\partial p_{,i}} = 0$$

gives

$$div\lambda = \lambda_{i,i} = 0$$

In order to quantize this, we must find out the Hamiltonian using the Legendre transformation:

$$\pi_i = \frac{\partial L}{\partial v_{i,t}} = \lambda_i$$

So the Hamiltonian density is

$$H = \pi_i v_{i,t} - L =$$

$$-\pi_i v_j v_{i,j} - \pi_i p_{,i}/\rho - \nu \pi_{i,j} v_{i,j} + \pi_i f_i + \mu v_{j,j}$$

The Hamilton equations are

$$\pi_{i,t} = -\frac{\delta H}{\delta v_i} = -\frac{\partial H}{\partial v_i} + \partial_j \frac{\partial L}{\partial v_{i,j}},$$

$$v_{i,t} = \frac{\delta H}{\delta \pi_i} = \frac{\partial H}{\partial \pi_i} - \partial_j \frac{\partial H}{\partial \pi_{i,j}},$$

$$\frac{\partial H}{\partial \mu} = 0$$

We leave it to the reader to check that these yield the correct equations of motion. Let $\pi_{\lambda i}$ denote the momentum density conjugate to λ_i and π_μ the momentum density conjugate to μ. We note that π_i is the momentum density conjugate to v_i. Also denote by π_p the momentum density conjugate to p. Then our constraint equations are

$$\pi_\mu = 0, \pi_{\lambda i} = 0, \pi_p = 0$$

We also have constraint equations

$$\chi = v_{i,i} = 0, M_i = \pi_i - \lambda_i = 0$$

We then form the Dirac bracket taking these constraints into account. For example, some of the standard Poisson brackets are

$$[v_i(t,r), \chi(t,r')] = 0, [\chi(t,r), \pi_i(t,r')] = i\partial_i \delta^3(r-r')$$

$$[\chi(t,r), M_i(t,r')] = [v_{j,j}(t,r), \pi_i(t,r')] = i\partial_i \delta^3(r-r')$$

etc. These formulas can be used to form the Dirac bracket for quantization

Reference: Steven Weinberg, "The quantum theory of fields", vol.1, Cambridge University Press.

3.15 Optimal control problems for fluid dynam-ics

The problem is to match the fluid velocity field $v(t,r)$ to a given/desired velocity field $v_d(t,r)$ over the time interval $[0,T]$ and space region $B \subset \mathbb{R}^3$. Assuming the fluid to be incompressible, we can write $div\, v = 0$. We also assume the desired velocity field v_d to be incompressible, ie, $div\, v_d = 0$. This means that we can derive v, v_d from stream fields ψ, ψ_d using

$$v = curl\, \psi, v_d = curl\, \psi_d$$

We may assume without loss of generality that $div\, \psi = div\, \psi_d = 0$ and then the problem of matching the velocity fields amounts to matching the stream vector fields $\psi(t,r)$ and $\psi_d(t,r)$. Taking the curl of the Navier-Stokes equation gives

$$\nabla^2 \psi_{,t} + \nabla \times (\nabla^2 \psi \times (curl\, \psi)) = \nu(\nabla^2)^2 \psi - g(t,r)$$

where

$$g = curl\, f$$

with f begin the control force field. We now regard g as the control force field and it should satisfy the constraint $div\, g = 0$. Thus, the objective is to minimize

$$S(g, \psi, \lambda, \mu) = \int_{[0,T] \times B} \| \psi(t,r) - \psi_d(t,r) \|^2 \, dt d^3 r$$

$$- \int_{[0,T] \times B} \lambda(t,r)^T (\nabla^2 \psi_{,t}(t,r) + \nabla \times (\nabla^2 \psi(t,r) \times (curl\, \psi(t,r))))$$

$$- \nu(\nabla^2)^2 \psi(t,r) - g(t,r)) dt d^3 r$$

$$- \int_{[0,T] \times B} \mu(t,r) div\, g(t,r) dt d^3 r$$

We leave it as an exercise to carry out the variation of S and derive the input and costate equations.

3.16 Hydrodynamic scaling limits for simple exclusion models

The lattice over which the exclusion process runs is $\mathbb{Z}_N^d$. The exclusion process is $\eta_t : \mathbb{Z}_N^d \to \{0,1\}$. $\eta_t(x)$ is one or zero according as the site $x \in \mathbb{Z}_N^d$ is occupied or not by a particle. If site x is occupied while the site y is not at time t, ie $\eta_t(x) = 1, \eta_t(y) = 0$, then there is a probability $p(x,y)dt$ of the particle at x jumping to y, otherwise not. Thus, we can define a family of independent Poisson processes $\{N_t(x,y) : t \geq 0\}, x, y \in \mathbb{Z}_N^d$ having rates $\lambda p(x,y)$ with $\sum_{y \in \mathbb{Z}_N^d} p(x,y) = 1$ and the exclusion process η_t then satisfies the stochastic differential equation

$$d\eta_t(x) = \sum_{y : y \neq x} \left(-\eta_t(x)(1 - \eta_t(y))dN_t(x,y) + \eta_t(y)(1 - \eta_t(x))dN_t(y,x) \right)$$

The process η_t is Markov and has the infinitesimal generator L where for $f : \{0,1\}^{\mathbb{Z}_N^d} \to \mathbb{R}$, we have

$$Lf(\eta) = \lambda \sum_{x \neq y} \eta(x)(1 - \eta(y))p(x,y)(f(\eta^{(x,y)}) - f(\eta))$$

where

$$\eta^{(x,y)} : \mathbb{Z}_N^d \to \{0,1\}$$

is the map defined by $\eta^{(x,y)}(z)$ equals $\eta(z)$ if $z \neq x, z \neq y$, equals $\eta(x)$ if $z = y$ and $\eta(y)$ if $z = x$. In other words, $\eta^{(x,y)}$ is the state that interchanges the state of the sites x and y and keeps the state of the other sites the same. We write

$$\rho(t, x/N) = c(N)\mathbb{E}(\eta_t(x)), x \in \mathbb{Z}_N^d$$

where $c(N)$ is a normalization constant. Then, in the limit $N \to \infty$, $\rho(t,.)$ converges to a function on the d-dimensional torus $\mathbb{T}^d = [0,1]^d$ and we call this function $\rho(t, \theta), \theta \in \mathbb{T}^d$ as the limiting density of the exclusion process. Varadhan and other researchers have derived nonlinear partial differential equations for $\rho(t, \theta)$ like the Burger's equation by imposing various conditions on the transition probabilities $p(x,y)$ using Large deviation theory (Reference:The Collected papers of S.R.S.Varadhan, Hindustan Book Agency). The basic idea is to start with a smooth function $J : \mathbb{T}^d \to \mathbb{R}$ and consider

$$\int_{[0,1]^d} J(\theta)\rho(t,\theta)d^d\theta$$

as the limit of

$$\sum_{x \in \mathbb{Z}_N^d} J(x/N)\rho(t, x/N)N^{-d}$$

as $N \to \infty$. We can easily using the sde for η_t write down

$$d\rho(t, x/N)/dt = c(N)\lambda\mathbb{E} \sum_{y : y \neq x} [(\eta_t(y)(1 - \eta_t(x))p(y,x) - \eta_t(x)(1 - \eta_t(y))p(x,y)]$$

3.17 Appendix: The complete fluid dynamical equations in orthogonal curvilinear coordinate systems specializing to cylindrical and spherical polar coordinates

Let (q_1, q_2, q_3) be an orthogonal curvilinear coordinate system. Let σ_{ab} be the stress tensor of pressure and viscosity in the cartesian system and $\tilde{\sigma}_{ij}$ in the curvilinear system. Then, we have

$$\tilde{\sigma}_{ij} = \sigma_{ab} e_{ia} e_{jb}$$

where (e_{i1}, e_{i2}, e_{i3}) is the unit vector along the q_i direction expressed w.r.t. the cartesian basis. Thus,

$$(e_{ia}, a = 1, 2, 3) = \nabla q_i / |\nabla q_i| = H_i^{-1} (\frac{\partial x_a}{\partial q_i})_{a=1}^3$$

with

$$H_i = |\frac{\partial r}{\partial q_i}| = \sum_{a=1}^3 (\frac{\partial x_a}{\partial q_i})^2$$

The total pressure and viscous force per unit volume along the e_i direction is given by

$$\tilde{s}_i = \sigma_{ab,b} e_{ia} = (\sigma_{ab} e_{ia})_{,b} - \sigma_{ab} e_{ia,b}$$

Now we note that since the matrix $((e_{ia}))$ is orthogonal, we have

$$\sigma_{ab} = \tilde{\sigma}_{ij} e_{ia} e_{jb}$$

and hence we can write

$$\sigma_{ab} e_{ia} = \tilde{\sigma}_{ij} e_{jb}$$

and so

$$(\sigma_{ab} e_{ia})_{,b} =$$

$$(\tilde{\sigma}_{ij} e_{jb})_{,b} = \tilde{\sigma}_{ij,k} q_{k,b} e_{jb} + \tilde{\sigma}_{ij} e_{jb,b}$$

$$= \tilde{\sigma}_{ij,k} (\nabla q_k, e_j) + \tilde{\sigma}_{ij} e_{jb,b}$$

Now,

$$(\nabla q_k, e_j) = |\nabla q_k|(e_k, e_j) = H_k^{-1} \delta_{kj}$$

Also,

$$e_{jb,b} = div e_j = (H_1 H_2 H_3)^{-1} \epsilon(jrs)(H_r H_s)_{,j}$$

by the formula for divergence in an orthogonal curvilinear coordinate system. Here, j, r, s denote curvilinear indices while a, b, c denote Cartesian indices. Thus,

$$(\sigma_{ab} e_{ia})_{,b} =$$

$$H_k^{-1}\tilde{\sigma}_{ik,k} + (H_1 H_2 H_3)^{-1}\tilde{\sigma}_{ij}\epsilon(jrs)(H_r H_s)_{,j}$$

Further,

$$\sigma_{ab}e_{ia,b} = \tilde{\sigma}_{rs}e_{ra}e_{sb}e_{ia,b}$$

Now,

$$e_{sb}e_{ia,b} = (e_s, \nabla)e_{ia} = H_s^{-1}e_{ia,s} = H_s^{-1}\frac{\partial e_{ia}}{\partial q_s}$$

on using the formula for the gradient in an orthogonal curvilinear coordinate system. Now,

$$e_{ra}e_{ia,s}$$

is zero if $r = a$ and if $r \neq a$, we can evaluate it as follows. For example, consider

$$e_{2a}e_{2a,1} = 0$$

because $e_{2a}e_{2a} = 1$ and hence $e_{2a}e_{2a,1} = 0$ which means that $e_{2,1}$ is orthogonal to e_2 and hence in the linear span of e_1 and e_3. We can thus write

$$e_{2,1} = c_1 e_1 + c_3 e_3$$

with

$$c_1 = (e_{2,1}, e_1), c_3 = (e_{2,1}, e_3)$$

Now,

$$(e_{2,1}, e_1) = -(e_2, e_{1,1})$$

Remark: From the formula for the curl in an orthogonal curvilinear coordinate system, we know that

$$curl(e_k/H_k) = 0, k = 1, 2, 3$$

Also from the formula for the divergence in an orthogonal curvilinear system, we know that

$$div(e_1/H_2 H_3) = div(e_2/H_3 H_1) = div(e_3/H_1 H_2) = 0$$

These equations give us $dive_k$ and $curle_k$ in the curvilinear system in terms of the $H_j's$ and their partial derivatives w.r.t the $q's$.

We now have

$$(e_{1,1}, e_2) = -(e_1, e_{2,1})$$
$$(e_1, e_{2,1}) = (e_1, (H_2^{-1}r_{,2})_{,1}) =$$
$$H_2^{-1}(e_1, r_{,21}) = H_2^{-1}H_1^{-1}(r_{,1}, r_{,21}) =$$
$$(2H_1 H_2)^{-1}(r_{,1}, r_{,1})_{,2} = (2H_1 H_2)^{-1}(H_1^2)_{,2}$$
$$= H_{1,2}/H_2$$
$$(e_{1,3}, e_2) = -(e_1, e_{2,3}) = -H_1^{-1}(r_{,1}, (H_2^{-1}r_{,2})_{,3}) = -(H_1 H_2)^{-1}(r_{,1}, r_{,23})$$
$$= (H_1 H_2)^{-1}(r_{,12}, r_{,3})$$

We now observe the following:

$$(r_{,1}, r_{,2}) = 0$$

$$(r_{,13}, r_{,2}) + (r_{,1}, r_{,23}) = 0$$

$$(r_{,1}, r_{,3}) = 0$$

$$(r_{,12}, r_{,3}) + (r_{,1}, r_{,23}) = 0$$

$$(r_{,2}, r_{,3}) = 0$$

$$(r_{,12}, r_{,3}) + (r_{,2}, r_{,13}) = 0$$

From these, we deduce that

$$(r_{,1}, r_{,23}) = 0$$

and likewise,

$$(r_{,12}, r_3) = 0, (r_{,13}, r_{,2}) = 0$$

Thus, we get

$$(e_{1,3}, e_2) = 0, (e_{1,2}, e_3) = 0, (e_{3,2}, e_1) = 0, (e_{2,1}, e_3) = 0$$

ie

$$(e_{r,s}, e_m) = 0$$

if r, s, m are all distinct. Thus, we can calculate $e_{r,s}$ as linear combinations of e_1, e_2, e_3 and hence evaluate $\tilde{s}_i$ in terms of the $\tilde{\sigma}'_{ij}s$, the H'_js and their partial derivatives w.r.t the q'_js. We leave this as an exercise to the reader. Finally, write down the Navier Stokes equations in the form

$$\rho((v, \nabla)v + v_{,t}) = \tilde{s}_i e_i + f$$

we need to express $(v, \nabla)v$ in the orthogonal curvilinear system and if we use the explicit form of $\tilde{s}_i e_i$ as $-\nabla p + \eta \nabla^2 v$, then we must also express $\nabla^2 v$ in the orthogonal curvilinear system. The gradient of p part is easy. It is

$$\nabla p = H_i^{-1} p_{,i} e_i$$

The other term is

$$\nabla^2 v = \nabla^2 (v_i e_i) =$$

$$(H_1 H_2 H_3)^{-1}((H_2 H_3 (v_i e_i)_{,1} / H_1)_{,1} + (H_3 H_1 (v_i e_i)_{,2} / H_2)_{,2} + (H_1 H_2 (v_i e_i)_{,3} / H_3)_{,3})$$

Since we have already calculated $e_{i,j}, 1 \leq i, j \leq 3$ as linear combinations of the e'_js, we leave it as an exercise to evaluate the curvilinear components of $\nabla^2 v$ in terms of the partial derivatives of the curvilinear components v_i of v w.r.t. the q'_js.

Chapter 4

Quantum robots in motion carrying Dirac current as quantum antennas

4.1 A short course in classical and quantum robotics with antenna theory applications

4.1.1 The Lagrangian for a rigid body

Let $B \subset \mathbb{R}^3$ denote the volume of the rigid body at time $t = 0$ and $R(t)(B)$ the same at time t. Thus, $R(t) \in SO(3)$. The kinetic energy of the body at time t is given by

$$K(t) = (\rho/2) \int_B |R'(t)r|^2 d^3r = (1/2)Tr(R'(t)JR'(t)^T)$$

where

$$J = \rho \int_B rr^T d^3r$$

is the moment of inertia matrix of the rigid body. Taking into account the gravitational potential and external torques, we can write down the Lagrangian as

$$L = K(t) - V(t)$$

where

$$V(t) = mgdR_{33}(t) + \tau(t)^T (\phi(t), \theta(t), \psi(t))^T$$

where

$$R(t) = R_z(\phi(t))R_x(\theta(t))R_z(\psi(t))$$

with ϕ, θ, ψ being the Euler angles.

4.1.2   The Hamiltonian for a rigid body

Study project:

[1] Express the velocity of each point of a rigid body in terms of the time derivative of the rotation matrix applied to the initial position of the point and hence integrate its norm square over the entire rigid body to obtain the kinetic energy of the rigid body as a quadratic function of the time derivative of the rotation matrix with the quadratic function being determined in terms of the moment of inertia matrix of the body.

[2] Express the gravitational potential of the rigid body as a linear function of the rotation matrix and hence determine the total Lagrangian of the body in terms of the instantaneous rotation matrix and its time derivative.

[3] By using the well known formula for the differential of the exponential map in Lie group theory, express the Lagrangian of the rigid body in terms of the standard three Lie algebra coordinates and their time derivatives.

[4] Perform the Legendre transformation on this Lagrangian to obtain the Hamiltonian of the rigid body in terms of the Lie algebra coordinates and their corresponding canonical momenta.

4.1.3   The Lagrangian and Hamiltonians of a d-link robot with 3-D rigid body links

Study project: Generalize the results of the previous subsection to a robot consisting of d three dimensional rigid links with each link pivoted at the top of the previous link. Describe the state of each link at a given time in terms of a rotation matrix and hence obtain the kinetic energy of the robot as a quadratic function of the time derivatives of d rotation matrices. Likewise, express the total potential energy the robot as a linear function of the d rotation matrices and set up the Lagrangian and Hamiltonian in terms of the Lie algebra coordinates of each rotation matrix using the differential of the exponential map.

4.1.4   The equations of motion of a d-link robot subject to gravitation, external forces and torques

Starting with the result of the previous subsection, add to the Lagrangian additional terms coming from the torques at each pivot provided by motors attached the pivot and hence set up the Euler-Lagrange equations of motion in Lie algebra coordinates. If an external force acts on each link causing the robot to acquire a translational velocity at its base, then describe the contribution of these forces to the Lagrangian in terms of the coordinates and velocities of the pivot of the first link.

4.1.5 Stochastic differential equations for a d-link robot in the presence of noise in the machine torques and noise in the human hand operator force

Add White Gaussian noise terms to the torque in the equations of motion of the previous subsection and express the resulting equations of motion as a system of 3d coupled stochastic differential equation by noting that White Gaussian noise times dt is the differential of Brownian motion.

4.1.6 Master-slave robots acting in teleoperation with feedback, a stochastic analysis

Describe the motion of two d link robots with the torque applied to the second one having an error feedback component coming from the difference between the Lie algebra coordinates and their velocites of the two robots. Explain using Lyapunov energy theory how these feedback torques can be used to achieve trajectory tracking.

4.2 A fluid of interacting robots

If the lattice site $x \in \mathbb{Z}^d$ is occupied by a robot at time t, we put $\eta_t(x) = 1$, otherwise $\eta_t(x) = 0$. The transition from site x to site y is accord with a Poisson process $N_t(x, y)$ having rate $p(x, y)$ and the transition takes place at time t iff $\eta_t(x) = 1$ and $\eta_t(y) = 0$. Thus, the process $\eta_t : \mathbb{Z}^d \to \{0, 1\}$ satisfies the sde

$$d\eta_t(x) = \sum_{y \neq x} \eta_t(y)(1 - \eta_t(x))dN_t(y, x) - \eta_t(x)(1 - \eta_t(y))dN_t(x, y)$$

Let X denote the space of all maps $\eta : \mathbb{Z}^d \to \{0, 1\}$, ie,

$$X = \{0, 1\}^{\mathbb{Z}^d}$$

The generator of the Markov process η_t described by the above sde is therefore given by

$$Lf(\eta) = \sum x \neq y p(x, y)\eta(x)(1 - \eta(y))(f(\eta^{(x,y)}) - f(\eta))$$

To describe fluid dynamics using this model, we introduce an empirical density $\tilde{\rho}_t$ by the equation

$$N^{-d} \sum_{y \in \mathbb{Z}_N^d} J(y/N)\tilde{\rho}_t(y/N) = N^{-d} \sum_{y \in \mathbb{Z}_N^d} J(y/N)\eta_t(y)$$

where

$$J : [0, 1]^d \to \mathbb{R}$$

is any function. We expect that as $N \to \infty$, this will converge to $\int_{[0,1]^d} J(\theta)\rho_t(\theta)d\theta$ where $\rho_t(\theta)$ satisfies an appropriate pde which is a nonlinear version of the standard heat equation. Note that

$$\tilde{\rho}_t(y/N) = \eta_t(y)$$

so we get

$$d(N^{-d} \sum_{y \in \mathbb{Z}_N^d} J(y/N)\rho_t(y/N)) = N^{-d} \sum_{x \neq y} J(y/N)(\eta_t(x)(1-\eta_t(y))dN_t(x,y)$$

$$-\eta_t(y)(1-\eta_t(x))dN_t(y,x))$$

so taking expectations on both sides, and denoting

$$\rho_t(x/N) = \mathbb{E}[\tilde{\rho}_t(x/N)] = \mathbb{E}[\eta_t(x)] = Pr(\eta_t(x) = 1)$$

we get

$$d/dt(N^{-d} \sum_{y \in \mathbb{Z}_N^d} J(y/N)\rho_t(y/N)) =$$

$$\sum_{x=y} J(y/N) - J(x/N))p(x,y)\mathbb{E}[\eta_t(x)(1 - \eta_t(y))]$$

4.3   Disturbance observer in a robot

The dynamics of the robot is

$$M(q)q'' + N(q,q') = \tau(t) + d(t)$$

where $q(t) \in \mathbb{R}^d$ and $M(q) \in \mathbb{R}^{d \times d}$ with $M(q) > 0$ for all $q \in [0, 2\pi)^d$. $d(t)$ is the disturbance to be estimated. Define

$$\hat{d}(t) = z(t) + p(q'(t))$$

where

$$z' = L(q,q')(N(q,q') - \tau - \hat{d})$$

Then,

$$d\hat{d}/dt = z' + p'(q')q''$$

$$= L(q,q')(N - \tau - \hat{d}) + p'(q')M(q)^{-1}(\tau + d - N))$$

Assume that

$$p'(q')M(q)^{-1} = L(q,q')$$

Then, we get

$$d\hat{d}/dt = L(q,q')(d - \hat{d})$$

Thus, if $L(q, q')$ is a positive definite matrix for all q, q', then we can expect that $d(t) - \hat{d}(t) \to 0$ as $t \to \infty$. In the special case when $L(q, q') = CM(q)^{-1}$ where C is a constant matrix, our disturbance observer reduces to

$$z' = C(N(q, q') - \tau - \hat{d}), \hat{d} = z + p(q')$$

We then require that

$$p'(q') = C$$

or equivalently,

$$p(q') = Cq'$$

which means that this disturbance observer is simply

$$z' = C(N(q, q') - \tau - \hat{d}), \hat{d} = z + Cq'$$

and this results in

$$d\hat{d}/dt = CM(q)^{-1}(d - \hat{d})$$

so we can expect convergence of $d - \hat{d}$ to zero provided that for all q, all the eigenvalues of the matrix $CM(q)^{-1}$ have negative real parts.

4.4 Robot connected to a spring mass with damping system

.

The original robot dynamical state equations are

$$X'(t) = \psi(t, X(t)) + G(t, X(t))\tau(t)$$

where $X(t) = [q(t)^T, q'^T(t)]^T \in \mathbb{R}^{2d}$ are the d-link robot angles and angular velocities. $\tau(t)$ is the external torque coming from the motors at the joints of the robot as well as from the environment. The end-effector position of the robot is

$$\eta(X(t)) \in \mathbb{R}^3$$

and this is connected to a spring mass system defined by a dynamical equation

$$x'(t) = v(t), v'(t) = -\gamma v(t) - kx(t) + f_e(t) - f_{rob}(t)$$

where $x(t) = \eta(X(t)) = \eta(q(t))$, $f_e(t)$ is the external force acting on the spring mass system coming from the environment and $-f_{rob}(t)$ is the back reaction of the robot end effector acting on the spring mass system. Thus,

$$f_{rob}(t) = f_e(t) - \gamma x'(t) - kx(t) - x''(t) = f_e(t) - \gamma \eta'(q(t))q'(t) - k\eta(q(t)) - (\eta'(q(t))q(t))'$$
$$= f_e(t) - \chi(X(t), X'(t))$$

is the net external force acting on the end effector of the robot and if $J(X(t)) = J(q(t))$ denotes the Jacobian matrix of the transformation $q(t) \to x(t)$ from the robot angles to the end-effector positions, then the net torque acting on the robot from its end effector is by D-Alembert's principle of virtual work, given by $J(X(t))^T f_{rob}(t)$ and thus, the robot dynamical equation when we take the computed torque with trajectory tracking error feedback torque $u_c(t)$ into account, is given by

$$X'(t) = \psi(t, X(t)) + G(t, X(t))(J(X(t))^T f_{rob}(t) + G(t, \hat{X}(t))^{-1}(K(t)(X_d(t)$$
$$-\hat{X}(t)) + X'_d(t) - \psi(t, \hat{X}(t)))$$

or equivalently,

$$X'(t) = \psi(t, X(t)) + G(t, X(t))J(X(t))^T(f_e(t) - \chi(\eta(X_d(t), X'_d(t))))$$
$$+G(t, X(t))G(t, \hat{X}(t))^{-1}(K(t)(X_d(t) - \hat{X}(t)) + X'_d(t) - \psi(t, \hat{X}(t)))$$

Here, the computed torque is

$$\tau_c(t) = G(t, \hat{X}(t))^{-1}(K(t)(X_d(t) - \hat{X}(t)) + X'_d(t) - \psi(t, \hat{X}(t)))$$

If we wish to eliminate the external force $f_e(t)$, ie, we regard this force as a disturbance to the robot dynamics, then we should subtract off its estimate $\hat{f}_e(t)$ from this dynamics thereby resulting in the modified dynamics

$$X'(t) = \psi(t, X(t)) + G(t, X(t))J(X(t))^T(f_e(t) - \hat{f}_e(t))$$
$$+G(t, X(t))G(t, \hat{X}(t))^{-1}(K(t)(X_d(t) - \hat{X}(t)) + X'_d(t) - \psi(t, \hat{X}(t))$$

It should be noted that $\hat{X}(t)$ is the observer output, ie, an estimate of the state $X(t)$ based on measurement data collected upto time t: $\{z(s) : s \le t\}$, where

$$dz(t) = h(t, X(t))dt + \sigma_v dV(t)$$

is the measurement model. The observer dynamics is a generalized version of the EKF:

$$d\hat{X}(t) = \psi(t, \hat{X}(t)) + L(t)(dz(t) - h(t, \hat{X}(t))dt)$$

where $L(t)$ is the output error feedback coefficient matrix. When tracking is good and the observer is also good, we have $X(t) \approx \hat{X}(t) \approx X_d(t)$, $f_e(t) \approx \hat{f}_e(t)$, in which case we get approximately, $X'(t) = X'_d(t)$ or equivalently, $X(t) \approx X_d(t)$ which demonstrates self-consistency. In practice, the external force/disturbance $f_e(t)$ will be functions of some unknown parameters $\theta(t)$ like the amplitude, frequency and phase of sinusoids and then we would represent it as $f_e(t, \theta(t))$ and its estimate as $f_e(t, \hat{\theta}(t))$ where the parameter estimates $\hat{\theta}(t)$ would be a part of an extended state vector estimate. It should be noted that in general, process noise is present in the robot state dynamics and hence the correct dynamical equation for in terms of computed torque and trajectory tracking error feedback without subtracting out the estimate of the external forces would be given by

$$X'(t) = \psi(t, X(t)) + G(t, X(t))J(X(t))^T(f_e(t) - \chi(\eta(X_d(t), X'_d(t))))$$
$$+G(t, X(t))G(t, \hat{X}(t))^{-1}(K(t)(X_d(t) - \hat{X}(t)) + X'_d(t) - \psi(t, \hat{X}(t))) + G(t, X(t))W(t)$$

where $W(t) = \sigma dB(t)/dt$ with $B(.)$ being vector valued Brownian motion. More precisely, this equation should be multiplied by dt and expressed in the form of an Ito stochastic differential equation.

Chapter 5

Design of quantum gates using electrons, positrons and photons, quantum information theory and quantum stochastic filtering

5.1 A short course in quantum gates, quantum computation and quantum information with antenna theory applications

5.1.1 one qubit quantum state and quantum gate: Examples including NOT gate, Phase gate, Hadamard gate

5.1.2 Multiple qubit quantum gates: CNOT gate, other controlled unitary gates, swap gate, quantum Fourier transform gate

5.1.3 Information/Von-Neumann entropy of a mixed quantum state

5.1.4 Examples of entropy computation in quantum systems. Entropy computation for noisy quantum evolutions

Assume that system 1 evolves according to the dynamics

$$\rho'(t) = -i[H_1, \rho(t)] + \theta_1(\rho(t))$$

77

and system 2 evolves according to

$$\sigma'(t) = -i[H_2, \sigma(t)] + \theta_2(\sigma(t))$$

where θ_1, θ_2 are Lindblad maps:

$$\theta_1(\rho) = (-1/2) \sum_{k=1}^{N} (L_k^* L_k \rho + \rho L_k^* L_k - 2L_k \rho L_k^*)$$

$$\theta_2(\sigma) = (-1/2) \sum_{k=1}^{M} (P_k^* P_k \sigma + \sigma P_k^* P_k - 2P_k \sigma P_k^*)$$

Then calculate the rate of change of the relative entropy between the two states at time t:

$$H(\rho(t), \sigma(t)) = Tr(\rho(t)(log(\rho(t)) - log(\sigma(t))))$$

hint: Use the formula for the differential of the exponential map in the form:

$$\frac{d}{dt} exp(Z(t)) = exp(Z(t)) g(ad(Z(t))^{-1}(Z'(t))$$

where

$$g(z) = \frac{z}{1 - exp(-z)}$$

Equivalently,

$$Z'(t) = g(ad(Z(t))(exp(-Z(t)). \frac{d}{dt} exp(Z(t)))$$

Take $\rho(t) = exp(Z(t))$ and deduce that

$$\frac{d}{dt} log(\rho(t)) = g(ad(log(\rho(t))))(\rho(t)^{-1} \rho'(t))$$

Now note that

$$g(z) = \frac{1}{1 - z/2! + z^2/3! + ...} = 1 + \sum_{k \geq 1} c[k] z^k$$

Note that $\rho(t)$ has eigenvalues in the range $[0, 1]$. So $log(\rho(t))$ has all eigenvalues in the range $(-\infty, 0]$. Hence, $ad(log(\rho(t))$ has eigenvalues in the range $\mathbb{R} = (-\infty, \infty)$. So in order that we be able to substitute $ad(log(\rho(t)))$ for z in the above equation, we require that the Taylor series for $g(z)$ be convergent for all $z \in \mathbb{R}$ which may not be the case. Note that $Tr(\rho'(t)) = 0$ and that for $k \geq 1$,

$$Tr(\rho(t) ad(log(\rho(t))^k (\rho(t)^{-1} \rho'(t))) = Tr(ad(log(\rho(t)))^k (\rho'(t))) = 0$$

since the commutator of two operators has trace zero (provided that the operators are bounded). Thus, we get

$$Tr(\rho(t) \frac{d}{dt} log(\rho(t))) = 0$$

and so

$$\frac{d}{dt}Tr(\rho(t).log(\rho(t))) = Tr(\rho'(t).log(\rho(t)))$$

$$= Tr(T(\rho(t)).log(\rho(t)))$$

where

$$T(X) = -i[H_1, X] + \theta_1(X)$$

Now,

$$Tr([H_1, \rho].log(\rho)) = Tr([H_1.log(\rho), \rho]) = 0$$

so

$$\frac{d}{dt}Tr(\rho(t).log(\rho(t))) = Tr(\theta_1(\rho(t)).log(\rho(t)))$$

5.1.5 Entropy of a quantum antenna interacting with a photon bath

Antenna in a quantum electromagnetic field: The antenna consists of N electrons and N positrons which start in a given pure state with prescribed momenta and spins for these particles. This system evolves under the interaction with a photon bath according to the interaction Hamiltonian

$$H_I(t) = \int J^\mu(x)A_\mu(x)d^3r = -e\int \psi(t,r)^*\alpha^\mu\psi(t,r)A_\mu(t,r)d^3r$$

where

$$\psi(t,r) = \sum_{k=1}^{N} a_k f_k(t,r) + b_k^* g_k(t,r)$$

with a_k, b_k denoting respectively the annihilation operator of the kth electron and the kth positron,

$$A_\mu(t,r) = \sum_{k=1}^{p} c_k h_k(t,r) + c_k^* \bar{h}_k(t,r)$$

with c_k denoting the annihilation operator of the kth photon in the bath. We write the initial state of the antenna and photon bath as

$$|\psi_i >= |e_1, ..., e_N, p_1, ..., p_N, \phi(u) >$$

where e_i is a one or a zero according as the ith electron is present or not and p_i is a one or a zero according as the ith positron is present or not while

$$u = (u_1, ..., u_p)^T \in \mathbb{C}^p$$

defines the coherent state of the photon bath. Specifically,

$$|\phi(u) >= \sum_{n} \mathbf{u}^n \mathbf{c}^{*n}|0 > /\mathbf{n}!$$

where

$$\mathbf{c}^{*\mathbf{n}} = c_1^{*n_1}...c_p^{*n_p}$$

The action of the operators $a_k, a_k^*, b_k, b_k^*, c_k, c_k^*$ on $|\psi_i>$ is as follows:

$$a_k|e_1, ..., e_N, p_1, ..., p_N, \phi(u)> = \delta(1 - e_k))|e_1, ..., 1 - e_k..., e_N, p_1, ...p_N, \phi(u)>$$

$$a_k^*|e_1, ..., e_N, p_1, ..., p_N, \phi(u)> = \delta(e_k)|e_1, ..., 1 - e_k, ...e_N, p_1, ..., p_N, \phi(u)>$$

and likewise for a_k, b_k^*,

$$c_k|e_1, ..., e_N, p_1, ..., p_N, \phi(u)> = u_k|e_1, ..., e_N, p_1, ..., p_N, \phi(u)>$$

The Hamiltonian of the system of electrons and positrons inside the quantum antenna is

$$H_A = \sum_{k=1}^{N}(a_k^* a_k - b_k^* b_k)$$

where the canonical anticommutation rules (CAR) are satisfied:

$$\{a_k, a_j^*\} = \delta_{kj}, \{b_k, b_j^*\} = \delta_{kj},$$

$$\{a_k, a_j\} = 0, \{b_k, b_j\} = 0, \{a_k, b_j\} = 0, \{a_k, b_j^*\} = 0,$$

$$\{a_k^*, a_j^*\} = 0, \{b_k, b_j^*\} = 0, \{a_k^*, b_j\} = 0$$

and the canonical commutation rules (CCR) are satisfied by the c_k, c_k^*:

$$[c_k, c_j^*] = \delta_{kj}, [c_k, c_j] = 0, [c_k^*, c_j^*] = 0$$

Exercise: Express $H_I(t)$ as a cubic functional of $\{c_k, c_k^*\}, \{a_k^*, b_k\}, \{a_k, b_k^*\}$ and hence design a perturbation theoretic technique for calculating the state at time t in the interaction picture:

$$|\psi(t)> = T\{exp(-i\int_0^t H_I(s)ds)\}|\psi_i>$$

Hence, calculate the state of the antenna at time t as

$$\rho_A(t) = Tr_B(|\psi(t)><\psi(t)|)$$

where Tr_B denotes partial trace over the photon state. Also calculate the entropy of the antenna at time t. Finally, using the expression

$$J^\mu(t, r) = -e\psi(t, r)^* \alpha^\mu \psi(t, r)$$

(in the interaction representation, the observables $\psi(t, r)$ evolve according to the unperturbed Hamiltonian of the electrons and positrons) calculate the far field electromagnetic field radiated out by the antenna and the moments of the associated Poynting vector in the state $|\psi(t)>$ or equivalently, in the state $\rho_A(t)$. We can express the evolution of $\rho_A(t)$ in terms of the Lindblad operators obtained by tracing out over the photon bath and hence evaluate the rate of its entropy change $-\frac{d}{dt}Tr(\rho(t).log(\rho(t)))$.

5.1.6 Entangled quantum states and the inequalities of John Bell:The impossibility of constructing a hidden variable theory in quantum mechanics

Let X_1, X_2, X_3 be three classical Bernoulli random variables on a given probability space, ie, they assume only the values ± 1. Then, it is easy to see that

$$X_1(X_2 - X_3) \leq 1 - X_2 X_3$$

and hence taking expectations,

$$\mathbb{E}(X_1 X_2) - \mathbb{E}(X_1 X_3) \leq 1 - \mathbb{E}(X_2 X_3)$$

Interchanging X_2, X_3 gives us therefore

$$|\mathbb{E}(X_1 X_2) - \mathbb{E}(X_1 X_3)| \leq 1 - \mathbb{E}(X_2 X_3)$$

This is called Bell's inequality. It is violated by quantum observables taking values ± 1 only. For example, let $\sigma_k, k = 1, 2, 3$ be the Pauli spin matrices. Their eigenvalues are ± 1. Now consider the mixed state

$$\rho = I_2/2$$

in the Hilbert space $\mathbb{C}^2$. Let a, b, c be unit vectors in $\mathbb{R}^3$ and define the observables

$$X_1 = (a, \sigma) = \sum_{k=1}^{3} a_k \sigma_k, X_2 = (b, \sigma), X_3 = (c, \sigma)$$

Then X_1, X_2, X_3 all have eigenvalues only ± 1 and are therefore quantum Bernoulli random variables. Further, their correlations in the state ρ are

$$r(k, j) = Tr(\rho X_k X_j)$$

and we find that

$$r(1, 2) = r(2, 1) = (a, b), r(2, 3) = r(3, 2) = (b, c), r(3, 1) = r(1, 3) = (a, c)$$

Now define θ_{kj} by

$$r(k, j) = cos(\theta_{kj})$$

Then if Bell's inequality is to be satisfied by these quantum correlations, we must have

$$|cos(\theta_{12}) - cos(\theta_{13})| \leq 1 - cos(\theta_{23})$$

Note that θ_{12} is the angle between the vectors a, b, θ_{23} is the angle between b, c and finally θ_{13} is the angle between a, c. Show that unit vectors a, b, c can be chosen so that this inequality is violated.

5.1.7 Communication using entangled states : Quantum teleportation and superdense coding

5.1.8 A property of quantum entropy

If A, B are two systems and $|e_i^A>, i = 1, 2, ..., n$ are orthonormal vectors in A's Hilbert space while $|e_i^B>, i = 1, 2, ..., n$ are orthonormal vectors in $B's$ Hilbert space, then consider the following pure state in the tensor product of $A's$ and $B's$ Hilbert space:

$$|\psi> = \sum_{i=1}^{n} \sqrt{p(i)}|e_i^A \otimes e_i^B>$$

where $p(i) \geq 0, \sum p(i) = 1$. Show that

$$<\psi|\psi> = 1$$

and that $A's$ state is

$$\rho_A = Tr_B(|\psi><\psi|) = \sum_i p(i)|e_i^A><e_i^A|$$

while $B's$ state is

$$\rho_B = Tr_A(|\psi><\psi|) = \sum_i p(i)|e_i^B><e_i^B|$$

Deduce that

$$S(\rho_A) = S(\rho_B)$$

where $S(.)$ denotes Von-Neumann entropy. If the joint system of A and B evolves according to the Hamiltonian

$$H = H_A + H_B + V_{AB}$$

where H_A acts in $A's$ Hilbert state only, H_B acts in $B's$ Hilbert space only and V_{AB} acts in the tensor product of the two spaces, then write down the evolution of the above state $|\psi>$ in the interaction picture assuming that $|e_i^A> ., i = 1, 2, ..., n$ are eigenstates of H_A and $|e_i^B>, i = 1, 2, ..., n$ are eigenstates of H_B.

5.1.9 Entanglement assisted quantum communication

Evaluation of the maximum rate of information transmission when the transmitter and receiver share an entangled state. [a] Let $|e_a>, a = 1, 2, ..., N$ be an onb for Alice's Hilbert space and $|f_a>, a = 1, 2, ..., N$ an onb for Bob's Hilbert space. Alice and Bob share the maximally entangled state

$$|\Phi> = N^{-1/2} \sum_{a=1}^{N} |e_a \otimes f_a>$$

Alice appends the state

$$|\psi> = \sum_{a=1}^{N} C(a)|e_a>$$

Now Alice appends this state to her share in the entangled state $|\Phi>$ that she shares with Bob. The resulting state of Alice and Bob is

$$|\chi> = N^{-1/2} \sum_a |\psi> |e_a> |f_a>$$

Now Alice applies a unitary operator W to her share of this state with W defined by the $N^2 \times N^2$ matrix

$$W(a,b|c,d)$$

defined by

$$W(|e_a> |e_b>) = \sum_{c,d} W(c,d|a,b)|c> |d>$$

Then the resulting state of Alice and Bob becomes

$$(W \otimes I_B)|\chi> = N^{-1/2} \sum W(|\psi> |e_a>)|f_a>$$

and since

$$W(|\psi> |e_a>) = \sum_b C(b) W(|e_b> |e_a>) =$$

$$= \sum_{b,c,d} W(c,d|a,b)|e_c> |e_d>$$

so that the resulting state of Alice and Bob can be written as

$$(W \otimes I_B)|\chi> =$$

$$N^{-1/2} \sum_{abcd} C(a) W(c,d|a,b)|e_c> |e_d> |f_a>$$

Now suppose that Alice applies the measurement $\{|e_a> |e_b>: 1 \le a,b \le N\}$ to her share of this state (ie the projections $|e_a> |e_b> < e_a| < e_b| = |e_a> < e_a| \otimes |e_b> < e_b| : 1 \le a,b \le N$). Then if $|e_c> |e_d>$ her outcome, the state of Bob after applying the collapse postulate becomes

$$|\eta(c,d)> = N^{-1/2} \sum_{a,b} C(a) W(c,d|a,b)|f_a>$$

Alice reports via classical communication, her measurement outcome (c,d) to Bob and hence Bob knows the numbers $C(a)\sum_b W(c,d|a,b), a = 1, 2, ..., N$ from which he gets to know the state $|\psi>$ ie the numbers $\{C(a):a=1,2,..., N\}$ that Alice wanted to transmit to him.

[b] Other descriptions of super-dense coding and quantum teleportation. Let $|e_a>, a = 1, 2, ..., d$ be an onb for $\mathbb{C}^d$. Assume that Alice and Bob share the epr state

$$\Phi_{AB} = d^{-1/2} \sum_{a=a}^{d} |e_a>^A |e_a>^B$$

For $0 \le \alpha \le d - 1$, let

$$\alpha = \sum_{k=0}^{r-1} \alpha[k] 2^k, \alpha[k] = 0, 1$$

be its binary expansion. Here, we are assuming that $d = 2^r$. Let X, Y, Z denote the 2×2 Pauli spin matrices and define

$$X(\alpha) = \otimes_{k=0}^{r-1} X^{\alpha[k]}, Y(\alpha) = \otimes_{k=0}^{r-1} Y^{\alpha[k]},$$

$$Z(\alpha) = \otimes_{k=0}^{r-1} Z^{\alpha[k]}$$

Then we have for $a, b, c, d = 0, 1$ that

$$\sum_{m=0,1} < m|Z^c X^d X^a Z^b|m >= 2\delta[a - d]\delta[b - c]$$

In fact if $a \ne d$ or $b \ne c$, then $Z^c X^d X^a Z^b$ is either proportional to X, Y or Z. The diagonal entries of X and Z are zero while the sum of the diagonal entries of Y is zero. On the other hand, if $a = d$ and $b = c$, then $Z^c X^d X^a Z^b = I_2$ and the sum of its diagonal entries is 2. This proves the claim. It follows from this identity that

$$\sum_{j=0}^{d-1} < j|Z(\beta')X(\alpha')X(\alpha)Z(\beta)|j >=$$

$$\Pi_{k=0}^{r-1} \sum_{j[k]=0,1} < j[k]|Z^{\beta'[k]} X^{\alpha'[k]} X^{\alpha[k]} Z^{\beta[k]}|j[k] >=$$

$$2^r \Pi_{k=0}^{r-1} \delta[\alpha[k] - \alpha'[k]]\delta[\beta[k] - \beta'[k]]$$
$$= d\delta[\alpha - \alpha']\delta[\beta - \beta']$$

(We denote $|e_a >$ by $|a - 1 >$ so that $a = 0, 1, ..., d - 1$). It follows from this observation that the vectors

$$|\Phi_{AB}^{\alpha,\beta}> = (X(\alpha)Z(\beta) \otimes I_d)|\Phi_{AB}>=$$

$$d^{-1/2} \sum_{j=0}^{d-1} (X(\alpha)Z(\beta)|j >^A)|j >^B, \alpha, \beta = 0, 1, ..., d - 1$$

forms an orthonormal basis for $\mathbb{C}^d \otimes \mathbb{C}^d = \mathbb{C}^{d^2}$. Note that if T is a linear transformation in $\mathbb{C}^d$ and if T^t denotes its transpose in the basis $|j >, j = 0, 1, ..., d - 1$, then

$$\sum_{j=0}^{d-1} (T|j >^A)|j >^B = \sum_{j=0}^{d-1} |j >^A T^t|j >^B$$

Now suppose Alice wishes to trasnmit a pair of numbers $(\mu, \nu), \mu, nu = 0, 1, ..., d-1$ to Bob. She Applies the gate $X(\mu)Z(\nu)$ to her share of q-dits in the shared state Φ_{AB}, so that the total state of Alice and Bob becomes

$$|\Phi_{AB}^{\mu,\nu}>= d^{-1/2}\sum_{j=0}^{d-1}(X(\alpha)Z(\beta)|j>^A)|j>^B$$

and then transmits her share of the state ($log_2(d)$ qubits) to Bob. Thus, Bob has the state $|\Phi_{AB}^{\mu,\nu}>$ and Bob makes a measurement with respect to the Bell basis defined above $\{|\Phi_{AB}^{\alpha,\beta}>, \alpha, \beta = 0, 1, ..., d-1\}$ and his measurement outcome is (μ, ν). Note that the PVM operators of Bob's measurement are $|\Phi_{AB}^{\alpha,\beta}><\Phi_{AB}^{\alpha,\beta}|, \alpha, \beta = 0, 1, ..., d-1$. Thus, by sharing $log_2 d$ qubits with Bob, Alice can transmit $2log_2(d)$ classical bits to Bob by transmitting only $log_2(d)$ qubits. This fact can be expressed by a resource inequality:

$$[qq, log_2(d)] + [q \rightarrow q, log_2(d)] \geq [c \rightarrow c, 2log_2(d)]$$

or more concisely as

$$[qq] + [q \rightarrow q] \geq [2c \rightarrow 2c]$$

5.1.10 Quantum neural networks (qnn), an example

The input states for training purpose are $|e_a>, a = 1, 2, ..., N$ and the corresponding desired output probability distributions are $P_Y(a, k), k = 1, 2, ..., N, a = 1, 2, ..., N$, ie $P_Y(a, k)$ is the desired probability of obtaining the output $|f_k>$ when the input is $|e_a>$. Here, $|e_a>, a = 1, 2, ..., N$ is an onb for the input Hilbert space $\mathcal{H}_i$ while $|f_k>, k = 1, 2, ..., N$ is an onb for the output Hilbert space $\mathcal{H}_o$. The qnn is a unitary matrix W of size $N \times N$ mapping $\mathcal{H}_i$ onto $\mathcal{H}_o$. We parametrize W by p Hermitian matrices $X_1, ..., X_p$, so that

$$W = W(\theta_1, ..., \theta_p) = W(\theta) = exp(i\sum_{k=1}^{p}\theta_k X_k)$$

The "weights" $\theta = (\theta_1, ..., \theta_p)^T$ are to be selected so that

$$|<f_k|W(\theta)|e_a>|^2 \approx P_Y(a, k), a, k = 1, 2, ..., N$$

This means that the weights θ have to be trained so that the "error energy"

$$E(\theta) = \sum_{k,a=1}^{N}|<f_k|W(\theta)|e_a>|^2 - P_Y(a, k)|^2$$

is a minimum. Such a training can be achieved for example using the gradient search algorithm. We can also talk about adaptive qnn's in, for example, the following way. Assume that the desired quantum system has Hamiltonian $H(t)$ so that the Schrodinger evolution is

$$|\psi'(t) >= -iH(t)|\psi(t) >, t \geq 0$$

$H(t)$ is to be estimated on a real time basis. Let us approximate the corresponding unitary evolution operator

$$U(t) = T\{exp(-i \int_0^t H(s)ds)\}$$

by $W(\theta(t))$. Then $\theta(t)$ is to be varied with time so that $W(\theta(t))|\psi(0) >$ follows $|\psi(t) >$. This can be achieved for example, using the gradient algorithm:

$$\theta(t + \Delta) = \theta(t) - \mu\nabla_\theta \mid\mid |\psi(t) > -W(\theta(t))|\psi(0) >\mid\mid^2$$

where μ is an adaptation constant.

5.1.11   Design of quantum gates using the interaction of electrons,positrons, photons and gravitons

The metric tensor of the gravitational field is

$$g_{\mu\nu}(x) = \eta_{\mu\nu} + h_{\mu\nu}(x)$$

where $\eta_{\mu\nu}$ is the Minkowski metric of flat space-time and $h_{\mu\nu}(x)$ is a small perturbation of flat space-time. The linearized Einstein field equations after choosing an appropriate coordinate system (harmonic coordinates) reduce to

$$\Box h_{\mu\nu}(x) = 0$$

the solution to which is given by a plane wave expansion

$$h_{\mu\nu}(x) = \int (d(K, \sigma)e_{\mu\nu}(K, \sigma)exp(-ik.x) + d(K, \sigma)^* \bar{e}_{\mu\nu}(K, \sigma)exp(ik.x))d^3K$$

where $k = (|K|, K), k.x = k_\mu x^\mu = |K|t - K.r$. The sum runs over σ taking five values $-2, -1, 0, 1, 2$. This follows from the constraints imposed by the four coordinate conditions

$$h^\mu_{\nu,\mu} - h_{,\nu}/2 = 0$$

required to obtain the wave equation from the linearized version of the Einstein field equations. These constraints lead to the condition that there are only five linearly independent linear combinations of the $e_{\mu\nu}(K)'s$ for a given value of the three vector K. For example, we can take a wave propagating along the z direction and write down these conditions to deduce this fact. This also means that gravitons have spin two. Now, we can write down the energy density of the gravitational field enclosed inside a finite volume. For doing so, we have to first write down the energy-momentum tensor of the gravitational field. That comes from Einstein field equations in the presence of matter and radiation. Suppose matter and em radiation have total energy momentum $T^{\mu\nu}$. The Einstein field equations are then

$$R^{\mu\nu} - (1/2)Rg^{\mu\nu} = -8\pi G T^{\mu\nu}$$

and the covariant divergence of both sides vanishes by virtue of the Bianchi identity:

$$(R^{\mu\nu} - (1/2)Rg^{\mu\nu})_{:\nu} = 0$$

Now define the Einstein tensor

$$G^{\mu\nu} = R^{\mu\nu} - (1/2)Rg^{\mu\nu}$$

and express it as

$$G^{\mu\nu} = G_{\mu\nu(1)} + G^{\mu\nu(2)}$$

where $G^{\mu\nu(1)}$ is linear in the $h_{\mu\nu}$ and its first and second order partial derivatives w.r.t to space-time. Then, $G^{\mu\nu(2)} = G^{\mu\nu} - G^{\mu\nu(1)}$. It is easy to show that the ordinary four divergence of $G^{\mu\nu(1)}$ vanishes:

$$G^{\mu\nu(1)}_{,\nu} = 0$$

and therefore, we get the conservation law

$$(T^{\mu\nu} + G^{\mu\nu(2)}/8\pi G)_{,\nu} = 0$$

which means that $\tau^{\mu\nu} = G^{\mu\nu(2)}/8\pi G$ must be interpreted as the pseudo-tensor of the gravitational field for this interpretation would guarantee that the total energy and momentum of matter, em radiation and gravitation is conserved. Now we can evaluate the energy density τ^{00} of the gravitational field upto quadratic orders in the $h_{\mu\nu}$ and hence the total energy of the gravitational field $\int \tau^{00} d^3 r$ upto quadratic orders in the coefficients $d(K, \sigma), d(K, \sigma)^*$. This would then ensure that upto second order, the Hamiltonian of the gravitational field is that of an ensemble of harmonic oscillators. The result of this calculation will yield the Hamiltonian of the gravitational field in the form

$$H_G = \int f(K, \sigma) d(K < \sigma)^* d(K, \sigma) d^3 K$$

with the Bosonic commutation relations

$$[d(K, \sigma), d(K', \sigma')^*] = \delta^3(K - K')\delta(\sigma, \sigma')$$

After discretizing this, we can write

$$H_G = \sum_{k=1}^{N_G} f_1[k]d[k]^*d[k]$$

Now the total unperturbed Hamiltonian of the electron-positron field (Dirac field) is given by

$$H_D = \sum_{k=1}^{N_D} E[k](a[k]^* a[k] - b[k]^* b[k])$$

and the unperturbed Hamiltonian of the photon field (Maxwell field) is given by

$$H_M = \sum_{k=1}^{M} f_2[k] c[k]^* c[k]$$

The interaction Hamiltonian between the Dirac field and the gravitational field is obtained using the spinor connection for the gravitational field. It is quadratic in the Dirac field but highly nonlinear in the gravitational field. It is represented by

$$H_{DG}(t) = -e \int V_a^\mu(x) Re(\psi(x)^* \alpha^a \Gamma_\mu(x) \psi(x)) \sqrt{-g(x)} d^3r$$

where $\alpha^a = \gamma^0 \gamma^a$ and $V_a^\mu(x)$ is the tetrad of the gravitational metric $g_{\mu\nu}(x)$. $\Gamma_\mu(x) = (-1/2) V^\nu)a(x) v_{b\nu:\mu}(x) J^{ab}$ is the spinor connection of the gravitational field where $J^{ab} = (1/4)[\gamma^a, \gamma^b]$ are the Lie algebra generators of the Dirac spinor representation of the Lorentz group. We can express H_{DG} as

$$H_{DG}(t) = \sum_{r,s} (F_{1rs}(t, d[m], d[m]^*, m = 1, 2, ..., N_G) a[r]^* a[s] +$$

$$+F_{2rs}(t, d[m], d[m]^*, m = 1, 2, ..., N_G) b[r] b[s]^* + F_{3rs}(t, d[m], d[m]^*,$$

$$m = 1, 2, ..., N_G) a[r]^* b[s]^*) + cc.$$

where cc denotes the adjoint of the previous terms. The interaction Hamiltonian between the Dirac field and the photon field is simpler. It is given by

$$H_{DEM}(t) = -e \int V_a^\mu(x) \psi(x)^* \alpha^a \psi(x) A_\mu(x) d^3r$$

We write

$$V_a^\mu(x) = \delta_a^\mu + U_a^\mu(x)$$

and then express

$$H_{DEM}(t) = H_{DEM0}(t) + H_{DEMG}(t)$$

where $H_{DEM0}(t)$ does not involve the gravitational tetrad V_a^μ, ie, V_a^μ is replaced by δ_a^μ while $H_{DEMG}(t)$ involves all the three fields $U_a^\mu, \psi(x), A_\mu(x)$. $H_{DEM}(t)$ can therefore be represented using the quantum Maxwell field representation

$$A_\mu(x) = \sum_k (c[k] \chi_{k\mu}(x) + c[k]^* \bar{\chi}_{k\mu}(x))$$

Thus, we can write

$$H_{DEM0}(t) = \sum_{kml} (g_1(t) a[k]^* a[m] c[l] + g_2(t) b[k] b[m]^* c[l] + g_3(t) a[k]^* b[m]^* c[l] + g_4(t) b[k] a[m] c[l]) + cc.$$

$$H_{DEMG}(t) = \sum_{kml}(L_1(t,d[m],d[m]^*,$$

$$m = 1,2,...,N_G))a[k]^*a[m]c[l] + L_2(t,d[m],d[m]^*, m = 1,2,...,N_G)b[k]b$$

Finally, we describe the interaction energy between the Maxwell photon field and the gravitational field. The corresponding Lagrangian density is proportional to $F_{\mu\nu}F^{\mu\nu}\sqrt{-g}$. This is quadratic in the EM potentials but highly nonlinear in the metric of space-time. If we subtract out the part from this not involving the metric perturbations $h_{\mu\nu}$, we get the Lagrangian density of the free electromagnetic field, ie, em field in flat space-time. Thus, the interaction Hamiltonian between the Maxwell field and the gravitational field can be expressed as

$$H_{GEM}(t) = \sum_{k,r}(G_{1kr}(t,d[m],d[m]^*, m = 1,2,...,N_G)c[k]c[r] + G_{2kr}(t,d[m],d[m]^*,$$

$$m = 1,2,...,N_G)c[k]^*c[r]) +$$

The total Hamiltonian is

$$H(t) = H_G + H_D + H_{EM} + H_{DG}(t) + H_{DEM0}(t) + H_{DEMG}(t)$$

It should be noted that in the above expression for H_G, we've taken only quadratic terms in the $d[m], d[m]^*$ into account. To be more accurate, we have to express the energy density of the gravitational field τ^{00} as a power series in $d[m], d[m]^*$, ie, H_G would be replaced by $H_{G0} + H_{G1}$ where H_{G0} is the H_G above, ie, quadratic in $d[m], d[m]^*$ while H_{G1} contains cubic and higher powers of $d[m], d[m]^*$.

5.1.12 The notion of measurement in the quantum theory

Let ρ be a mixed state in a Hilbert space $\mathcal{H}$ and let $|\psi>$ be a pure state in another Hilbert space $\mathcal{K}$. Then $\rho \otimes |\psi><\psi|$ is a mixed state in $\mathcal{H} \otimes \mathcal{K}$. Let U be a unitary operator in this tensor product space and consider the state

$$T(\rho) = Tr_{\mathcal{K}}(U(\rho \otimes |\psi><\psi|)U^*)$$

We can write

$$U = \sum_{m=1}^{N} V_m \otimes W_m$$

where V_m, W_m are linear operators respectively in the spaces $\mathcal{H}$ and $\mathcal{K}$. Then, we find that

$$T(\rho) = \sum_{m,n=1}^{N} V_m\rho V_n jTr(W_m|\psi><\psi|W_n^*)$$

$$= \sum_{m,n=1}^{N} <\psi|W_n^*W_m|\psi> V_m\rho V_n^*$$

Now $((< \psi | W_n^* W_m | \psi >))$ is clearly an $N \times N$ positive definite complex matrix and hence, it has an eigendecompostion

$$\sum_{r=1}^{N} \lambda[r] |e_r >< e_r|, \lambda[r] \geq 0, < e_r | e_s >= \delta_{rs}$$

or equivalently in terms of components,

$$< \psi | W_n^* W_m | \psi >= \sum_{r=1}^{N} \lambda[r] \bar{e}_r[n] e_r[m]$$

From this, we get

$$T(\rho) = \sum_{m,n,r=1}^{N} \lambda[r] \bar{e}_r[n] e_r[m] V_m \rho V_n^*$$

$$= \sum_{r=1}^{N} E_r \rho E_r^*$$

where E_r is a linear operator in $\mathcal{H}$ defined by

$$E_r = \sqrt{\lambda[r]} \sum_{m=1}^{N} e_r[m] V_m$$

Clearly, the condition $Tr(\rho) = 1$ implies $Tr(T(\rho)) = 1$ and hence

$$\sum_{r=1}^{N} E_r^* E_r = I_{\mathcal{H}}$$

5.2   The Baker-Campbell-Hausdorff formula. A, B are n x n matrices

The aim is to define the matrix $C(t)$ by

$$exp(C(t)) = exp(tA).exp(tB)$$

or equivalently,

$$C(t) = log(exp(tA).exp(tB))$$

and obtain a Taylor expansion for $C(t)$ with matrix coefficients. We note that by the formula for the differential of the exponential map,

$$\frac{d}{dt}(exp(tA).exp(tB)) = exp(tA)(A+B)exp(tB)$$

$$= exp(C(t))((I-exp(-ad(C(t))/ad(C(t)))(C'(t))$$

or equivalently,

$$exp(-t.ad(B))(A) + B = exp(-tB)(A+B).exp(tB) = g(ad(C(t))^{-1}(C'(t))$$

where

$$g(z) = \frac{z}{1 - exp(-z)}$$

Thus,

$$C'(t) = g(ad(C(t))(exp(-t.ad(B))(A) + B) --- (1)$$

By writing

$$C(t) = \sum_{m \geq 0} C_m t^m$$

and substituting this into the differential equation (1), we can successively determine the coefficients $C_m, m = 0, 1, ...$ Note that $C_0 = I$. We will of course require the Taylor expansion of $g(z)$.

5.3 Yang-Mills radiation field (an approxima-tion)

$\tau_a, a = 1, 2, ..., N$ are the Hermitian generators of the gauge group $G \subset U(N)$ and $C(abc)$ are the associated structure constants:

$$[\tau_a, \tau_b] = C(abc)i\tau_c$$

The gauge field is

$$A_\mu(x) = A_\mu^a(x)\tau_a$$

and the covariant derivative is

$$\nabla_\mu = \partial_\mu + ieA_\mu$$

The field tensor $F_{\mu\nu}(x)$ is defined by the curvature of this connection:

$$ieieF_{\mu\nu} = [\nabla_\mu, \nabla_\nu] = ie(A_{\nu,\mu} - A_{\mu,\nu}) - e^2[A_\mu, A_\nu]$$

$$= [ie(A_{\nu,\mu}^a - A_{\mu,\nu}^a) - e^2 C(bca)A_\mu^b A_\nu^c]i\tau_a$$

so writing

$$F_{\mu\nu} = F_{\mu\nu}^a \tau_a$$

we get

$$F_{\mu\nu}^a = A_{\nu,\mu}^a - A_{\mu,\nu}^a - eC(bca)A_\mu^b A_\nu^c$$

The field equations are derived from the action principle

$$\delta \int F_{\mu\nu}^a F^{a\mu\nu} d^4x = 0$$

This gives

$$F^{a\mu\nu}_{,\mu} + eC(acb)A^c_\nu F^{b\mu\nu} = 0$$

or equivalently,

$$A^{a,\nu}_{\nu,\mu} - A^{a,\nu}_{\mu,\nu} - eC(bca)(A^b_\mu A^c_\nu)^{,\nu}$$

$$+e^2 C(acb)C(pqb)A^{c\nu}A^p_\mu A^q_\nu = 0$$

This contains linear, quadratic and cubic terms in the gauge potentials $A^a_\mu(x)$. We impose a Lorentz gauge condition

$$A^{a,\mu}_\mu = 0$$

or equivalently,

$$A^{a\mu}_{,\mu} = 0$$

Then, the field equations simplify to

$$A^{a,\nu}_{\mu,\nu} - eC(bca)A^c_\nu A^{b,\nu}_\mu$$

$$+e^2 C(acb)C(pqb)A^{c\nu}A^p_\mu A^q_\nu = 0 ---(1)$$

We solve this approximately upto $O(e^2)$ using perturbation series taking e as the perturbation parameter. Specifically,

$$A^a_\mu = A^{a(0)}_\mu + eA^{(a)(1)}_\mu + e^2 A^{a(2)}_\mu + ...$$

Substituting this into (1) and equating coefficients of e^0, e^1, e^2 respectively gives

$$\Box A^{a(0)}_\mu(x) = 0, ---(2)$$

$$\Box A^{a(1)}_\mu(x) =$$

$$-C(bca)A^{c(0)}_\nu A^{b(0),\nu}_\mu$$

$$+e^2 C(acb)C(pqb)A^{c\nu(0)}A^{p(0)}_\mu A^{q(0)}_\nu = 0 ---(3)$$

$$\Box A^{a(2)}_\mu =$$

$$C(bca)(A^{c(0)}_\nu A^{b(1),\nu}_\mu + A^{c(1)}_\nu A^{b(0),\nu}_\mu$$

$$-C(acb)C(pqb)A^{c\nu(0)}A^{p(0)}_\mu A^{q(0)}_\nu$$

5.4 Belavkin filter applied to estimating the spin of an electron in an external magnetic field. We assume that the magnetic field is $B_0(t) \in \mathbb{R}^3$

The spin magnetic moment of the electron is

$$\mu = geh\sigma/8\pi m = a\sigma$$

say, where

$$\sigma = (\sigma_1, \sigma_2, \sigma_3)$$

are the three Pauli spin matrices and we define

$$\sigma_0 = I_2$$

The HP Schrodinger equation is written down taking the system Hilbert space as

$$\mathfrak{h} = \mathbb{C}^2$$

and the noise bath space as

$$\Gamma_s(L^2(\mathbb{R}_+))$$

It is given by

$$dU(t) = (-(iH(t) + P)dt + L_1 dA + L_2 dA^* + Sd\Lambda)U(t)$$

where

$$H(t), P, L_1, L_2, S \in \mathbb{C}^{2 \times 2} = \mathcal{B}(\mathfrak{h}), H(t) = (\mu, B_0(t)) = a(\sigma, B_0(t))$$

The star unital homomorphism associated with $U(t)$ is given by

$$j_t(X) = U(t)^* X U(t)$$

where $X \in \mathfrak{h} \otimes \Gamma_s(L^2(\mathbb{R}_+))$. The Belavkin filter measurement is taken as a mixture of quantum Brownian motion and the quantum Poisson process:

$$Y_o(t) = U(t)^* Y_i(t) U(t), Y_i(t) = A(t) + A(t)^* + c\Lambda(t) = B(t) + c\Lambda(t)$$

where

$$B(t) = A(t) + A(t)^*$$

is classical Brownian motion. The quantum Ito table is

$$dA.dA = 0, dA.dA^* = dt, dA.d\Lambda = dA, dA^*dA = 0, dA^*dA^* = 0, dA^*d\Lambda = 0,$$

$$d\Lambda.dA = 0, d\Lambda.dA^* = dA^*, d\Lambda.d\Lambda = d\Lambda$$

Using this table, we compute

$$(dY_i(t))^2 = dt + c.dB + c^2 d\Lambda,$$

and in general, for $n \geq 1$ we write

$$(dY_i(t))^n = a[n]dt + b[n]dB(t) + d[n]d\Lambda(t)$$

Then, we get

$$(dY_i(t))^{n+1} = (dB + cd\Lambda)(b[n]dB + d[n]d\Lambda) =$$

$$b[n]dt + cd[n]d\Lambda + d[n]dA + cb[n]dA^* = a[n+1]dt + b[n+1]dB + d[n+1]d\Lambda$$

from which we infer the recursions

$$a[n+1] = b[n], b[n+1] = d[n] = cb[n], d[n+1] = cd[n], n \geq 1$$

with the initial conditions

$$a[1] = 0, b[1] = 1, d[1] = c$$

The solution to this recursion is

$$b[n] = c^{n-1}, d[n] = c^n, a[n+1] = c^{n-1}, n \geq 1$$

Thus, for $n \geq 2$, we get

$$(dY_i(t))^n = c^{n-2}dt + c^{n-1}dB(t) + c^n d\Lambda(t)$$

$$= c^{n-2}dt + c^{n-1}(dB + cd\Lambda) = c^{n-2}dt + c^{n-1}dY_o, n \geq 2$$

We now have for a system observable X,

$$dj_t(X) = j_t(\theta_{0t}(X))dt + j_t(\theta_1(X))dA(t) + j_t(\theta_2(X))dA(t)^* + j_t(\theta_3(X))d\Lambda(t)$$

where

$$\theta_{0t}(X) = i[H(t), X] + L_2^* X L_2 - P^* X - XP$$

$$\theta_1(X) = L_2^* X + XL_1 + L_2^* XS, \theta_2(X) = L_1^* X + XL_2 + S^* XL_2,$$

$$\theta_3(X) = S^* XS + S^* X + XS$$

Also,

$$dY_o(t) = dY_i(t) + dU(t)^* dY_i(t)U(t) + U(t)^* dY_i(t)dU(t)$$

$$dA.dY_i = dt + cdA, dA^*.dY_i = cdA^*, d\Lambda.dY_i = dA^* + cd\Lambda,$$

$$dY_i dA = cdA, dY_i dA^* = dt + cdA^*, dY_i d\Lambda = dA + cd\Lambda$$

We therefore deduce that

$$dY_o(t) = dY_i(t) + j_t(cL_1^* + cL_2 + S^*)dA^* + j_t(cL_2^* + cL_1 + S)dA$$

$$+cj_t(S^* + S)d\Lambda$$

$$= j_t(cL_2^* + cL_1 + S + 1)dA + j_t(cL_1^* + cL_2 + S^* + 1)dA^* + cj_t(S^* + S + 1)d\Lambda$$

Problem: Calculate $(dY_o(t))n, n = 2, 3, \ldots$ using recursion formulas.

hint: Let

$$(dY_o(t))^n = j_t(P_0[n])dt + j_t(P_1[n])dA(t) + j_t(P_2[n])dA(t)^* + j_t(P_3[n])d\Lambda(t), n \geq 1$$

where $P_k[n], k = 0, 1, 2, 3$ are 2×2 matrices, ie, system matrices. Then using the Homomorphism property of j_t and quantum Ito's formula, we deduce that

$$j_t(P_0[n+1])dt + j_t(P_1[n+1])dA(t) + j_t(P_2[n+1])dA(t)^* + j_t(P_3[n+1])d\Lambda(t) =$$

$$(dY_o(t))^{n+1} =$$

$$(j_t(cL_2^* + cL_1 + S + 1)dA + j_t(cL_1^* + cL_2 + S^* + 1)dA^* + cj_t(S + S^* + 1)d\Lambda)$$

$$\times (j_t(P_0[n])dt + j_t(P_1[n])dA(t) + j_t(P_2[n])dA(t)^* + j_t(P_3[n])d\Lambda(t))$$

$$= j_t((cL_2^* + cL_1 + S + 1)P_2[n])dt + j_t((cL_2^* + cL_1 + S^* + 1)P_3[n])dA$$

$$+ j_t(c(S + S^* + 1)P_2[n])dA^*$$

$$+ j_t(c(S + S^* + 1)P_3[n])d\Lambda$$

and therefore

$$P_0[n+1] = (cL_2^* + cL_1 + S + 1)P_2[n], P_1[n+1] = (cL_2^* + cL_1 + S^* + 1)P_3[n],$$

$$P_2[n+1] = c(S + S^* + 1)P_2[n], P_3[n+1] = c(S + S^* + 1)P_3[n]$$

The initial conditions are

$$P_0[1] = 0, P_1[1] = cL_2^* + cL_1 + S + 1, P_2[1] = cL_1^* + cL_2 + S^* + 1, P_3[1] = c(S + S^* + 1)$$

The solution to the above recursion is therefore

$$P_3[n] = (c(S + S^* + 1))^n, P_1[n+1] = (cL_2^* + cL_1 + S^* + 1)(c(S + S^* + 1))^n,$$

$$P_2[n] = (c(S + S^* + 1))^{n-1}(cL_1^* + cL_2 + S^* + 1),$$

$$P_0[n+1] = (cL_2^* + cL_1 + S + 1)(c(S + S^* + 1))^{n-1}(cL_1^* + cL_2 + S^* + 1), n \geq 1$$

We can solve for $dA, dA^*, d\Lambda$ in terms of $dt, dY_o, (dY_o)^2$ and $(dY_o)^3$ with coefficients being j_t of some system operators. Specifically, we consider the three linear equations

$$(dY_o)^n - j_t(P_0[n])dt = j_t(P_1[n])dA + j_t(P_2[n])dA^* + j_t(P_3[n])d\Lambda, n = 1, 2, 3$$

and use $j_t(X)^{-1}j_t(Y) = j_t(X^{-1}Y)$ and analogous relations to solve for $dA, dA^*, d\Lambda$ as linear combinations of $dt, (dY_o)^n, n = 1, 2, 3$ with coefficients being j_t of some functions of the system operators $P_k[n], k = 0, 1, 2, 3, n = 1, 2, 3$. The generalized Belavkin filtering equation can be expressed as

$$d\pi_t(X) = F_t(X)dt + \sum_{k \geq 1} G_{kt}(X)(dY_o(t))^k$$

where

$$F_t(X), G_{kt}(X) \in \eta_t = \sigma(Y_o(s) : s \leq t)$$

We note that by definition,

$$\pi_t(X) = \mathbb{E}[j_t(X)|\eta_t]$$

and hence (the orthogonality principle)

$$\mathbb{E}[(j_t(X) - \pi_t(X))C_t] = 0, C_t \in \eta_t$$

Defining $C_t \in \eta_t$ by

$$dC_t = \sum_{k \geq 1} f_k(t)(dY_o(t))^k C_t, C_0 = 1$$

we get by applying quantum Ito's formula to the above equation and using the arbitrariness of the complex functions $f_k(t)$ that

$$\mathbb{E}(dj_t(X) - d\pi_t(X)|\eta_t) = 0 - - - (1),$$

$$\mathbb{E}[(dj_t(X) - d\pi_t(X))(dY_o(t))^k|\eta_t] + \mathbb{E}[(j_t(X) - \pi_t(X))(dY_o(t))^k|\eta_t] = 0, j \geq 1 - - - (2)$$

Thus, assuming that expectations are in the state $|f \otimes \phi(u) >$ where $|\phi(u) >= exp(-|u|^2/2)|e(u) >$ with $|e(u) >$ the standard exponential vector and $f \in \mathbb{C}^2, |f| = 1$, we get

$$\pi_t(\theta_{0t}(X) + u(t)\theta_1(X) + \bar{u}(t)\theta_2(X) + |u(t)|^2\theta_3(X)) =$$

$$F_t(X) + \sum_{k \geq 1} G_{kt}(X)\pi_t(P_0[k] + u(t)P_1[k] + \bar{u}(t)P_2[k] + |u(t)|^2P_3[k]) - - - (3)$$

and

$$\mathbb{E}(j_t(\theta_1(X))dA(t)(dY_o(t))^k|\eta_t) + \mathbb{E}(j_t(\theta_2(X))dA(t)^*(dY_o(t))^k|\eta_t)$$

$$+\mathbb{E}(j_t(\theta_3(X))d\Lambda(t)(dY_o(t))^k|\eta_t)$$

$$-F_t(X)\pi_t(P_0[k] + u(t)P_1[k] + \bar{u}(t)P_2[k] + |u(t)|^2P_3[k])dt$$

$$-\sum_{m \geq 1} G_{mt}(X)\pi_t(P_0[k+m] + u(t)P_1[k+m] + \bar{u}(t)P_2[k+m] + |u(t)|^2P_3[k+m])dt$$

$$+\pi_t(X(P_0[k] + u(t)P_1[k] + \bar{u}(t)P_2[k] + |u(t)|^2P_3[k]))dt$$

$$-\pi_t(X)\pi_t(P_0[k] + u(t)P_1[k] + \bar{u}(t)P_2[k] + |u(t)|^2P_3[k])dt = 0 - - - (4a)$$

Now,

$$dA(t).(dY_o(t))^k = j_t(P_2[k])dt + j_t(P_3[k])dA(t)$$

so that

$$\mathbb{E}(j_t(\theta_1(X))dA(t).(dY_o(t))^k|\eta_t) = \pi_t(\theta_1(X)(P_2[k]) + u(t)P_3[k]))dt$$

$$dA(t)^*(dY_o(t))^k = 0$$

$$d\Lambda(t)(dY_o(t))^k = j_t(P_2[k])dA(t)^* + j_t(P_3[k])d\Lambda(t)$$

and hence

$$\mathbb{E}(j_t(\theta_3(X))d\Lambda(t)(dY_o(t))^k|\eta_t) = \pi_t(\theta_3(X)(\bar{u}(t)P_2[k] + |u(t)|^2 P_3[k]))dt$$

Hence, (4a) can be expressed as

$$\pi_t(\theta_1(X)(P_2[k] + u(t)P_3[k]))$$

$$+\pi_t(\theta_3(X)(\bar{u}(t)P_2[k] + |u(t)|^2 P_3[k]))$$

$$+\pi_t(X(P_0[k] + u(t)P_1[k] + \bar{u}(t)P_2[k] + |u(t)|^2 P_3[k]))$$

$$-\pi_t(X)\pi_t(P_0[k] + u(t)P_1[k] + \bar{u}(t)P_2[k] + |u(t)|^2 P_3[k]) =$$

$$F_t(X)\pi_t(P_0[k] + u(t)P_1[k] + \bar{u}(t)P_2[k] + |u(t)|^2 P_3[k])$$

$$+\sum_{m\geq 1} G_{mt}(X)\pi_t(P_0[k+m]+u(t)P_1[k+m]+\bar{u}(t)P_2[k+m]+|u(t)|^2 P_3[k+m]), k \geq 1 ---(4b)$$

(3) and (4b) are to be solved for $F_t(X), G_{kt}(X), k \geq 1$ giving the desired generalized Belavkin filter. Eliminating $F_t(X)$ using (3), we can express the filter as

$$d\pi_t(X) =$$

$$\pi_t(\theta_{0t}(X) + u(t)\theta_1(X) + \bar{u}(t)\theta_2(X) + |u(t)|^2\theta_3(X))dt$$

$$+\sum_{k\geq 1} G_{kt}(X)((dY_o(t))^k - \pi_t(P_0[k] + u(t)P_1[k] + \bar{u}(t)P_2[k] + |u(t)|^2 P_3[k])dt)$$

where $G_{kt}(X), k \geq 1$ satisfy

$$\pi_t(\theta_1(X)(P_2[k] + u(t)P_3[k])) + \pi_t(\theta_3(X)(\bar{u}(t)P_2[k] + |u(t)|^2 P_3[k]))$$

$$+\pi_t(X(P_0[k] + u(t)P_1[k] + \bar{u}(t)P_2[k] + |u(t)|^2 P_3[k]))$$

$$-\pi_t(X)\pi_t(P_0[k] + u(t)P_1[k] + \bar{u}(t)P_2[k] + |u(t)|^2 P_3[k]) =$$

$$(\pi_t(\theta_{0t}(X)+u(t)\theta_1(X)+\bar{u}(t)\theta_2(X)+|u(t)|^2\theta_3(X))$$

$$-\sum_{m\geq 1} G_{mt}(X)\pi_t(P_0[m]+u(t)P_1[m]+\bar{u}(t)P_2[m]+|u(t)|^2 P_3[m]))$$

$$\times \pi_t(P_0[k] + u(t)P_1[k] + \bar{u}(t)P_2[k] + |u(t)|^2 P_3[k])$$

$$+\sum_{m\geq 1} G_{mt}(X)\pi_t(P_0[k+m]+u(t)P_1[k+m]+\bar{u}(t)P_2[k+m]+|u(t)|^2 P_3[k+m]), k \geq 1 ---(4b)$$

or equivalently,

$$\pi_t(\theta_1(X)(P_2[k] + u(t)P_3[k])) + \pi_t(\theta_3(X)(\bar{u}(t)P_2[k] + |u(t)|^2 P_3[k]))$$

$$+\pi_t(X(P_0[k] + u(t)P_1[k] + \bar{u}(t)P_2[k] + |u(t)|^2 P_3[k]))$$

$$-\pi_t(X)\pi_t(P_0[k] + u(t)P_1[k] + \bar{u}(t)P_2[k] + |u(t)|^2 P_3[k]) =$$

$$(\pi_t(\theta_{0t}(X) + u(t)\theta_1(X) + \bar{u}(t)\theta_2(X) + |u(t)|^2\theta_3(X)).\pi_t(P_0[k] + u(t)P_1[k] + \bar{u}(t)P_2[k] + |u(t)|^2P_3[k])$$

$$+ \sum_{m \geq 1} G_{mt}(X)[\pi_t(P_0[k+m] + u(t)P_1[k+m] + \bar{u}(t)P_2[k+m] + |u(t)|^2P_3[k+m])$$

$$-\pi_t(P_0[m] + u(t)P_1[m] + \bar{u}(t)P_2[m] + |u(t)|^2P_3[m])\pi_t(P_0[k] + u(t)P_1[k] + \bar{u}(t)P_2[k] + |u(t)|^2P_3[k])] --- (5)$$

Chapter 6

Pattern classification for image fields in motion using Lorentz group representations

6.1 SL(2,C), SL(2,R) and image processing

For dealing with images having rapid time variations, special relativity must be taken into account while dealing with measurements of the images in different inertial frames. Specifically, if (t, x, y, z) is a coordinate system in the frame K and the image field in this coordinate system is $f(t, x, y, z) = f(t, r)$, then in a frame K' that is moving relative to K with a uniform velocity after rotation, the coordinates (t', r') are related to the coordinates (t, r) in K by a Lorentz transformation

$$(t', r')^T = L(t, r)^T$$

L will be a proper orthochronous Lorentz transformation. Writing this as

$$x'_r = L_{rs} x_s, r = 0, 1, 2, 3$$

with summation over the repeated index $s = 0, 1, 2, 3$ being implied where $x_0 = t, x_1 = x, x_2 = y, x_3 = z$ and likewise for $x'_r, r = 0, 1, 2, 3$, we have

$$x_0'^2 - x_1'^2 - x_2'^2 - x_3'^2 = x_0^2 - x_1^2 - x_2^2 - x_3^2$$

and hence we deduce that

$$L_{00}^2 - L_{10}^2 - L_{20}^2 - L_{30}^2 = 1$$

so that

$$L_{00} = \pm\sqrt{1 + L_{10}^2 + L_{20}^2 + L_{30}^2}$$

In particular,

$$|L_{00}| \geq 1$$

Note that since L^T is also a Lorentz transformation, we have

$$L_{00}^2 - L_{01}^2 - L_{02}^2 - L_{03}^2 = 1$$

and hence

$$L_{00} = \pm\sqrt{1 + L_{01}^2 + L_{02}^2 + L_{03}^2}$$

If $L_{00} \geq 1$, then we say that L is an orthochronous Lorentz transformation. In this case, suppose $x_0^2 - x_1^2 - x_2^2 - x_3^2 > 0$ and $x_0 > 0$. Then,

$$x_0' = L_{00}x_0 + L_{01}x_1 + L_{02}x_2 + L_{03}x_3$$

$$= \sqrt{1 + L_{01}^2 + L_{02}^2 + L_{03}^2}x_0 + L_{01}x_1 + L_{02}x_2 + L_{03}x_3$$

$$\geq \sqrt{1 + L_{01}^2 + L_{02}^2 + L_{03}^2}x_0 - \sqrt{L_{01}^2 + L_{02}^2 + L_{03}^2}\sqrt{x_1^2 + x_2^2 + x_3^2}$$

$$\geq \sqrt{L_{01}^2 + L_{02}^2 + L_{03}^2}(x_0 - \sqrt{x_1^2 + x_2^2 + x_3^2}) > 0$$

This property characterizes orthchronocity of a Lorentz transformation. A Lorentz transformation has the property that det $L = \pm 1$ If $detL = 1$, we say that L is proper. All proper orthochronous Lorentz transformations form a group and this group is generated by proper rotations and boosts of the space-time manifold. This group is denoted by $G_0 = \{L : L \in G : detL = 1, L_{00} \geq 1\}$ where G is the group of all Lorentz transformations, ie all L such that $L^T \eta L = \eta$ where $\eta = diag[1, -1, -1, -1]$ is the Minkowski metric. G_0 is the connected component of G containing the identity transformation. $SL(2, \mathbb{C})$ is the covering group of G_0. In fact for any $L \in G_0$, there exists a unique pair $\pm A$ with $A \in SL(2, \mathbb{C})$ such that

$$A\Phi(t,r)A^* = \Phi(L(t,r))$$

where

$$\Phi(t,r) = \begin{pmatrix} t + z & x - iy \\ x + iy & t - z \end{pmatrix}$$

We see therefore that G_0 is isomorphic to the group $SL(2, \mathbb{C})/\{\pm I\}$. If we restrict to $SL(2, \mathbb{R})$, then we obtain all Lorentz transformations in the space $(t, x, 0, z)$, ie, rotations and boosts in the $x - z$ plane. Thus, we can use the characters of the discrete series to do pattern recognition for planar image field moving at relativistic speeds.

Exercise: Let $A \in SL(2, \mathbb{R})$. Assume that L is the Lorentz transformation defined by

$$\Phi(L(t,r)) = A\Phi(t,r)A^*$$

Show that L fixes y. In fact this is obvious since y occurs with a factor of i in $\Phi(t,r)$ while t, x, z appear with factors ± 1. Hence if A is a real $SL(2, \mathbb{C})$ matrix, it would not touch the imaginary components $\pm iy$ in $\Phi(t,r)$.

The discrete series for $SL(2, \mathbb{R})$: Let H, X, Y be the standard generators of $SL(2, \mathbb{C})$, ie,

$$H = \begin{pmatrix} 1 & 0 \\ 0 & -1 \end{pmatrix},$$

$$X = \begin{pmatrix} 0 & 1 \\ 0 & 0 \end{pmatrix},$$

$$Y = \begin{pmatrix} 0 & 0 \\ 1 & 0 \end{pmatrix}$$

These satisfy the commutation relations

$$[H, X] = 2X, [H, Y] = -2Y, [X, Y] = H$$

The irreducible representations of $SL(2, \mathbb{C})$ are obtained as follows. Suppose first that there is a highest weight vector v_0 in this representation π. Then,

$$X.v_0 = 0, H.v_0 = \lambda_0 v_0$$

where by $X.v_0$, we mean $\pi(X).v_0$ etc with π denoting the irreducible representation of the Lie algebra $\mathfrak{sl}(2, \mathbb{R})$. Then the commutation relations imply on defining

$$v_s = Y^s v_0, s = 0, 1, 2, \ldots$$

that

$$H.v_s = (\lambda_0 - 2s)v_s, Y v_s = v_{s+1}, s \geq 0$$

and since the representation is irreducible, we must have

$$X v_s = \alpha(s)v_{s-1}$$

for some scalars $\alpha(s)$. We find that

$$\alpha(s)v_s = \alpha(s)Y v_{s-1} = Y X v_s = [Y, X]v_s + XY v_s$$
$$= -H v_s + X v_{s+1} = -(\lambda_0 - 2s)v_s + \alpha(s+1)v_s$$

or

$$\alpha(s) = \alpha(s+1) - (\lambda_0 - 2s)$$

With the initial condition

$$\alpha(0) = 0$$

we get

$$\alpha(s) = \sum_{k=0}^{s-1}(\lambda_0 - 2k) = \lambda_0 s - s(s-1) = s(\lambda_0 - s + 1)$$

We thus have an infinite dimensional representation of $SL(2, \mathbb{R})$ spanned by the vectors $\{v_s : s \geq 0\}$ with the Lie algebra actions

$$H v_s = (\lambda_0 - 2s)v_s, X v_s = s(\lambda_0 - s + 1)v_{s-1}, Y v_s = v_{s+1}, s \geq 0$$

Suppose that for some $s_0 = 1, 2, ...$, we have $\lambda_0 - s_0 + 1 = 0$, ie, $\lambda_0 = s_0 - 1$. Then, $span\{v_s : s \geq s_0\}$ would be an invariant subspace for the representation and hence the representation would not be irreducible. This means that the representation is irreducible iff λ_0 is not a non-negative integer. Suppose $\lambda_0 = -m$ for some $m \geq 1$. Then we get an irreducible representation of $SL(2, \mathbb{R})$ and this representation is in the discrete series. Likewise, we can start with a lowest weight vector and generate another infinite dimensional irreducible representation in the discrete series.

Now define the $sl(2, \mathbb{C})$ elements

$$H' = i(X - Y), X' = (H - i(X + Y))/2, Y' = (H + i(X + Y))/2$$

Then,

$$[H', X'] = i[X - Y, H]/2 + [X, Y] = -iX - iY + H = 2X',$$

$$[H', Y'] = i[X - Y, H]/2 - [X, Y] = -iX - iY - 2H = -2Y'$$

$$[X', Y'] = i[H, X + Y]/2 = i(X - Y) = H'$$

Hence, $\{H', X', Y'\}$ satisfy the same commutation relations as $\{H, X, Y\}$ and therefore the Lie algebra spanned by $\{H', X', Y'\}$ is isomorphic to $sl(2, \mathbb{C})$. More precisely, the map $\{H, X, Y\} \rightarrow \{H', X', Y'\}$ is a Lie algebra automorphism which is easily proved to be an inner automorphism. It follows that if π is the same irreducible representation of $sl(2, \mathbb{C})$ corresponding to $\lambda_0 = -m$ as defined above, then the character of the corresponding Lie group representation evaluated at $exp(tH')$ is the same as its character evaluated at $exp(tH)$. Thus, the character of this representation evaluated at $exp(\theta(X - Y))$ is the same as the character of this representation evaluated at $exp(-i\theta H)$. Now as seen above, the character of the above discrete series representation for $\lambda_0 = -m$ evaluated at $exp(tH)$ is given by

$$\chi_m(exp(tH)) = \sum_{s \geq 0} exp(-t(m + 2s)) = \frac{exp(-mt)}{1 - exp(-2t)}$$

$$= \frac{exp(-(m-1)t)}{exp(t) - exp(-t)}$$

and hence, it follows by the above argument, that the character of this same representation evaluated at $u(\theta) = exp(\theta(X - Y))$ is given by

$$\chi_m(u(\theta)) = \frac{exp(i(m-1)\theta)}{exp(-i\theta) - exp(i\theta)}$$

Remark: $\mathbb{R}.H$ and $\mathbb{R}.(X - Y)$ are two non-conjugate Cartan subalgebras of $sl(2, \mathbb{R})$. The first generates a noncompact subgroup L of $SL(2, \mathbb{R})$ consisting

of matrices $\begin{pmatrix} a & 0 \\ 0 & 1/a \end{pmatrix}$, $a > 0$ while the second generates a compact subgroup
$B = SO(2)$ of $SL(2, \mathbb{R})$ consisting of matrices

$$u(\theta) = \begin{pmatrix} cos(\theta) & -sin(\theta) \\ sin(\theta) & cos(\theta) \end{pmatrix}, \theta \in [0, 2\pi)$$

We have the singular value decomposition of any $g \in SL(2, \mathbb{R})$:

$$g = u(\theta_1)a(t).u(\theta_2)$$

where

$$u(\theta) = exp(\theta(X - Y)) = \begin{pmatrix} cos(\theta) & -sin(\theta) \\ sin(\theta) & cos(\theta) \end{pmatrix},$$

$$a(t) = exp(tH) = \begin{pmatrix} e^t & 0 \\ 0 & e^{-t} \end{pmatrix}$$

The Iwasawa decomposition of $G = SL(2, \mathbb{R})$ is

$$G = KMAN$$

where $K = SO(2)$, $M = \{\pm I\}$, $A = exp(\mathbb{R}.H)$ and $N = exp(\mathbb{R}.X)$. This is essentially the QR decomposition obtained via a Gram-Schmidt orthonormalization of the columns of G. We

Image processing using discrete series characters: Given any $g \in G = SL(2, \mathbb{R})$ (or equivalently, a proper orthochronous Lorentz transformation in the xz plane), we have the basic result that g is conjugate in $G = SL(2, \mathbb{R})$ either to an element of the elliptic Cartan subgroup $B = exp(\mathbb{R}.(X - Y))$ or to an element of the hyperbolic Cartan subgroup $L = exp(\mathbb{R}.H)$. Let f_1, f_2 be two functions on $SL(2, \mathbb{R})$ and let χ_m denote the character of the above discrete series representation. Then, we consider

$$F_m(f_1, f_2) = \int_{G \times G} f_1(g)f_2(h)\chi_m(gh^{-1})dgdh$$

We have that

$$F_m(f_1 ox, f_2 ox) = \int_{G \times G} f_1(xg)f_2(xh)\chi_m(gh^{-1})dgdh$$

$$= \int f_1(g)f_2(h)\chi_m(x^{-1}gh^{-1}x)dgdh = \int f_1(g)f_2(h)\chi_m(gh^{-1})dgdh = F_m(f_1, f_2)$$

for all $x \in G$. Thus, $(f_1, f_2) \to F_m(f_1, f_2)$ is a G-invariant function defined on all image pairs where each image is a function on $G = SL(2, \mathbb{R})$.

Chapter 7

Optimization problems in classical and quantum stochastics and information with antenna design applications

7.1 A course in optimization techniques

7.1.1 Linear optimization problems using least squares methods

The orthogonal projection theorem in a finite dimensional Hilbert space.

7.1.2 Minimum mean squares estimation

Conditional expectation of L^1 random variables using the Radon-Nikodym derivative; Conditional expectation using the orthogonal projection theorem for $L^2(P)$ random variables and the density of $L^2(P)$ in $L^1(P)$. These are applications of the orthogonal projection theorem in an infinite dimensional Hilbert space.

7.1.3 Orthogonal projection theorem in infinite dimensional Hilbert spaces using the Apollonius theorem-existence and uniqueness theorems and properties of the orthogonal projection operator

Study project: Read about the proof of the existence and uniqueness of the best approximant of a vector from a Hilbert space in a closed subspace of it or more generally in a closed convex subset of it. The existence proof is based on constructing a sequence in the subspace/convex subset whose distance from the given vector converges to the minimum distance and then using Apollonius's theorem to show that this sequence is Cauchy and hence converges. Likewise the uniqueness proof is based on assuming two best approximants to the given vector and then using Apollonius' theorem to show that these two are equal. Linearity of the orthogonal projection map that takes any given vector to its best approximant in the closed subspace is established using the orthogonality priniciple which states that the approximation error is orthogonal to the subspace. This fact is established by noting that the approximation error norm square is a minimum for all vectors in the subspace. This minimum condition is expressed by setting to zero the derivative of the error norm square w.r.t a one parameter family of curves in the subspace passing through the best approximant. This orthogonality principle is also used to establish the self-adjointness of the orthogonal projection operator.

7.1.4 Orthogonal projection theorem for closed convex subsets of an infinite dimensional Hilbert space. Proofs of existence and uniqueness

Study project:The Apollonius theorem works even if the closed subspace is replaced by a closed convex subset since it involves only three vector in the subset, the first two being any two vectors in the sequence whose distance from the given vector converges to the minimum and the third being the average value of these two vectors which by convexity of the subset, is also an element of the convex set. However now we cannot talk about linearity or selfadjointness of the orthogonality operator.

7.1.5 The variational derivatives of a function defined on infinite dimensional Banach and Hilbert spaces. The Frechet and Gateaux derivatives

7.1.6 The variational principle of Lagrange on spaces of twice differentiable functions

Study project: The Euler-Lagrange equations for functions of one variable and for functions of several variables. Application to classical mechanics and classical field theory. Learn about Noether's theorem which states that if the Lagrangian is invariant under a Lie group of transformations, then we can derive a conservation law both in mechanics and in classical field theory.

7.1.7 The Euler-Lagrange variational principle combined with the Feynman path integral (sum over histories) to verify the transition from quantum mechanics to classical mechanics in the limit $h \to 0$ via the method of stationary phase

Study project:The Feynman path integral (FPI) between two space-time points is the sum of a phase factor over different paths in time passing between the two points. The phase factor equals the action integral along the path divided by Planck's constant. This path integral solves the Schrodinger equation ie it is precisely the Schrodinger evolution kernel as first pointed out by Richard Feynman in his PhD thesis. The contribution to the FPI from each path is interpreted as being the quantum mechanical amplitude for the particle to go from the first point to the second in the given time interval. Thus, the FPI can be used as the starting point for formulating quantum mechanics. Moreover, in this formulation, the analogy between classical and quantum mechanics becomes apparent at once. Indeed, when Planck's constant goes to zero, then the phase becomes infinite and hence rapidly oscillates in the vicinity of any trajectory between the two points except the classical trajectory causing phase cancellations in the FPI around any trajectory except the classical one. The action functional and hence the phase is stationary around the classical trajectory by the Euler-Lagrange theorem and hence no phase cancellations occur around this trajectory. This means that only the contribution from the classical trajectory contributes to the quantum mechanical transition amplitude in the limit as Planck's constant converges to zero or more precisely when the ratio of the action integral to Planck's constant becomes very large as is the case for macroscopic bodies.

7.1.8 Large deviation theory as an exercise in optimization

The theorems of Sanov, Cramer, Gartner-Ellis, Bryc and Varadhan. Calculating the rate function for various sequences of random variables and random processes. Schilder's rate function for Brownian motion. The variational principle of Varadhan, for evaluating

$$lim_{\epsilon \to 0} \mathbb{E}[exp(\phi(Z_\epsilon)/\epsilon)]$$

when $Z_\epsilon, \epsilon \to 0$ satisfies the LDP with a known rate function $I(z)$.

7.1.9 Nonlinear filtering theory as an optimization problem

Calculating $p(x_t|Y_t)$ on a real time basis as a solution to a stochastic pde driven by the measurement process $z(t)$ with $Y_t = \{z(s) : s \leq t\}$. Calculating

$$\hat{\phi}(t|t) = \mathbb{E}(\phi(x(t)|Y_t)$$

as the optimal mmse of $\phi(x(t))$ given measurements upto time t. Approximations leading to the EKF. Deriving the EKF using optimization w.r.t the Kalman gain matrix both in discrete and in continuous time.

7.1.10 Variational principles applied to the Brachistochrone problem

(curve having minimum time of descent in uniform and non-uniform gravitational fields) and applied to the catenary problem of determining the shape of the hanging chain problem with applications to transmission line theory.

[10] Entropy maximization in physics for deriving the Maxwell-Boltzmann, Fermi-Dirac and Bose-Einstein statistics.

7.1.11 Optimization in filter design problems

Designing a rational transfer function that has minimum L^p distance from a given transfer function.

7.1.12 Channel capacity calculation as an optimization problem in cc and cq Shannon problems

In Cq problems, the problem is to maximize

$$H(\sum_{x \in A} p(x)\rho(x)) - \sum_{x \in A} p(x)H(\rho(x))$$

w.r.t $(p(x), \rho(x)), x \in A$ where $p(.)$ is a probability distribution on A and $\rho(x)$ is a density matrix in a Hilbert space $\mathcal{H}$ for each $x \in A$.

7.1.13 Another exercise in quantum channel capacity calculation as an optimization problem

ρ_i is the input state, $\rho_o = \sum_{k=1}^{p} E_k \rho_i E_k^*$ is the output state of a quantum noisy channel. Here, $\sum_{k=1}^{p} E_k^* E_k = I$. If A is a finite alphabet and $\rho_i : A \rightarrow S(\mathcal{H})$ is a mapping from A into the space of density operators in a Hilbert space $\mathcal{H}$, then then choose a POVM $\{M_k : 1 \leq k \leq q\}$ in $\mathcal{H}$, ie, $0 \leq M_k \leq I, \sum_{k=1}^{q} M_k = I$

so that the output probability distribution $q(k|x)$ given that the input source alphabet is $x \in A$ is given by

$$q(k|x) = Tr(\rho_o(x)M_k), \rho_o(x) = \sum_{k=1}^{p} E_k \rho_i(x) E_k^*$$

Calculate the capacity of this classical channel $q(.|.)$ and then maximize this capacity over all POVM's $\{M_k\}$.

7.1.14 Optimization in fluid dynamics

construction of the optimal stirring forces for the fluid velocity field to match an given velocity field with applications to optical flow problems.

The fluid equations are

$$(v, \nabla)v + v_{,t} = -\nabla p/\rho + \nu\nabla^2 v + f(t, r)$$

and

$$div\, v(t, r) = 0$$

Thus,

$$v = \nabla \times \psi, div\,\psi = 0$$

and

$$\nabla \times v = \Omega = -\nabla^2\psi$$

Taking the curl of the Navier-Stokes equation gives us a pde for ψ, ie, with $\nabla \times f = g$, we have

$$\nabla \times (\nabla^2\psi \times (\nabla \times \psi)) + \nabla^2\psi_{,t} = \nu(\nabla^2)^2\psi + g$$

or equivalently, with $\Delta = \nabla^2$,

$$\psi_{,t} = \nu\Delta\psi - \Delta^{-1}(\nabla \times (\Delta\psi \times (\nabla \times \psi))) + \Delta^{-1}g - - - (a)$$

Now design the forcing function g such that $div\, g = 0$ and subject to the equation of motion constraint (a),

$$\int L(\psi(t, r), \psi_d(t, r), g(t, r))d^3r dt$$

is a minimum where $\psi_d(t, r)$ is the desired stream function vector field.

7.1.15 Optimization in electromagnetics

Design of an antenna to match a given radiation pattern. Suppose $P_d(\omega, \hat{r})$ is the desired power flow per unit solid angle at a large distance r from the origin. Let $J(\omega, r)$ denote the current density of the source. The far field magnetic vector potential is

$$A(\omega, r) = (\mu/4\pi r)exp(-jkr) \int_S J(\omega, r')exp(jk\hat{r}.r')d^3r'$$

where S stands for the source region (centred around the origin). The far field magnetic field is

$$H(\omega, r) = \nabla \times A(\omega, r) = (exp(-jkr))/4\pi r)(-jk)\hat{r} \times \int_S J(\omega, r')exp(jk\hat{r}.r')d^3r'$$

and hence the power flow per unit solid angle in the radiation zone is given by

$$P(\omega, \hat{r}) = (\eta/2)r^2|H(\omega, r)|^2 =$$

$$(\eta/16\pi^2)|\int_S J(\omega, r')exp(jk\hat{r}.r')d^3r'|^2$$

Let $\sigma(\omega, r)$ denote the conductivity of the source. Then, the power dissipated in the source at frequency ω is given by

$$\int_S (2\sigma(\omega, r')^{-1}|J(\omega, r')|^2d^3r'$$

The optimization problem is to choose the current field $J(\omega, r'), r' \in S$ such that for a given power dissipation at frequency ω, $P(\omega, \hat{r})$ is as close as possible to $P_d(\omega, \hat{r})$ for all $\hat{r} \in S^2$.

7.1.16 Quantum gate design using optimization techniques. Discussion of the gravitational search algorithm (GSA)

7.1.17 Solving nonlinear least squares problems using perturbation theory

Minimizing

$$E(\theta) = \| y - \sum_{k=1}^{p} A_k(\theta^{\otimes k}) \|^2$$

w.r.t $\theta \in \mathbb{R}^p$.

7.1.18 Bellman-Hamilton-Jacobi dynamic programming for solving deterministic and stochastic optimal control problems

The state is a Markov process $X(t) \in \mathbb{R}^d$ whose generator K depends on the input $u(t) \in \mathbb{R}^p$ which is to be controlled so that

$$S[X, u, \lambda] = \mathbb{E} \int_0^T L(X(t), u(t), t)dt + \lambda^T \mathbb{E}\psi(X(T))$$

is extremized. This problem amounts to minimizing the average cost over the fuel input $u(.)$ for a fixed time duration T subject to the constraint that at time T, the average value of some function of the state is known. This is therefore a minimum fuel problem. The fuel input $u(t)$ at time t is allowed to depend only on the state $X(t)$ at time t, ie $u(t) = f(t, X(t))$ where f is a non-random function. Alternately, we could also regard the time duration T as a variable over which the average cost is to be minimized. For example, if $L = 1$, this becomes a minimum time problem.

7.1.19 Optimization problems for state diffusion problems

The Hudson-Parthasarathy noisy Schrodinger equation reads

$$dU(t) = (-(iH + P)dt + L_1 dA + L_2 dA^* + Sd\Lambda)U(t)$$

where H, P, L_1, L_2, S are system operators satisfying certain relations that guarantee $U(t)$ to be unitary in $\mathfrak{h} \otimes \Gamma_s(L_2(\mathbb{R}_+))$. We choose a pure state $|f \otimes \phi(u) >$ in this space where $|\phi(u) >= exp(-|u|^2/2)|e(u) >$ is coherent state of the bath and define the pure state

$$|\psi(t) >= U(t)|f \otimes \phi(u) >\in \mathfrak{h} \otimes \Gamma_s(L^2(\mathbb{R}_+))$$

Then $|\psi(t) >$ satisfies

$$d|\psi(t) >= (-(iH+P)|\psi(t) > +u(t)(L_1-L_2)|\psi(t) >)dt+L_2 dB(t)|\psi(t) > \\ +S(dA^* dA/dt)|\psi(t) >$$

where

$$B(t) = A(t) + A(t)^*$$

is classical Brownian motion. We have used the well known relation

$$d\Lambda = dA^* dA/dt$$

which follows from

$$< e(u)|dA^* dA/dt|e(v) >=< dAe(u)|dAe(v) > /dt \\ = \bar{u}(t)v(t)dt < e(u)|e(v) >=< e(u)|d\Lambda|e(v) >$$

We now observe that

$$(dA^* dA/dt)|\psi(t) >= U(t)(dA^* dA/dt)|f \otimes \phi(u) >=$$

$$u(t)U(t)dA^*|f \otimes \phi(u) >= u(t)U(t)(dB - dA)|f \otimes \phi(u) >$$
$$= u(t)dB(t)|\psi(t) > -u(t)^2 dt|\psi(t) >$$

Thus, the HP equation when applied to the above pure state results in a "state diffusion" (in the sense of Gisin and Percival):

$$d|\psi(t) >= (-(iH+P)|\psi(t) > +u(t)(L_1-L_2)|\psi(t) >)dt+L_2 dB(t)|\psi(t) >$$
$$+S(u(t)dB(t)-u^2(t)dt)|\psi(t) >$$
$$= (-(iH + P) + u(t)L - u^2(t)S)dt + (L_2 + u(t)S)dB(t))|\psi(t) >$$

It should be noted that this is a classical stochastic differential equation and the state vector $|\psi(t) >$ can now be regarded as a classical random process adapted to the classical Brownian motion $B(.)$ and taking values in the system Hilbert space $\mathfrak{h}$. It should also be noted that in this interpretation, if $< .,. >=< .,. >_{\mathfrak{h}}$ is the inner product in $\mathfrak{h}$, then $< \psi(t)|\psi(t) >= 1$, rather $< \psi(t)|\psi(t) >$ is a random process adapted to the Brownian motion $B(.)$ and the condition

$$< \psi(t)|\psi(t) >_{\mathfrak{h}\otimes\Gamma_s(L^2(\mathbb{R}_+))}= 1$$

translates to

$$\mathbb{E}[< \psi(t)|\psi(t) >] = 1$$

or equivalently,

$$\int < \psi(t)|\psi(t) > dP(B) = 1$$

In other words, averaging w.r.t. the state of the bath is equivalent to averaging w.r.t. the probability law of the classical Brownian motion $B(.)$. If $|\psi(t) >$ is to be regarded as the system state at time t, then it must be normalized w.r.t the system inner product. The system state at time t then becomes

$$|\chi(t) >=< \psi(t)|\psi(t) >^{-1/2} |\psi(t) >= |\psi(t) > / \| \psi(t) \|$$

In fact, this is the state to which the system state collapses after making a measurement of the bath over time (t, ∞) without noting the outcome. We seek now to determine using the classical Ito formula, the dynamics of $|\chi(t) >$. We first write down the dynamics of $|\psi(t) >$ as

$$d|\psi(t) >= (A_1(t)dt + A_2(t)dB(t))|\psi(t) >$$

where

$$A_1(t) = -(iH + P) + u(t)L - u^2(t)S, A_2(t) = L_2 + u(t)S$$

Then

$$d(\| \psi(t) \|^{-1}) = d < \psi(t)|\psi(t) >^{-1/2}=$$
$$(-1/2) \| \psi(t) \|^{-3} (d < \psi(t)|\psi(t) >) + (3/8) \| \psi(t) \|^{-5} (d < \psi(t)|\psi(t) >)^2$$

and

$$d < \psi(t)|\psi(t) >= 2Re(< \psi(t)|d\psi(t) >)+ < d\psi(t)|d\psi(t) >$$
$$= 2Re(< \psi(t)|A_1|\psi(t) >)dt + 2Re(< \psi(t)|A_2|\psi(t) >)dB(t)$$

$$+ < \psi(t)|A_2^* A_2|\psi(t) > dt$$

$$(d < \psi(t)|\psi(t) >)^2 = 4(Re(< \psi(t)|A_2|\psi(t) >))^2 dt$$

So,

$$d|\chi(t) >= d(\| \psi(t) \|^{-1})|\psi(t) > + \| \psi(t) \|^{-1} d|\psi(t) >$$

$$+ d(\| \psi(t) \|^{-1}).d|\psi(t) >$$

Now the above equations imply

$$d(\| \psi(t) \|^{-1}) =$$

$$(-1/2) \| \psi(t) \|^{-3} (d < \psi(t)|\psi(t) >) + (3/8) \| \psi(t) \|^{-5} (d < \psi(t)|\psi(t) >)^2$$

$$= (-1/2) \| \psi(t) \|^{-3} [2Re(< \psi(t)|A_1|\psi(t) >)dt + 2Re(< \psi(t)|A_2|\psi(t) >)dB(t)$$

$$+ < \psi(t)|A_2^* A_2|\psi(t) > dt] + (3/2) \| \psi(t) \|^{-5} (Re(< \psi(t)|A_2|\psi(t) >))^2 dt$$

Thus,

$$d|\chi(t) >=$$

$$= (-1/2) \| \psi(t) \|^{-3} [2Re(< \psi(t)|A_1|\psi(t) >)dt + 2Re(< \psi(t)|A_2|\psi(t) >)dB(t)$$

$$+ < \psi(t)|A_2^* A_2|\psi(t) > dt]|\psi(t) > + (3/2) \| \psi(t) \|^{-5} (Re(< \psi(t)|A_2|\psi(t) >))^2 dt|\psi(t) >$$

$$+ \| \psi(t) \|^{-1} (A_1|\psi(t) > dt + A_2|\psi(t) > dB(t))$$

$$- \| \psi(t) \|^{-3} Re(< \psi(t)|A_2|\psi(t) >)A_2|\psi(t) > dt$$

$$= dt[-Re(< \chi(t)|A_1|\chi(t) > -(1/2) < \chi(t)|A_2^* A_2|\chi(t) >$$

$$+ (3/2)(Re(< \chi(t)|A_2|\chi(t) >))^2 + A_1]|\chi(t) >$$

$$+ dB(t)[-Re(< \chi(t)|A_2|\chi(t) >) + A_2]|\chi(t) >$$

Thus, after making a measurement on the environment over the time interval (t, ∞), the state of the system collapses to the normalized state $|\chi(t) >$ which satisfies a nonlinear diffusion equation.

7.2 Group theoretical techniques in optimization theory

7.2.1 Statement of the basic problems

A group G acts on a manifold $\mathcal{M}$. The image field is a map $f_1 : \mathcal{M} \to \mathbb{R}$. The image field f after transformation by the group element $g \in G$ and after having been corrupted by a noise field $w(x), x \in \mathcal{M}$, ie $w : \mathcal{M} \to L^2(\Omega, \mathcal{F}, P)$ is given by

$$f_2(x) = f_1(g^{-1}x) + w(x), x \in \mathcal{M}$$

The aim is to estimate the group transformation element g from measurements of f_1 and f_2 and to obtain the mean square error in the estimation error. More generally, we may be given a whole set of image field pairs $(f_{a1}, f_{a2}), a = 1, 2, ..., K$ on $\mathcal{M}$ such that for a fixed $g \in G$, we have

$$f_{a2}(x) = f_{a1}(g^{-1}x) + w_a(x), a = 1, 2, ..., K$$

Then from measurements of these pairs, g must be estimated. The least squares method of doing this according to

$$\hat{g} = argmin_{h \in G} \sum_{a=1}^{K} \int_{\mathcal{M}} w_a(x)|f_{a2}(x) - f_{a1}(h^{-1}x)|^2 d\mu(x)$$

where μ is a G-invariant measure on $\mathcal{M}$ and w_a is a non-negative weight function on $\mathcal{M}$ involves a search and is computationally very expensive. Techniques based on group representation theory considerably simplify the problem to a linear problem. For example suppose we decompose

$$L^2(\mathcal{M}, \mu) = \bigoplus_{k=1}^{\infty} \mathcal{H}_k$$

as an orthogonal direct sum, where $\mathcal{H}_k$ is a Hilbert subspace of $L^2(\mathcal{M}, \mu)$ and is invariant and irreducible under G, ie, $f \in \mathcal{H}_k$ implies $foh^{-1} \in \mathcal{H}_k$ for all $h \in G, k = 1, 2, ...$ and further, $\mathcal{H}_k$ contains no proper G-invariant subspace. Let π_k denote the corresponding representation. Specifically, choose an onb $\{e_{km} : m = 1, 2, ..., d_k\}$ for $\mathcal{H}_k$ and let

$$e_{km}(h^{-1}x) = \sum_{m'=1}^{d_k} [\pi_k(h)]_{m'm} e_{km'}(x)$$

Since μ is a G-invariant measure on $\mathcal{M}$, it follows that the representation $\pi_k : G \to GL_(\mathbb{C}, d_k)$ is in fact a unitary representation, ie $\pi_k : G \to U(\mathbb{C}, d_k)$. In fact, we get from the above that

$$\delta_{m,n} = < e_{km}, e_{kn} > = < e_{km}oh^{-1}, e_{kn}oh^{-1} > =$$

$$\sum_{m',n'} [\bar{\pi}_k(h)]_{m'm}[\pi_k(h)]_{n'n} < e_{km'}, e_{kn'} >$$

$$= \sum_{m',n'} [\bar{\pi}_k(h)]_{m'm}[\pi_k(h)]_{n'n}\delta_{m',n'}$$

$$= \sum_{m'} [\bar{\pi}_k(h)]_{m'm}[\pi_k(h)]_{m'n} = [\pi_k(h)^*\pi_k(h)]_{mn}$$

which proves the unitarity of $\pi_k(h)$. Now the equation

$$f_2(x) = f_1(g^{-1}x) + w(x)$$

implies

$$< e_{km}, f_2 >=< e_{km}, f_1 og^{-1} > + < e_{km}, w >=$$

$$< e_{km} og, f_1 > + < e_{km}, w >= \sum_{m'} [\bar{\pi}_k(g^{-1})]_{m'm} < e_{km'}, f_1 > + < e_{km}, w >$$

$$= \sum_{m'} [\pi_k(g)]_{mm'} < e_{km'}, f_1 > + < e_{km}, w >$$

or equivalently, denoting by $\mathbf{f}_1[k]$ the $d_k \times 1$ complex vector $((< e_{km}, f_1 >))_{m=1}^{d_k}$ and likewise for $\mathbf{f}_2[k]$ and $\mathbf{w}[k]$, we can express the above equation in matrix form as

$$\mathbf{f}_2[k] = \pi_k(g)\mathbf{f}_1[k] + \mathbf{w}[k]$$

and in the case when we have several such image pairs,

$$\mathbf{f}_{2a}[k] = \pi_k(g)\mathbf{f}_{1a}[k] + \mathbf{w}_a[k]$$

so that for each k we can estimate the matrix $\pi_k(g)$ by minimizing

$$E_k(\mathbf{X}) = \sum_{a=1}^{K} w_a \parallel \mathbf{f}_{2a}[k] - \mathbf{X}\mathbf{f}_{1a}[k] \parallel^2$$

w.r.t $\mathbf{X}$ and by doing so for several such k, a good estimate of g may be obtained.

7.2.2 Some aspects of the root space decomposition of a semisimple Lie algebra

Use Weyl' character formula for compact semisimple groups to construct invariants for image pairs and thereby do pattern classification/pattern recognition. Let G be a compact complex semisimple group and let $\mathfrak{g}$ denote its Lie algebra. Let $\mathfrak{h}$ be a Cartan subalgebra of $\mathfrak{g}$, note that since the group is complex, all Cartan subalgebras are conjugate to each other. Corresponding to this Cartan subalgebra, let

$$\mathfrak{g} = \mathfrak{h} \oplus \bigoplus_{\alpha \in \Delta} \mathfrak{g}_\alpha$$

be its root space decomposition. Here, Δ is the set of all roots of $\mathfrak{g}$. Let Δ_+ denote the set of positive roots. Then, the root space decomposition can also be expressed as

$$\mathfrak{g} = \mathfrak{h} \oplus \bigoplus_{\alpha \in \Delta_+} \mathfrak{g}_\alpha \oplus \mathfrak{g}_{-\alpha}$$

Note that $dim\mathfrak{g}_\alpha = 1, \alpha \in \Delta$. Let λ be a dominant integral weight, ie, $\lambda \in \langle^*$ and $\lambda(H_\alpha)$ is a positive integer for every simple root α. Note that corresponding to the set Δ of roots, we have a set $P \subset \Delta_+$ of roots such that given any $\alpha \in \Delta_+$, there exist integers $n(\alpha, \beta) \geq 0, \beta \in P$ such that

$$\alpha = \sum_{\beta \in P} n(\alpha, \beta)\beta$$

and further there does not exist any proper subset of P with this property. In other words, every root is either a positive integer linear combination of the simple roots P or a negative integer linear combination of the simple roots P. Define

$$\rho = (1/2) \sum_{\alpha \in \Delta_+} \alpha$$

In the above notation, H_α is a normalized version of $[X_\alpha, X_{-\alpha}]$ where $X_\alpha \in \mathfrak{g}_\alpha$ with $\alpha \in \Delta_+$ and the normalization is carried out in such a way so that $\{H_\alpha, X_\alpha, X_{-\alpha}\}$ are the standard generators of a Lie algebra isomorphic to $\mathfrak{sl}(2, \mathbb{C})$, ie,

$$[H_\alpha, X_\alpha] = 2X_\alpha, [H_\alpha, X_{-\alpha}] = -2X_{-\alpha},$$

$$[X_\alpha, X_{-\alpha}] = H_\alpha$$

This means that

$$\alpha(H_\alpha) = 2$$

Corresponding to the dominant integral weight λ, there is an irreducible representation π_λ of $\mathfrak{g}$ whose character χ_λ on the Cartan subgroup $T = exp(\mathfrak{h})$ is given by

$$\chi_\lambda(t) = \frac{\sum_{s \in W} \epsilon(s) exp(s(\rho + \lambda))}{\sum_{s \in W} \epsilon(s) exp(s\rho)}$$

where W is the Weyl group of $\mathfrak{h}$, ie, if N_T denotes the normalizer of $\mathfrak{h}$ in G, ie the set of all $g \in G$ such that

$$Ad(g)(\mathfrak{h}) \subset \mathfrak{h}$$

then the Weyl group W is N_T/T. Note that $T = exp(\mathfrak{h})$ is the centralizer of $\mathfrak{h}$ in G, ie $t \in T$ iff $Ad(t)(H) = H \forall H \in \mathfrak{h}$. The Weyl group acts on $\mathfrak{h}$ through the adjoint action. More precisely, for each $s \in W$, there is an element $g \in N_T$ such that

$$Ad(g)(H) = s.H, H \in \mathfrak{h}$$

and conversely, given any $g \in N_T$ we can find a unique element $s \in W$ such that

$$Ad(g)(H) = s.H, H \in \mathfrak{h}$$

For each simple root α, ie, $\alpha \in P$, there is an element $s_\alpha \in W$ such that s_α is a reflection about the plane $\{H \in \mathfrak{h} : \alpha(H) = 0\}$. Equivalently,

$$s_\alpha.H = H - \alpha(H)H_\alpha$$

Any element in W is expressible as a product of the $s_\alpha, \alpha \in P$. We note that

$$\alpha(s_\alpha.H) = \alpha(H) - 2\alpha(H) = -\alpha(H)$$

and further,

$$s_\alpha^2.H = s_\alpha.H - \alpha(H)s_\alpha.H_\alpha$$

$$= H - \alpha(H)H_\alpha - \alpha(H)(H_\alpha - 2H_\alpha)$$

$$= H$$

ie,

$$s_\alpha^2 = 1$$

which confirms the fact that s_α is a reflection. $s_\alpha, \alpha \in P$ are therefore called simple reflections. If $s \in W$ is expressible as a product of an even number of simple reflections, then we set $\epsilon(s) = -1$, otherwise $\epsilon(s) = 1$. Note that the Weyl group, by duality acts on $\mathfrak{h}^*$, ie, if $s \in W$ and $\mu \in \mathfrak{h}^*$, then

$$(s.\mu)(H) = \mu(s^{-1}.H)$$

It follows that if $\alpha \in P$, then

$$(s_\alpha.\lambda)(H) = \lambda(s_\alpha.H) =$$

$$\lambda(H - \alpha(H).H_\alpha) = \lambda(H) - \lambda(H_\alpha)\alpha(H)$$

$$= \lambda - <\lambda, \alpha > \alpha(H)$$

ie

$$s_\alpha.\lambda - \lambda - <\lambda, \alpha > \alpha$$

where

$$<\lambda, \alpha> = \lambda(H_\alpha)$$

for any $\lambda \in \mathfrak{h}^*$. Note that here λ is a dominant integral weight and hence $<\lambda, \alpha>$ is a non-negative integer for all $\alpha \in P$. Then, it follows that since any $s \in W$ is a product of simple reflections and since if $\alpha, \beta \in P$, then $<\alpha, \beta>$ is a negative integer, ie, $\alpha(H_\beta)$ is a negative integer, we have

$$s.\lambda = \lambda - \sum_{\alpha \in P}^{p} n(\lambda, \alpha)\alpha$$

where $n(\lambda, \alpha)$ are non-negative integers.

7.2.3 Weyl's integration formula and Weyl's character formula

Let G be a compact semisimple group and T a Cartan subgroup. Consider the mapping $\psi : G/T \times T \to G$ defined by

$$\psi(g.T, h) = Ad(g)(h) = ghg^{-1}$$

Note that this map is well defined since T is an Abelian subgroup. We compute the differential of this map: for $Z \in \mathfrak{g}$ infinitesimal and $H \in \mathfrak{t}$ infinitesimal with $\mathfrak{t}$ being the Lie algebra of T (T is a maximal torus in G), we have

$$g(1 + Z)h(1 + H)(1 - Z)g^{-1} = (g + gZ)(h + hH)(g^{-1} - Zg^{-1})$$

$$= ghg^{-1} + ghHg^{-1} + gZhg^{-1} - ghZg^{-1}$$

$$= ghg^{-1} + ghg^{-1}Ad(g)(H) + ghg^{-1}Ad(g)Ad(h^{-1})(Z) - ghg^{-1}Ad(g)(Z)$$

It follows from this expression that the differential of ψ at $(gT, h$ is given by

$$d\psi_{(gT,h)}(Z, H) = Ad(g)(H + Ad(h)^{-1}(Z) - Z)$$

Choosing successively Z to be the root vectors $X_\alpha, \alpha \in \Delta$ and H to run over a basis for $\mathfrak{h}$, we easily see that the Jacobian determinant of the map ψ at (gT, h) is given by

$$\Pi_{\alpha \in \Delta}(exp(\alpha(log(h))) - 1)$$

$$= |\Pi_{\alpha \in \Delta_+}(exp(\alpha(log(h))/2) - exp(-\alpha(log(h))/2))|^2$$

$$= |\Delta(h)|^2$$

where

$$\Delta(h) = \Pi_{\alpha \in \Delta_+}(exp(\alpha(log(h))/2) - exp(-\alpha(log(h))/2))$$

It follows that

$$\int_G f(g)dg = |W|^{-1} \int_{G/T \times T} f(xhx^{-1})|\Delta(h)|^2 dx dh$$

where $|W|$ is the number of elements in the Weyl group. We note that χ_λ as defined above when restricted to the torus T is a finite Fourier series with integer coefficients. This is one of the requirements for any character on T because T is an Abelian group and hence since every representation of T is a direct sum of irreducible representations of T, it follows by taking the trace that every character on T is an integer linear combination of irreducible characters and any irreducible character on T is simply of the form $t_1^{n_1}...t_l^{n_l}$ where $l = dimT$, $t_1, ..., t_l$ are complex numbers of unit magnitude and $n_1, ..., n_l$ are non-negative integers. Note that by the definition of a torus, there exist linearly independent elements $H_1, ..., H_l \in \mathfrak{h}$ such that $exp(i2\pi H_k) = 1, k = 1, 2, ..., l$ and that every element $t \in T$ can be expressed as

$$t = exp(2\pi i(n_1\theta_1 H_1 + .. + n_l\theta_l H_l))$$

where $n_1, ..., n_l$ are non-negative integers and $\theta_1, ..., \theta_k \in [0, 1)$. We can thus define $t_j = exp(2\pi i\theta_j H_j)$, identify it with the complex number $exp(2\pi i\theta_j)$ and note that t can then be identified with $t_1^{n_1}...t_l^{n_l}$. We now note that if $s_1, s_2 \in W$ are such that $s_1.(\lambda + \rho) = s_2.(\lambda + \rho)$, then

$$\lambda + \rho = s.(\lambda + \rho), s = s_1^{-1}s_s$$

and we can write

$$\lambda - s.\lambda = \sum_{\alpha \in P} n(\alpha)\alpha, n(\alpha) \in \mathbb{Z}_+$$

On the other hand,

$$\rho - s.\rho = (1/2) \sum_{\alpha \in \Delta_+} \alpha - s.\alpha \geq 0$$

Thus, we must necessarily have $s = 1$, ie, $s_1 = s_2$. It follows that each weight $s.(\lambda + \rho), s \in W$ appears exactly once in the sum $\sum_{s \in W} \epsilon(s) exp(s.(\lambda + \rho))$ from which, we deduce on using the fact that $s.\lambda$ assumes only integer values on $H_\alpha, \alpha \in P$ and that the same is true for ρ that

$$\int_T | \sum_{s \in W} \epsilon(s) exp(s.(\lambda + \rho)(log(h)))|^2 dh$$

$$= |W|$$

Thus,

$$\int_G |\chi_\lambda(g)|^2 dg = |W|^{-1} \int_{G/T \times T} |\chi_\lambda(xhx^{-1})|^2 |\Delta(h)|^2 dx dh$$

$$= 1$$

since

$$\chi_\lambda(xhx^{-1}) = \chi_\lambda(h) = \Delta(h)^{-1} \sum_{s \in W} \epsilon(s).exp(s.(\lambda + \rho))$$

Thus, the function χ_λ on G defined by (a) χ_λ is a class function and

$$\chi_\lambda(h) = \frac{\sum_{s \in W} \epsilon(s).exp(s.(\lambda + \rho)(log(h)))}{\sum_{s \in W} \epsilon(s) exp(s.\rho(log(h)))}, h \in T$$

satisfies all of the following properties:

[a] χ_λ restricted to T is a finite Fourier series with integer coefficients.
[b] $\int_G |\chi_\lambda|^2 dg = 1$
[c] $\chi_\lambda(s.h) = \chi_\lambda(h), h \in T, s \in W$

These three conditions guarantee that χ_λ is an irreducible character of G whenever λ is a dominant integral weight.

7.2.4 Irreducible representations of $SU(2)$ and hence by Weyl's unitarian trick of SL(2,C) using [a] Wigner's bivariate polynomials and [b] The Lie algebraic method

7.3 Feynman's diagrammatic approach to computation of the scattering amplitudes of electrons, positrons and photons

[a] Compton scattering: An electron of four momentum p and spin σ denoted by a straight line absorbs a photon of four momentum k and helicity s, represented by a wavy line, then travels along a straight line to a point where it has

a momentum $p' + k'$, it emits a photon of momentum k' and helicity s' at this point and then acquires a final momentum of p' and a spin σ'. The scattering amplitude for this process is to be written down. By conservation of four momentum, $p + k = p' + k'$ and by the diagrammatic technique, this amplitude is proportional to

$$e_\nu(k', s')^* \bar{u}(p', \sigma') \gamma^\nu S(p+k) \gamma^\mu u(p, \sigma) e_\mu(k, s)$$

where

$$S(p) = (\gamma.p - m - i\epsilon)^{-1}$$

is the electron propagator, $e_\mu(k, s)$ is the standard photon wave function in the momentum domain when it corresponds to an external em four potential, $u(p, \sigma)$ is the electron Dirac wave function in the momentum domain. By $\bar{u}$, we mean $u^{*T}\gamma^0$.

[b] The electron self-energy: An electron of four momentum p and spin σ propagates through space, it then emits a photon of four momentum k, thereby acquiring a four momentum of $p - k$, it then propagates further and absorbs the emitted photon after which it acquires back its momentum p. The photon is thus represented by a wavy line which starts at some point on the electron line and ends at some further point. According to the standard Feynman rule, the correction to the electron propagator caused by this photon emission is given by upto a proportionality constant by

$$\Sigma(p) = \int D_{\mu\nu}(k) \gamma^\nu S(p - k) \gamma^\mu d^4k$$

where

$$D_{\mu\nu}(k) = \frac{\eta_{\mu\nu}}{k^2 - i\epsilon}$$

is the photon propagator. Thus the correction to the electron propagator can be expressed upto a proportionality constant as

$$\Sigma(p) = \int (k^2 - i\epsilon)^{-1} \gamma_\mu S(p - k) \gamma^\mu d^4k$$

The corrected electron propagator due to such self energy terms is obtained by putting several such loops in the path of the electron and summing up all these diagrams. The resulting corrected electron propagator is thus

$$S_T(p) = S(p) + S(p)\Sigma(p)S(p) + S(p)\Sigma(p)S(p)\Sigma(p)S(p) + ...$$

$$= (S(p)^{-1} - \Sigma(p))^{-1} = (\gamma.p - m - \Sigma(p))^{-1}$$

Chapter 8

Quantum waveguides and cavity resonators

8.1 Quantum waveguides

Consider a waveguide of arbitrary cross section with axis parallel to the z axis. Introduce an orthogonal curvilinear coordinate system (q_1, q_2) in the xy plane so that $q_1 = 0$ is the boundary of the guide. Let $H_k, k = 1, 2$ denote the Lame's coefficients. Denote by $-\gamma$ the differential operator $\partial/\partial z$ and by $j\omega$ the operator $\partial/\partial t$. The Maxwell curl equations give

$$E_\perp(t, x, y, z) = \sum_n ((-\gamma/h_n^2)\nabla_\perp(E_z[n, t, z]u_n(q)) - (j\omega\mu/k_n^2)\nabla_\perp H_z[n, t.z]v_n(q)\times\hat{z})$$

$$E_z(t, x, y, z) = \sum_n E_z[n, t, z]u_n(q)$$

$$H_\perp = \sum_n (-\gamma/k_n^2)\nabla_\perp(H_z[n, t, z]v_n(q)) + (j\omega\epsilon/h_n^2)(\nabla_\perp E_z[n, t, z]u_n(q)) \times \hat{z}$$

$$H_z(t, x, y, z) = \sum_n H_z[n, t, z]v_n(q)$$

where $E_z[n, t, z], H_z[n, t, z]$ satisfy the eigen-equations

$$(\partial_z^2 + h_n^2 - \mu\epsilon\partial_t^2)E_z[n, t, z] = 0,$$

$$(\partial_z^2 + k_n^2 - \mu\epsilon\partial_t^2)H_z[n, t, z] = 0$$

Note that γ^2 stands for the operator ∂_z^2 and ∂_t^2 is the operator $-\omega^2$. The different eigenvalues h_n^2 and k_n^2 for the E_z and H_z components arise because of the different boundary conditions satisfied by these. These eigenvalues are determined by the solution to the boundary value problems

$$(\nabla_\perp^2 + h_n^2)u_n(q) = 0, u_n(q_1 = 0, q_2) = 0,$$

$$(\nabla_\perp^2 + k_n^2)v_n(q) = 0, \partial v_n(q_1 = 0, q_2)/\partial q_1 = 0$$

These boundary conditions correspond to the fact that the tangential components of the electric field and the normal component of the magnetic field vanishes on the boundary. We may assume that $\{u_n\}$ and $\{v_n\}$ are respectively orthonormal bases for $L^2(S)$ with the Dirichlet and Neumann boundary conditions where S is the cross section of the guide. Note that q_1 is directed along the normal to the guide wall while q_2 is along the tangent to the guide wall at the boundary. We also note that the area measure on S is $dS = dxdy = H_1 H_2 dq_1 dq_2$. We next calculate the Lagrangian density of the fields within the guide the density taken w.r.t. the z variable. It is given by

$$\mathcal{L}(E_z[n,t,z], H_z[n,t,z], E_{z,z}[n,t,z], H_{z,z}[n,t,z], E_{z,t}[n,t,z], H_{z,t}[n,t,z]) =$$

$$\int_S [(\epsilon/2)((E_z(t,x,y,z))^2 + |E_\perp(t,x,y,z)|^2) - (\mu/2)((H_z(t,x,y,z))^2$$

$$+ |H_\perp(t,x,y,z)|^2)]dS$$

The various terms are evaluated as follows:

$$\int_S E_z(t,x,y,z)^2 dS = \sum_n E_z[n,t,z]^2$$

$$\int_S H_z(t,x,y,z)^2 dS = \sum_n H_z[n,t,z]^2$$

$$\int_S |E_\perp(t,x,y,z)|^2 dS = \sum_n (c[n]h_n^{-4}E_{z,z}[n,t,z]^2 + d[n]\mu^2 k_n^{-4} H_{z,t}[n,t,z]^2)$$

where

$$c[n] = \int_S |\nabla_\perp u_n(q)|^2 dS, \quad d[n] = \int_S |\nabla_\perp v_n(q)|^2 dS$$

where we have used the fact that

$$\int_S (\nabla_\perp u_n(q), \nabla_\perp v_m(q) \times \hat{z}) dS =$$

$$\int_S (u_{n,1}(q)v_{m,2}(q) - u_{n,2}(q)v_{m,1}(q)) dq_1 dq_2 = 0$$

using integration by parts and the fact that $u_n(q)$ and $v_{m,1}(q)$ vanish on the boundary of S, ie when $q_1 = 0$. Likewise,

$$\int_S |H_\perp(t,x,y,z)|^2 dS =$$

$$\sum_n (d[n]k_n^{-4}H_{z,z}[n,t,z]^2 + c[n]\epsilon^2 h_n^{-4} E_{z,t}[n,t,z]^2)$$

The Lagrangian density of the waveguide confined electromagnetic field is therefore

$$\mathcal{L}(E_z[n,t,z], H_z[n,t,z], E_{z,z}[n,t,z], H_{z,z}[n,t,z], E_{z,t}[n,t,z], H_{z,t}[n,t,z], n = 1,2,...) =$$

$$\sum_n (\epsilon/2) E_z[n,t,z]^2 - (\mu/2) H_z[n,t,z]^2)$$

$$+ \sum_n (\epsilon/2)(c[n]h_n^{-4} E_{z,z}[n,t,z]^2 + d[n]\mu^2 k_n^{-4} H_{z,t}[n,t,z]^2)$$

$$- \sum_n (\mu/2)(d[n]k_n^{-4} H_{z,z}[n,t,z]^2 + c[n]\epsilon^2 h_n^{-4} E_{z,t}[n,t,z]^2)$$

To quantize this confined em field, we assume that the position fields are $E_z[n,z,t], H_z[n,z,t], n = 1,2,\ldots$ and the corresponding conjugate momentum fields are

$$\pi_E[n,z,t] = \frac{\partial \mathcal{L}}{\partial E_{z,t}[n,z,t]} =$$

$$-\mu\epsilon^2 c[n]h_n^{-4} E_{z,t}[n,z,t],$$

$$\pi_H[n,z,t] = \frac{\partial \mathcal{L}}{\partial H_z[n,z,t]} =$$

$$\epsilon\mu^2 d[n]k_n^{-4} H_{z,t}[n,z,t]$$

and hence by applying the Legendre transformation, the Hamiltonian density is

$$\mathcal{H}(E_z[n,z,t], H_z[n,z,t], \pi_E[n,z,t], \pi_H[n,z,t], n = 1,2,\ldots] =$$

$$\sum_n (\pi_E[n,z,t]E_z[n,z,t] + \pi_H[n,z,t]H[n,z,t]) - \mathcal{L} =$$

$$\sum_n ((-h_n^4/2\mu\epsilon^2 c[n])\pi_E[n,z,t]^2 + (k_n^4/2\epsilon\mu^2 d[n])\pi_H[n,z,t]^2)$$

$$+ \sum_n (\mu/2)H_z[n,t,z]^2 - (\epsilon/2)E_z[n,t,z]^2)$$

$$- \sum_n (\epsilon/2)(c[n]h_n^{-4} E_{z,z}[n,t,z]^2)$$

$$+ \sum_n (\mu/2)(d[n]k_n^{-4} H_{z,z}[n,t,z]^2)$$

This Hamiltonian density of the confined em field is therefore the sum of Hamiltonians of a discrete conutably infinite set of Harmonic oscillators some of which have negative masses. More precisely since the z coordinate variable is also involved, this Hamiltonian density is the sum of Hamiltonian densities of a countably infinite set of one dimensional strings, some of which have negative linear mass densities. The magnetic field modes have positive mass densities while the electric field modes have negative mass densities. We therefore rearrange these as

$$\mathcal{H} =$$

$$\sum_n [(k_n^4/2\epsilon\mu^2 d[n])\pi_H[n,z,t]^2) + (\mu/2)H_z[n,t,z]^2 + (\mu/2)(d[n]k_n^{-4} H_{z,z}[n,t,z]^2)$$

$$-[\sum_n (h_n^4/2\mu\epsilon^2 c[n])\pi_E[n,z,t]^2 + (\epsilon/2)E_z[n,t,z]^2 + (\epsilon/2)c[n]h_n^{-4}E_{z,z}[n,t,z]^2]$$

The canonical commutation reltions are

$$[E_z[n,z,t], \pi_E[n',z',t]] = i\delta_{n,n'}\delta(z-z'),$$

$$[H_z[n,z,t], \pi_H[n',z',t]] = i\delta_{n,n'}\delta(z-z')$$

Hence, we can formally set up the Schrodinger equation for the wave function of the confined field within a waveguide of length d as

$$(\int_0^d \mathcal{H}(E_z[n,z], H_z[n,z], -i\delta/\delta E_z[n,z], -i\delta/\delta H_z[n,z],$$

$$n = 1,2,...)dz)\psi_t(E_z[m,z], H_z[m,z], m = 1,2,...)$$

$$= i\partial_t\psi_t(E_z[m,\xi], H_z[m,\xi], m = 1,2,...0 \leq \xi \leq d)$$

Chapter 9

Classical and quantum filtering and control based on Hudson-Parthasarathy calculus, and filter design methods

9.1 Belavkin filter and Luc-Bouten control for electron spin estimation and quantum Fourier transformed state estimation when corrupted by quantum noise

QFT is realized by Schrodinger evolution for a fixed time duration under a designed system Hamiltonian but while implementing the evolution, it gets corrupted by quantum noise; the resulting state evolving according to the noisy HP -Schrodinger equation has to be estimated as well as the Lindblad noise appearing the HP equation has to be reduced.

Belavkin quantum filter for estimating the spin of an electron as an example in quantum computation. The spin of an electron along any given direction can have only two eigenvalues $\pm 1/2$. We may represent such a particle at time t as $c_1(t)|1> +c_0(t)|0>$ where $c_1(t), c_0(t)$ are complex random processes with the constraint $|c_1(t)|^2 + |c_0(t)|^2 = 1, t \geq 0$ where in the state $|1>$ the spin component along the given direction has eigenvalue $+1/2$ and in the state $|0>$, it has the eigenvalue $-1/2$. This is a one qubit state in the language of quantum computation and we may wish to transmit this state through a one qubit quantum channel. We may on the other hand allow this state to evolve according to a noisy Schrodinger equation, take non-demolition measurements on the noisy bath passed through the system and from these measurements, estimate the evolved spin using the Belavkin filter. In such a case, the estimated state of the electron at time t is a 2×2 positive definite matrix with unit trace measurable w.r.t the Abelian Von-Neumann algebra of measurements upto time t. Such an estimated density can also be viewed as a 2×2 random matrix with some probability distribution when the Bath state is fixed, for example when the bath is in a coherent state.

When the spin of the electron interacts with a real valued nonrandom time varying magnetic field $B(t)$, its interaction energy is $H(t) = ge(\sigma, B(t))/4m$ where $g \approx 2$ and σ is the triplet of the Pauli spin matrices. In addition, when the electron interacts with a noisy bath modeled in the Hudson-Parthasarathy formalism as a Boson Fock space $\Gamma_s(L^2(\mathbb{R}_+))$, its dynamics in described by the Hudson-Parthasarathy noisy Schrodinger evolution equation

$$dU(t) = (-i(H(t) + P)dt + L_1 dA(t) + L_2 dA(t)^* + Sd\Lambda(t))U(t)$$

where $A(t), A(t)^*, \Lambda(t)$ are operators in the Boson Fock space and P, L_1, L_2, S are like $H(t)$ system space operators, ie 2×2 complex matrices. Let X be a system space observable, ie, a 2×2 Hermitian matrix like for example the spin $(\sigma, \hat{n})/2$ along a fixed direction $\hat{n}$ The Hudson-Parthasarathy noisy Schrodinger equation can be regarded as a time varying single qubit quantum channel which transforms the initial state of the electron $|\psi_s(0) >= c_1(0)|1 > +c_0(0)|0 >$ into a mixed state $\rho_s(t)$ at time t defined by

$$\rho_s(t) = Tr_2(U(t)(\rho_s(0) \otimes |\phi(u) >< \phi(u)|)U(t)^*)$$

where

$$\rho_s(0) = |\psi_s(0) >< \psi_s(0)|$$

and $|\phi(u) >= exp(- \parallel u \parallel^2 /2)|e(u) >$ is the state of the bath, ie, the bath is in a coherent state. After transmission of the system state over the noisy HP channel, the Belavkin filter at the receiver end, decodes the system state based on non-demolition measurements. Thus, this whole setup can be regarded as a prototype single qubit quantum communication system with a receiver which operates on a real time basis. We may consider more general cases like for example suppose H is an $N \times N$ Hermitian matrix such that $U_0(T) = exp(-iTH)$ is the quantum Fourier transform matrix. Here $N = 2^r$ and we wish the system state at time 0 specified as $|\psi_s(0) >= \sum_{k=0}^{N-1} c_k(0)|k >$ to be quantum Fourier transformed at time T to the state

$$U_0(T)|\psi_s(0) >= \sum_{k,m=0}^{N-1} N^{-1/2}exp(-i2\pi kn/N)c_k(0)|n >$$

However owing to quantum noise coming from the bath some error is introduced and the Belavkin filter tries to estimate this state and if we follow this up with the Luc-Bouten algorithm of quantum control, we can remove part of the HP noise and obtain thereby a more reliable estimate of the quantum Fourier transformed state.

9.2   General Quantum filtering and control

(Ref:Naman Garg, Ph.D thesis, NSIT).

In this chapter, we propose to solve three kinds of related problems related to quantum filtering and control. The first deals with designing standard finite matrix based MATLAB simulation algorithms for the Hudson-Parthasarathy noisy Schrodinger equation which is today accepted as the standard technique for describing the unitary evolution of system and noisy bath with the system getting noisy inputs from the bath in the form of the three kinds of fundamental quantum noise processes:Creation, annihilation and conservation processes. The second deals with simulating the well known Belavkin quantum filter for obtaining estimates of system states observables on a real time basis with the measurements being non-demolition measurements on the bath noise that passes through the system where the system evolves according to the noisy Hudson-Parthasarathy-Schrodinger equation. Our basic method of simulating these quantum stochastic differential equations is based on the action of noise operators on coherent vectors and constructing truncated orthonormal bases using coherent vectors for the bath space. The Belavkin filter is the quantum non-commutative generalization of the classical Kushner-Kallianpur nonlinear filter and can be used to estimate quantum processes like the spin of the electron interacting with a magnetic field. The third kind of problem we deal with concerns design and implementation of real time quantum control algorithms using a sequence of control unitaries acting on the Belavkin filtered state to obtain to lessen the amount of Lindblad noise in the Hudson-Parthasarathy evolved state as well as to track a given state trajectory. These control algorithms have their roots in Luc-Bouten's famous thesis on filtering and control in quantum optics. The final problem dealt in this work involves computing the evolution of the Von-Neumann entropy of, (a) the Hudson-Parthasarathy noisy Schrodinger system state after tracing out over the bath variables, (b) the Belavkin filtered state and (c) the state after applying the control algorithm. These entropy computations involve Lie algebraic techniques and are important in physics for validating the second law of thermodynamics on monotonic entropy increase and also in modern quantum communication involving evaluating the amount of information transmitted through a noisy quantum channel.

9.3   Some topics in quantum filtering theory

[i] Historical survey of classical and quantum filtering theory
  [ii] Mathematical details of quantum filtering theory.
  [a] Schrodinger equation for state evolution in noiseless quantum dynamics.

The Schrodinger equation for the state vector $|\psi(t)>$ taking values in a Hilbert space $\mathcal{H}$ is given by

$$id/dt|\psi(t) >= H(t)|\psi(t) >$$

where $H(t)$ is a Hermitian operator. This guarantees unitarity of the evolution, ie,

$$< \psi(t)|\psi(t) >=< \psi(0)|\psi(0) >, t \geq 0$$

which implies conservation of probabilities. For example, if we use the position representation PAM Dirac), then the wave function would be $< r|\psi(t) >= \psi(t,r), r \in \mathbb{R}^3$ and the above equation would read as

$$\int |\psi(t,r)|^2 d^3r = 1, t \geq 0$$

provided that

$$\int |\psi(0,r)|^2 d^3r = 1$$

This means that the particle cannot with positive probability escape away to ∞ in a bound state. Schrodinger discovered the form of the Hamiltonian operator $H(t)$ for a particle moving in $\mathbb{R}^3$ with a kinetic energy of $p^2/2m$ and a potential energy of $V(t,r)$. The total energy is then $p^2/2m + V = E$. According to Planck's quantum hypothesis, the frequency of a wave associated with this energy is $\omega = 2\pi E/h$ where h is Planck's constant and according to De-Broglie's matter wave duality principle, the wave-vector associated to such a wave is $k = 2\pi p/h$ and the associated plane wave to this particle should be

$$\psi(t,r) = A.exp(-i(\omega t - k.r))$$

After doing this, Schrodinger observed that for such a plane wave,

$$2\pi ih\partial\psi(t,r)/\partial t = (h\omega/2\pi)\psi(t,r) = E\psi(t,r)$$

and

$$(-ih/2\pi)\nabla\psi(t,r) = (hk/2\pi)\psi(t,r) = p\psi(t,r)$$

Schrodinger postulated based on this intuitive notion, that even when the wave function is square integrable, E should be regarded as an energy operator $2\pi ih\partial/\partial t$ and p as a momentum operator $(-ih/2\pi)\nabla$. Then, Newton's relation between energy and momentum given by

$$E - p^2/2m - V = 0$$

should be interpreted as a wave equation

$$(E - p^2/2m - V)\psi(t,r) = 0$$

or equivalently,

$$(ih/2\pi)\partial\psi(t,r)/\partial t - (h^2/8\pi^2 m)\nabla^2\psi(t,r) - V(t,r)\psi(t,r) = 0$$

which is precisely the Schrodinger wave equation. Schrodinger's genius was to state that this wave equation is valid for all nonrelativistic quantum phenomena even when its solution $\psi(t, r)$ does not represent a plane wave. Schrodinger postulated that the bound states of atoms and molecules should correspond only to square integrable solutions and it became clear to him that such solutions can be expressed by separating the space and time variables as

$$\psi(t, r) = \sum_{n=1}^{\infty} exp(-iE_n t)\psi_n(r)$$

where the "energy eigenvalues" E_n are real numbers satisfying the eigenvalue problem, namely the stationary Schrodinger equation:

$$E_n\psi_n(r) - (h^2/8\pi^2 m)\nabla^2\psi_n(r) - V(r)\psi_n(r) = 0$$

in the special case when the potential V does not depend explicitly on time. The eigenfunctions $\psi_n(r)$ should be square integrable and this condition causes the energy eigenvalue spectrum of bound states to be discrete. He explicitly thus determined the eigenfunctions and energy spectrum of the Hydrogen atom and the quantum harmonic oscillator corresponding to potentials $V = -e^2/r$ and $V = Kr^2/2$. Later on Dirac recognized that we can also have unbounded state solutions as in scattering theory in which an projectile arrives from ∞, interacts with a repulsive potential and gets scattered to ∞. Such states are not normalizable and are characterized by the energy spectrum being continuous. It was Max Born who suggested that for bound states $|\psi(t, r)|^2$ should be interpreted as the probability density of the particle being in a unit volume around r at time t after normalizing the wave function to have a unit integral for its modulus square. It should be mentioned here that the time dependent Schrodinger equation stated above can be cast in form of a unitary operator evolution equation :

$$|\psi(t) >= U(t)|\psi(0) >, iU'(t) = H(t)U(t)$$

where $U(t)$ is a unitary operator in the Hilbert space of states in which the Hamiltonian operator $H(t)$ is defined. If the energy operator $H(t) = H$ is time independent, then its formal solution is

$$U(t) = exp(-itH)$$

where the exponential of an unbounded operator must be defined via its resolvent (Kato):
$$exp(-itH) = lim_{n\to\infty}(I + itH/n)^{-n}$$

rather than as
$$lim_{n\to\infty}(I - itH/n)^n$$

the reason being that the resolvent of an unbounded operator is bounded over the resolvent set while the positive integer powers of an unbounded operator will generally have smaller and smaller domains. In the case when $H(t)$ is time dependent as in quantum electrodynamics, we have to express the solution $U(t)$ as a Dyson series

$$U(t) = I + \sum_{n \geq 1} (-i)^n \int_{0 < t_n < ... < t_1 < t} H(t_1)...H(t_n)dt_1...dt_n$$

It is the unitary evolution operator formalism which is the most useful while formulating the noisy Schrodinger equations of Hudson and Parthasarathy because we shall be replacing the Schrodinger equation by equations of the form

$$dU(t) = (-iH(t)dt + \sum_k L_k dN_k(t))U(t)$$

where L_k are system operators and $dN_k(t)$ are noise operator differentials. These are non-random families of operators in an extended Hilbert space which plays the role of a noise bath coupled to the system. Although this equation does not contain any random terms, when we impose the condition that the bath is initially in a given state, like a coherent state, then system observables X after time t will evolve to system$\otimes$ noise space observables $U(t)^*(X \otimes I)U(t)$ which will display different kinds of statistical behaviour in different initial system$\otimes$ noise states.

[b] Noisy Schrodinger equations.
   When the potential contains randomly fluctuating terms, we can model it as a space-time random field $V(t,r)$ and then develop approximate perturbation theoretic techniques for calculating the statistical properties of the wave function. However, there are certain problems where the potential fluctuates so rapidly that we are compelled to add white noise terms. An example of such a Schrodinger equation is

$$dU(t) = (-iH(t)dt - iV(t)dB(t))U(t)$$

where $B(t)$ is Brownian motion and the equation is interpreted as an Ito stochastic differential equation. Here $H(t), V(t)$ are non-random Hermitian operators in the system Hilbert space. However, the Ito interpretation prevents $U(t)$ from being unitary:

$$d(U(t)^*U(t)) = dU^*.U + U^*.dU + dU^*.dU = U(t)^*V(t)^2U(t)dt \neq 0$$

To rectify this situation, we add a non-Hermitian term to $H(t)$, namely an Ito correction term to get our noisy Schrodinger evolution as

$$dU(t) = (-(iH(t) + V(t)^2/2)dt - iV(t)dB(t))U(t)$$

and then it is easily seen that

$$d(U^*U) = 0$$

which means that $U(0)^*U(0) = I$ implies $U(t)^*U(t) = I \forall t \geq 0$. Such an equation can be simulated in discrete time as

$$U[t+\Delta] = (I+i(H(t)\Delta+V(t)\sqrt{\Delta}w[t+1])/2)^{-1}.(I-i(H(t)\Delta+V(t)\sqrt{\Delta}w[t+1])/2)U[t], t = 0, \Delta, 2\Delta, ...$$

where $w[t], t = 0, \Delta, 2\Delta, ...$ are iid $N(0,1)$ random variables. Such a simulation guarantees unitary evolution rather than a simulation of the form

$$U[t + \Delta] = (I - i(H(t)\Delta + V(t)\sqrt{\Delta}w[t + 1]))U[t]$$

Now suppose $\rho(0)$ is the initial state of the system. Then after time t, the system state becomes after averaging over the bath, ie, w.r.t the probability distribution of the Brownian motion

$$\rho(t) = \mathbb{E}(U(t)\rho(0)U(t)^*) = \int U(t)\rho(0)U(t)^* dP(B)$$

It is easy to show by using the Ito's formula that $\rho(t)$ satisfies the GKSL/master equation

$$\rho'(t) = i[H(t), \rho(t)] - (1/2)(V(t)^2\rho(t) + \rho(t)V^2(t) - 2V(t)\rho V(t))$$

However, this master equation is not the most general kind of master equation owing to the fact that $V(t)$ is restricted to be a Hermitian operator. The general form of the GKSL equation is a variant of

$$\rho'(t) = i[H(t), \rho(t)] - (1/2)(L(t)^*L(t)\rho(t) + \rho(t)L(t)^*L(t) - 2L(t)\rho(t)L(t)^*)$$

where $L(t)$ is not necessarily Hermitian. More generally, we could have

$$\rho'(t) = i[H(t), \rho(t)] - (1/2)\sum_k (L_k(t)^*L_k(t)\rho(t) + \rho(t)L_k(t)^*L_k(t) - 2L_k(t)\rho(t)L_k(t)^*)$$

It is not hard to see that this general GKSL can be used to describe the motion of a damped Harmonic oscillator while the original classical Brownian motion based GKSL will not yield the equations of motion of a damped harmonic oscillator. No matter what kind of classical noise we add to the Schrodinger equation, we cannot after classical probabilistic averaging reproduce the damped harmonic oscillator equation. Somehow the number of degrees of freedom in classical noise is not sufficient to get quantum equations for damped motions. Some drastic change has to be made in the structure of the noise added. It was only after the advent of the Hudson-Parthasarathy noisy Schrodinger equation that such a description became possible. The HP theory yields two kinds of quantum Brownian motions $A(t)$ and $A(t)^*$ which do not commute and satisfy the quantum Ito law $dA.dA^* = dt, dA^*dA = 0$. The HP equation is a dilated form of the general GKSL equation, ie, it yields a unitary evolution in a larger Hilbert space namely the tensor product of the system Hilbert space and the bath space which is now a Boson Fock space and when averaged out w.r.t the bath state, it yields the most general form of the GKSL equation. The bath state here plays the role of the probability distribution of the Brownian motion process in the case of classical noise considered earlier but in this situation, we are able to describe a wider class of quantum systems. It should also be noted that maintaining the classical scenario, we can have even classical Poisson processes as perturbations to the Hamiltonian and yet we would not be able to realize the full GKSL degrees of freedom that are possible only by considering quantum Brownian motions (creation and annihilation processes) and quantum Poisson processes (Conservation processes) introduced by Hudson and Parthasarathy. Specifically using the classical Ito rule for independent Brownian motions $B_i(t), i = 1, 2, ..., p$ and independent Poisson processes $N_i(t), i = 1, 2, ..., q$:

$$dB_i dB_j = \delta_{ij} dt, dN_i dN_j = \delta_{ij} dN_i, dB_i dN_j = 0$$

we can construct a stochastic Schrodinger equation

$$dU(t) = (-(iH(t) + P(t))dt - i \sum_{k=1}^{p} V_k(t)dB_k(t) + \sum W_k(t)dN_k(t))U(t)$$

where in order to maintain unitarity of the evolution, ie, $d(U^*U) = 0$, we would require that

$$W_k(T)^* + W_k(t) + W_k(t)^* W_k(t) = 0, V_k(t)^* = V_k(t), P(t) = \sum_{k=1}^{p} V_k(t)^2/2$$

and yet this equation would not yield all possible open quantum systems, ie, all possible kinds of GKSL generators.

[c] The basic noise processes of Hudson and Parthasarathy using Boson Fock space.

We shall present one version of the Hudson-Parthasarathy theory which has used in this thesis. Here, the system Hilbert space is $\mathfrak{h} = \mathbb{C}^p$ with the standard inner product while the bath space is the Boson-Fock space $\Gamma_s(L^2(\mathbb{R}_+) \otimes \mathbb{C}^d)$. We write

$$\mathcal{H} = L^2(\mathbb{R}_+) \otimes \mathbb{C}^d$$

so that the Bath space is

$$\Gamma_s(\mathcal{H}) = \mathbb{C} \oplus \bigoplus_{n \geq 1} \mathcal{H}^{\otimes_s n}$$

where $\mathcal{H}^{\otimes_s n}$ is the n-fold symmetric tensor product of $\mathcal{H}$. For $u \in \mathcal{H}$, define for $u \in \mathcal{H}$ the exponential vector

$$|e(u)> = 1 \oplus \bigoplus_{n \geq 1} u^{\otimes n}/\sqrt{n!}$$

Then an easy computation shows that any symmetric tensor in $\Gamma_s(\mathcal{H})$ is expressible as a limit of finite linear combinations of the vectors $|e(u)>, u \in \mathcal{H}$. For example

$$u^{\otimes n} = (n!)^{-1/2} \frac{d^n}{dt^n} |e(tu) > |_{t=0}$$

Other examples include

$$u \otimes v + v \otimes u = (1/2)((u+v) \otimes (u+v) - (u-v) \otimes (u-v))$$

$$= (u+v) \otimes (u+v) - u \otimes u - v \otimes v$$

One way to realize this Boson Fock space is via the Harmonic oscillator algebra: Let $a(n), n = 1, 2, ...$ be operators in independent copies of $L^2(\mathbb{R})$ satisfying the canonical commutation relations

$$[a(n), a(m)^*] = \delta_{nm}$$

Each pair $(a(n), a(n)^*)$ determines an independent quantum harmonic oscillator. These operators can be expressed using the canonical position and momentum operators in $L^2(\mathbb{R})$. Let $|0>$ be the vacuum state, ie, the state in which all the operators $a(n)^*a(n)$ have zero eigenvalue and for complex numbers $u[n] \in \mathbb{C}, n = 1, 2, ...$, consider the state

$$|f(u) >= \sum_n \Pi_n u[n]^{m(n)} a(n)^{*m(n)} |0 > /\Pi_n m(n)!$$

$$= \sum_n \Pi_n u[n]^{m(n)} \otimes_n |m(n) > /\Pi_n \sqrt{m(n)!}$$

Then an easy computation shows that

$$< f(u)|f(v) >= exp(< u, v >)$$

where

$$< u, v >= \sum_{n,j} \bar{u}[n] v[n]$$

We now choose an onb $\{\psi_n(r) = (\psi_{n1}(r), ..., \psi_{nd}(r))^T : n = 1, 2, ...\}$ for $L^2(\mathbb{R}) \otimes \mathbb{C}^d$ and define

$$u(r) = \sum_n u[n] \psi_n(r) \in \mathbb{L}^2(\mathbb{R}) \otimes \mathbb{C}^d$$

Then, the map $|e(u) >] \to |f(u) >$ defines a Hilbert space isomorphism. In fact, this mapping is uniquely identified by the mapping

$$u^{\otimes N} \to \sum_{m(1)+m(2)+...=N} \sqrt{N!} \Pi_n u[n]^{m(n)} \Pi_n a(n)^{*m(n)} |0 > /\Pi_n m(n)!$$

$$= \sum_{m(1)+...+..=N} \sqrt{N!} u[1]^{m(1)} u[2]^{m(2)} ... |m(1), m(2), ... > /\sqrt{m(1)!m(2)!...}$$

which preserves inner products. The exponential vectors $|e(u) >, u \in \mathcal{H}$ span a dense linear manifold of $\Gamma_s(\mathcal{H})$ and hence the action of operators on exponential vectors is sufficient to determine their action on their domain in the entire Boson Fock spaces. Keeping this in mind, for $u \in \mathcal{H}$, we define

$$a(u)|e(v) >=< u, v > |e(v) >,$$

and then it is easy to verify using

$$< e(u), e(v) >= exp(< u, v >)$$

that the adjoint of $a(u)$ is given by

$$a(u)^* | e(v) >= \frac{d}{dt} |e(v + tu) > |_{t=0}$$

One then also easily verifies by considering the action on exponential vectors that

$$[a(u), a(v)^*] =< u, v > I, [a(u), a(v)] = 0, [a(u)^*, a(v)^*] = 0, u, v \in \mathcal{H}$$

In other words, the operator fields $a(u), a(u)^*, u \in \mathcal{H}$ define an algebra of harmonic oscillators. More precisely, choosing an orthonormal basis $|e_i >, i = 1, 2, ...$ of $\mathcal{H}$, we can write for $u \in \mathcal{H}$,

$$a(u) = \sum_i \bar{u}_i a_i, a(u)^* = \sum_i u_i a_i^*, u_i =< e_i, u >$$

and then $(a_i, a_i^*), i = 1, 2, ...$ define a countably infinite sequence of one dimensional quantum harmonic oscillators in the sense that the canonical commutation relations

$$[a_i, a_j^*] = \delta_{ij}, [a_i, a_j] = 0, [a_i^*, a_j^*] = 0$$

are satisfied. As in conventional quantum mechanics, we can also define a conservation/number operator Λ_i by

$$\Lambda_i = a_i^* a_i, i = 1, 2, ...$$

and then, we can obtain an onb for $\Gamma_s(\mathcal{H})$ as

$$|n_1, n_2, ... >$$

where

$$\Lambda_i |n_1, n_2, ... >= n_i |n_1, n_2, ... >$$

and

$$a_i |n_1, n_2, ... >= \sqrt{n_i} |n_1, ..., n_{i-1}, n_i - 1, n_{i+1}, ... >$$
$$a_i^* |n_1, n_2, ... >= \sqrt{n_i + 1} |n_1, n_2, ... >$$

It is then easily seen that the exponential vector can be represented by

$$|e(u) >= \sum_{n_1, n_2, ...} u_1^{n_1} u_2^{n_2} ... |n_1, n_2, ... > / \sqrt{n_1! n_2! ...}$$

and that the above actions of a_i, a_i^* on $|n_1, n_2, ... >$ are equivalent to

$$a(v) | e(u >=< v, u > | e(u) >, a(v)^* | e(u) >= \sum_i v_i \frac{\partial}{\partial u_i} | e(u) >= \frac{d}{dt} | e(u + tv) > |_{t=0}$$

or equivalently,

$$a_i|e(u)>=u_i|e(u)>,a_i^*|e(u)>=\frac{\partial}{\partial u_i}|e(u)>=\frac{d}{dt}|e(u+te_i)>|_{t=0}$$

An easy calculation using the above identities shows further that

$$<e(u)|\Lambda_i|e(v)>=\bar{u}_i v_i<e(u)|e(v)>$$

Based on this idea, Hudson and Parthasarathy [1984] introduced apart from the creation and annihilation operator fields $a(u), a(u)^*, u \in \mathcal{H}$, a conservation operator field $\lambda(Q)$ where Q is a Hermitian operator in $\mathcal{H}$ by the rule

$$exp(it\lambda(Q))|e(u)>=|e(exp(itQ)u)>,t \in \mathbb{R}$$

or equivalently,

$$i\lambda(Q)|e(u)>=\frac{d}{dt}|e(exp(itQ)u>|_{t=0}$$

It is easy to see then that

$$<e(v)|\lambda(Q)|e(u)>=<v|Q|u><e(v)|e(u)>$$

and in particular if

$$Q=|e_i><e_i|,$$

then

$$<v|Q|u>=\bar{v}_i u_i$$

so that

$$\lambda(|e_i><e_i|)=\Lambda_i$$

More generally, if Q has the spectral representation

$$Q=\sum_i |e_i>q_i<e_i|$$

then

$$\lambda(Q)=\sum_i q_i\Lambda_i=\sum_i q_i a_i^* a_i$$

Most of this formalism was well known even before the seminal paper of Hudson and Parthasarathy [1984]. The crucial observation of Hudson and Parthasarathy was to introduce quantum processes, ie creation, annihilation and conservation processes by bringing time into the picture in the form of a continuous unfolding of the tensor product appearing in Boson Fock space. Without going into the mathematics, the idea here is to consider the indicator function $\chi_{[0,t]}$ in $L^2(\mathbb{R}_+)$ and then choose an orthonormal basis $|f_i>, i=1,2,...,d$ for $\mathbb{C}^d$ so that

$$\chi_{[0,t]}|f_i>\in \mathcal{H}=L^2(\mathbb{R}_+)\otimes \mathbb{C}^d$$

Then Hudson and Parthasarathy define

$$A_i(t) = a(\chi_{[0,t]}|f_i >), A_i(t)^* = a(\chi_{[0,t]}|f_i >)^*,$$

$$\Lambda_i(t) = \lambda(\chi_{[0,t]}|f_i >< f_i|)$$

where $i = 1, 2, ..., d$ and where $\chi_{[0,t]}|f_i >< f_i| = \chi_{[0,t]} \otimes |f_i >< f_i|$ is a Hermitian operator in $\mathcal{H}$ with $\chi_{[0,t]}$ acting as a multiplication. The action of these operators on the exponential vectors is then easily deduced to be given by

$$A_i(t)|e(u) >= (\int_0^t u_i(s)ds)|e(u) > - - -(a),$$

$$< e(v)|A_i(t)^*|e(u) >= (\int_0^t \bar{v}_i(s)ds) < e(v)|e(u) > - - -(b),$$

$$< e(v)|\Lambda_i(t)|e(u) >= (\int_0^t \bar{v}_i(s)u_i(s)ds) < e(v)|e(u) > - - -(c)$$

and more generally,

$$< e(v)|\lambda(Q_t)|e(u) >= (\int_0^t < v(s)|Q|u(s) > ds) < e(v)|e(u) >$$

where

$$Q_t = \chi_{[0,t]}Q$$

and $u, v \in \mathcal{H} = L^2(\mathbb{R}_+) \otimes \mathbb{C}^d$ is specified in coordinate form as

$$u(t) = [u_1(t), ..., u_d(t)]^T, t \geq 0, u_k(.) \in \mathbb{L}^2(\mathbb{R}_+)$$

Since in quantum electrodynamics, the annihilation operator fields are proportional to the magnetic vector potential in the spatial Fourier domain and since the electric field in the spatial frequency domain is also proportional to the magnetic vector potential provided that we adopt the Coulomb gauge ($div A = 0$) which means that the electric scalar potential A^0 becomes a matter field and not any combination of the field part, we can interpret the equation (a) as giving the total complex amplitude, ie, amplitude and phase of the photons in the exponential/coherent state $|e(u) >$ upto time t. Likewise, the equation (c) determines the number of accumulated photons in the ith a coherent state upto time t:

$$< e(u)|\Lambda_i(t)|e(u) >= (\int_0^t |u_i(t)|^2 dt) < e(u)|e(u) >$$

The equation (b) tells us the total complex amplitude of the photons in the ith mode that have been absorbed into the coherent state $|e(v) >$. The processes $A_i(t), A_i(t)^*, \Lambda_i(t), i = 1, 2, ..., d$ are known as the fundamental noise processes in the Hudson-Parthasarathy calculus. They are not random functions, they are simply families of linear operators in the Boson Fock space, but when we restrict our bath space to be in a given state like a coherent state, then these families of operators display statistics like the classical Brownian motion and Poisson processes as special commutative cases. The important point to be noted is that these operator valued processes are non-commutative unlike the classical case and hence can exhibit more general kinds of statistical behaviour than those displayed by classical stochastic processes. The non-commutativity of these processes is precisely the fact that the most general GKSL equations can be dilated into unitary evolutions using quantum noisy Schrodinger equations, ie, the Hudson-Parthasarathy noisy Schrodinger equation. This is not possible using only classical stochastic noise. The reason behind this is the quantum Ito formula discussed next which can be regarded as an Ito formula taking into account Heisenberg's uncertainty principle arising from the non-commutativity of the operator valued processes.

[d] Quantum Ito's formula and the quantum noisy HP-Schrodinger equation. If $z_1, z_2 \in \mathbb{C}^d$, then $\chi_{0,t]} z_k \in \mathcal{H} = L^2(\mathbb{R}_+) \otimes \mathbb{C}^d, k = 1, 2$ and it is not hard to show that

$$da(\chi_{[0,t]} z_1).da(\chi_{[[0,t]} z_2)^* =< z_1, z_2 > dt$$

and on the other hand all the other products of these two differentials are zero, ie, $o(dt)$. To verify the above formula, we observe that the Harmonic oscillator commutation relations imply

$$[a(\chi_{[0,t]} z_1), a(\chi_{[0,s]} z_2)^*] =< \chi_{[0,t]}, \chi_{[0,s]} >< z_1, z_2 >$$

$$= min(t, s) < z_1, z_2 >= (s\theta(t - s) + t\theta(s - t)) < z_1, z_2 >$$

and thus taking the differentials w.r.t t and using $d\theta(t - s) = \delta(t - s)dt$ and $x\delta(x) = 0$ gives us

$$[da(\chi_{[0,t]} z_1), a(\chi_{[0,s]} z_2)^*] = ((s - t)\delta(t - s) + \theta(s - t)dt) < z_1, z_2 >$$

$$= \theta(s - t)dt < z_1, z_2 >$$

Now take the differential on both sides w.r.t s and use the identity $\delta(s - t)ds|_{s=t}dt = dt$ to get the result

$$[da(\chi_{[0,t]} z_1), da(\chi_{[0,t]} z_2)^*] =< z_1, z_2 > dt$$

Now it is clear that

$$da(\chi_{[0,t]} z_2)^*.da(\chi_{[0,t]} z_1) = 0$$

as follows by taking the matrix element of the lhs w.r.t $|e(u) >$ and $|e(v) >$:

$$< e(u)|da(\chi_{[0,t]} z_2)^* da(\chi_{[0,t]} z_1)|e(v) >=$$

$$< da(\chi_{[0,t]} z_2)e(u)|da(\chi_{[0,t]} z_1)e(v) >=$$

$$< z_2, u(t) >^* dt. < z_1, v(t) > dt. < e(u), e(v) >= O(dt^2)$$

In other words, we have

$$da(\chi_{[0,t]} z_2)^* da(\chi_{[0,t]} z_1) = 0$$

We have thus proved the required quantum Ito formula (choosing $z_1 = f_i, z_2 = f_j \in \mathbb{C}^d$)

$$dA_i(t).dA_j(t)^* = \delta_{ij}dt$$

and all the other products of these two differentials vanish. It should be noted that in deducing this quantum Ito formula, we have made use of Bosonic commutation relations and this is why one often says that quantum Ito formula can be traced to the Heisenberg uncertainty principle. The other quantum Ito formula are obtained as follows. First observe that

$$< e(v)|dA_i^*(t)dA_j(t)|e(u) >= u_j(t)\bar{v}_i(t)dt^2 < e(v)|e(u) >$$

and since

$$< e(v)|d\lambda(Q_t)|e(u) >= d < e(v)|\lambda(Q_t)|e(u) >$$

$$=< e(v)|e(u) > d \int_0^t < v(s)|Q|u(s) > ds$$

$$=< e(v)|e(u) >< v(t)|Q|u(t) > dt$$

it follows that

$$dA_i(t)^*.dA_j(t)/dt = d\lambda(\chi_{[0,t]}||f_i >< f_j|)$$

In particular,

$$dA_i(t)^*.dA_i(t)/dt = d\Lambda_i(t)$$

and hence, application of the quantum Ito formula derived earlier gives us

$$dA_j(t)d\Lambda_i(t) = \delta_{ij}dA_i(t), d\Lambda_i(t).dA_j(t) = \delta_{ij}dA_i(t)^*,$$

$$d\Lambda_i(t).d\Lambda_j(t) = \delta_{ij}d\Lambda_i(t)$$

These are known as the quantum Ito formulae and are fundamental in studying the effect of noise on quantum systems as well as in formulating a quantum filtering theory based on non-demolition measurements. The processes $B_i(t) = A_i(t) + A_i(t)^*$ have all the properties of classical Brownian motion in the vacuum coherent state. First, they commute with each other:

$$[B_i(t), B_j(s)] = 0 \forall t, s, i, j$$

second, for $f(t) = (f_i(t)) \in L^2(\mathbb{R}_+) \otimes \mathbb{C}^d$,

$$< e(u)|exp(\sum_{i=1}^N int_0^T f_i(t)dB_i(t))|e(u) >=$$

$$< e(u)|exp(a(f\chi_{[0,T]}) + a(f\chi_{[0,T]})^*)|e(u) >=$$

[e] Matrix elements of noise operators in Boson Fock space relative to coherent vectors.

Given an HP qsde of the form

$$dU(t) = (-(iH + P)dt \sum_i (L_i dA_i(t) + M_i dA_i(t)^* + S_i d\Lambda_i(t))U(t)$$

we can simulate this approximately as follows: We first choose a set $u_1, ..., u_N$ of linearly independent vectors in $\mathcal{H} = L^2(\mathbb{R}_+) \otimes \mathbb{C}^d$ and then apply the Gram-Schmidt orthonormalization to the corresponding exponential vectors $|e(u_k) >$ $, k = 1, 2, ..., N$. Denote the resulting orthonormal vectors by $|\xi_k >, k = 1, 2, ..., N$. Thus,

$$|\xi_k >= \sum_{m=1}^{k} c(k, m)|e(u_m) >, 1 \le k \le N,$$

with inverse

$$|e(u_k) >= \sum_{m=1}^{k} d(k, m)|\xi_m >, 1 \le k \le N$$

and

$$< \xi_k|\xi_m >= \delta_{km}$$

Since the exponential vectors span a dense linear manifold of the Boson Fock space [KRP book], it follows that if N is large, the vectors $|\xi_k >, k = 1, 2, ..., N$ will span almost completely the Boson Fock space. Choosing then an orthonormal basis $|\eta_k >, k = 1, 2, ..., p$ for the system Hilbert space $\mathfrak{h}$, we get an orthonormal set $|\eta_r \otimes \xi_k >, 1 \le k \le N, 1 \le r \le p$ in $\mathfrak{h} \otimes \Gamma_s(\mathcal{H})$=system space$\otimes$ bath space. We can then take matrix elements on both sides of the above HP equation relative to this basis and derive a sequence of linear deterministic equations for the truncated matrix elements of the unitary evolution operator $U(t)$. The advantage of using orthonormal sets for computing matrix elements is due to the composition law:

$$< \eta_r \otimes \xi_k|AB|\eta_s \otimes \xi_m >\approx$$

$$\sum_{q=1}^{N} \sum_{l=1}^{p} < \eta_r \otimes \xi_k|A|\eta_l \otimes \xi_q >< \eta_l \otimes \xi_q|B|\eta_s \otimes \xi_m >$$

This equation is exact iff A, B are operators defined in the truncated system $\otimes$ noise space. It should be noted that this method can be combined with the fact that if L is a system operator, then

$$LdA_k(t)U(t) = (L \otimes I)U(t)dA_k(t) = (L \otimes I)U(t) \otimes dA_k(t)$$

and likewise with $A_k(t)$ replaced by $A_k(t)^*$ or with $\Lambda_k(t)$, the reason being that $A_k(t), A_k(t)^*, \Lambda_k(t)$ act in $\Gamma_s(L^2([0, t] \otimes \mathbb{C}^d)$ while $dA_k(t), dA_k(t)^*, d\Lambda_k(t)$ act in $\Gamma_s(L^2(t, t + dt])$ and we have the Hilbert space isomorphism

$$\Gamma_s(\mathcal{H}_1 \oplus \mathcal{H}_2) = \Gamma_s(\mathcal{H}_1) \otimes \Gamma_s(\mathcal{H}_2)$$

as follows by choosing $w_k, v_k \in \mathcal{H}_k, k = 1, 2$ and noting that

$$< e(w_1 \oplus v_1)|e(w_2 \oplus v_2) >= exp(< w_1 \oplus v_1|w_2 \oplus v_2 >) =$$

$$exp(< w_1|w_2 > + < v_1|v_2 >) =< e(w_1)|e(w_2) >< e(v_1)|e(v_2) >$$

$$=< e(w_1) \otimes e(v_1)|e(w_2) \otimes e(v_2) >$$

In other words for $u, v \in \mathcal{H}$ with $< u|v >= 0$, we can under this isomorphism identify $|e(u + v) >$ with $|e(u) \otimes e(v) >$. In view of this, we have the Hilbert space isomorphism

$$\Gamma_s(\chi_{[0,t+dt]}\mathcal{H}) = \Gamma_s(\chi_{[0,t]}\mathcal{H}) \otimes \Gamma_s(\chi_{(t,t+dt]}\mathcal{H})$$

To see why $dA_k(t), dA_k(t)^*, d\Lambda_k(t)$ act in $\Gamma_s(\chi_{(t,t+dt]}\mathcal{H})$, we use the isomorphism identity just established in the form

$$e(\chi_{[0,T]}u) = e(\chi_{[0,t]}u) \otimes e(\chi_{(t,T]}, 0 < t < T$$

in the sense that

$$< e(\chi_{[0,T]}u)|e(v\chi_{[0,T]}) >= exp(\int_0^T < u(s)|v(s) > ds)$$

$$= exp(\int_0^t < u(s)|v(s) > ds).exp(\int_t^T < u(s)|v(s) >)$$

$$=< e(u\chi_{[0,t]}) \otimes e(u\chi_{(t,T]})|e(v(\chi_{[0,t]}) \otimes e(v\chi_{(t,T]}) >$$

We can thus write for $s < t$,

$$< e(v)|A_k(t) - A_k(s)|e(u) >=< e(v)| \int_s^t u_k(\tau)d\tau|e(u) >$$

$$= (\int_s^t u_k(\tau)d\tau) < e(v)|e(u) >$$

on the one hand and on the other hand,

$$|e(u) >= |e(\chi_{[0,s]})u) \otimes e(\chi_{s,t]}) \otimes e(\chi_{(t,\infty)}) >$$

with

$$< e(\chi_{(s,t]}v)|(A_k(t) - A_k(s)|e(\chi_{(s,t]})u >$$

$$= (\int_s^t u_k(\tau)d\tau) < e(\chi_{(s,t]}v)|e(\chi_{(s,t]}u) >$$

which is in view of the above isomorphism, the same as the previous equation since the isomorphism implies

$$|e(u) >= |e(\chi_{(0,s]}u) \otimes e(\chi_{(s,t]}u) \otimes e(\chi_{(t,\infty)}) >$$

and further, if I is any interval in $\mathbb{R}$ non-overlapping with $(s,t]$, then

$$< e(\chi_I.v)|A_k(t) - A_k(s)|e(\chi_I.u) >= (\int_{I \cap (s,t]} u_k(\tau)d\tau) < e(\chi_I)v)|e(\chi_I.u) >= 0$$

It should be noted that in view of these discussions, the isomorphism means that

$$< e(v)|A_k(t) - A_k(s)|e(u) >=$$

$$< e(v\chi_{(0,s]}) \otimes e(v.\chi_{(s,t]}) \otimes e(v.\chi_{(t,\infty)})|A_k(t) - A_k(s)|$$

$$.e(u\chi_{(0,s]}) \otimes e(u.\chi_{(s,t]}) \otimes e(u.\chi_{(t,\infty)}) >$$

$$=< e(v.\chi_{(0,s]})|e(u\chi_{(0,s]}) >< e(v.\chi_{(s,t]})|A_k(t)$$

$$- A_k(s)|e(u.\chi_{(s,t]}) > . < e(v.\chi_{(t,\infty)})|e(u.\chi_{(t,\infty)}) >$$

and likewise for the other noise operators $A_k(t)^*$ and $\Lambda_k(t)$. This is the reason why we can say that if $M(t)$ is any one of these noise operators, then for $s < t$, we can say that $M(t) - M(s)$ acts in the Hilbert space $\Gamma_s(\chi_{(s,t]}\mathcal{H})$ when we consider the isomorphism

$$\Gamma_s(\mathcal{H}) = \Gamma_s(\chi_{(0,s]}\mathcal{H} \oplus \chi_{(s,t]}\mathcal{H} \oplus \chi_{(t,\infty)}\mathcal{H})$$

$$= \Gamma_s(\chi_{(0,s]}\mathcal{H}) \otimes \Gamma_s(\chi_{(s,t]}\mathcal{H}) \otimes \Gamma_s(\chi_{(t,\infty)}\mathcal{H})$$

It is precisely this property of the noise operators, ie, quantum independent increment property that makes the HP equation easy to simulate using matrix elements w.r.t. bath states built out of finite linear combinations of exponential/coherent vectors. For example, the matrix elements of the terms

$$LdA(t)U(t), MdA(t)^*U(t), Sd\Lambda(t)U(t)$$

can be evaluated by noting that L, M, S act in $\mathfrak{h}$, the system Hilbert space, $U(t)$ acts in

$$\mathfrak{h} \otimes \Gamma_s(L^2[0,t] \otimes \mathbb{C}^d) = \mathfrak{h} \otimes \Gamma_s(\chi_{[0,t]}\mathcal{H})$$

while $dA(t), d(t)^*, d\Lambda(t)$ act in $\Gamma_s(\chi_{(t,t+dt]}\mathcal{H})$: Let dM denote any one of $dA_k, dA_k^*, d\Lambda$. Then,

$$< \eta_r \otimes \xi_k|LdM(t)U(t)|\eta_s \otimes \xi_l >=$$

$$\sum_{m,q,m',q'} < \eta_r \otimes \xi_k|L|\eta_q \otimes \xi_m >< \eta_q \otimes \xi_m|U(t)|\eta_{q'} \otimes \xi_{m'} >< \eta_{q'} \otimes \xi_{m'}|dM(t)|\eta_s \otimes \xi_l >$$

$$= \sum_{m,q,m',q'} < \eta_r|L|\eta_q > \delta_{km} < \eta_q \otimes \xi_m|U(t)|\eta_{q'} \otimes \xi_{m'} > \delta_{q's} < \xi_{m'}|dM(t)|\xi_l >$$

$$= \sum_{q,m'} < \eta_r|L|\eta_q >< \eta_q \otimes \xi_k|U(t)|\eta_s \otimes \xi_{m'} >< \xi_{m'}|dM(t)|\xi_l >$$

and if $dM(t) = dA_k(t)$, we get

$$< \xi_m|dM(t)|\xi_l >= \sum_{a,b} \bar{c}(m,a)c(l,b) < e(u_a)|dA_k(t)|e(u_b) >=$$

$$= \sum_{a,b} \bar{c}(m,a)c(l,b)u_{bk}(t) < e(u_a)|e(u_b) > dt$$

while if $dM(t) = dA_k(t)^*$, then

$$< \xi_m|dM(t)|\xi_l >= \sum_{a,b} \bar{c}(m,a)c(l,b) < e(u_a)|dA_k(t)^*|e(u_b) >=$$

$$= \sum_{a,b} \bar{c}(m,a)c(l,b)\bar{u}_{ak}(t) < e(u_a)|e(u_b) > dt$$

and finally if $dM(t) = d\Lambda_k(t)$, then

$$< \xi_m|dM(t)|\xi_l >= \sum_{a,b} \bar{c}(m,a)c(l,b) < e(u_a)|d\Lambda_k(t)|e(u_b) >=$$

$$= \sum_{a,b} \bar{c}(m,a)c(l,b)\bar{u}_{ak}(t)u_{bk}(t) < e(u_a)|e(u_b) > dt$$

These formulas enable simulation of the HP qsde in the form of deterministic difference equations after time has been discretized.

[f] The GKSL equation–a derivation based on the HP-Schrodinger equation.

The HP equation is a dilated version of the GKSL equation. This means that the GKSL equation does not describe unitary evolution of the system. In fact, it describes the evolution of only the mixed system state in the presence of a noisy bath. It does not describe the evolution of a pure system state. If the system state is initially pure, then under interaction with the bath, it becomes mixed after sometime. This leads us to suspect that the system Hilbert space can be enlarged, ie, dilated to include the bath Hilbert space in such a way that the overall evolution of system and bath is described by a unitary evolution operator which when applied to an initally pure state on the system and bath and then averaged out over the bath via a partial trace will yield the GKSL. The answer to this is provided by the HP-noisy Schrodinger equation. To see this, we start with the HP equation

$$dU(t) = (-(iH + P)dt + \sum_k (L_k dA_k + M_k dA_k^* + S_k d\Lambda_k))U(t)$$

and take a system observable X. Define

$$j_t(X) = U(t)^*XU(t) = U(t)^*(X \otimes I)U(t)$$

Then $j_t : \mathcal{B}(\mathfrak{h}) \rightarrow \mathcal{B}(\mathfrak{h} \otimes \Gamma_s(\chi_{[0,t]}\mathcal{H})$ is a $*$ unital homomorphism, ie,

$$j_t(c_1X + c_2Y) = c_1j_t(X) + c_2j_t(Y), j_t(XY) = j_t(X)j_t(Y), j_t(X^*) = j_t(X)^*$$

Note that the system operators H, P, L_k, M_k, S_k are chosen so that quantum Ito's formula ensures that $U(t)$ is unitary for all t. Another application of quantum Ito's formula yields

$$dj_t(X) = j_t(\theta_0(X))dt + \sum_k (j_t(\theta_{1k}(X))dA_k(t) + j_t(\theta_{2k}(X)dA_k(t)^*j_t(\theta_{3k}(X))d\Lambda_k(t)$$

where the $\theta_0, \theta_{1k}, \theta_{2k}, \theta_{3k}$ are linear maps from $\mathcal{B}(\mathfrak{h})$ to itself expressible in term of the system operators $H, P, L_k, M_k, S_k, k = 1, 2, ..., d$. Now, assume that the initial state of the system $\otimes$ bath is

$$\rho(0) = \rho_s(0) \otimes |\phi(u) >< \phi(u)|, |\phi(u) >= exp(- \parallel u \parallel^2 /2)|e(u) >$$

Then after time t, the state of the system will be

$$\rho_s(t) = Tr_2(U(t)\rho(0)U(t)^*)$$

Thus, if X is a system observable, we have that its average at time t is given by

$$Tr(\rho_s(t)X) = Tr(U(t)\rho(0)U(t)^*(X \otimes I))$$

$$= Tr(\rho(0)U(t)^*(X \otimes I)U(t)) = Tr(\rho(0)j_t(X))$$

and its differential is given by

$$dt.Tr(\rho'_s(t)X) = Tr(\rho(0)dj_t(X)) =$$

$$Tr(\rho(0)j_t(\theta_0(X)))dt + \sum_k Tr(\rho(0)j_t(\theta_{1k}(X))dA_k(t)) +$$

$$Tr(\rho(0)j_t(\theta_{2k}(X))dA_k(t)^*) + Tr(j_t(\theta_{3k}(X))d\Lambda_k(t))$$

Now, it easily follows that

$$Tr(\rho(0)j_t(\theta_0(X))) = Tr(\rho_s(t)\theta_0(X)),$$

and from the quantum independent increment property of the fundamental noise processes,

$$Tr(\rho(0)j_t(\theta_{1k}(X))dA_k(t)) = Tr((\rho_s(0) \otimes dA_k(t)|\phi(u) >< \phi(u)|)j_t(\theta_{1k}(X)))$$

$$= u_k(t)dt.Tr((\rho_s(0) \otimes |\phi(u) >< \phi(u)|)j_t(\theta_{1k}(X)))$$

$$= u_k(t)dt.Tr(\rho(0)j_t(\theta_{1k}(X))) = u_k(t)dt.Tr(\rho_s(t)\theta_{1k}(X)),$$

$$Tr(\rho(0)j_t(\theta_{2k}(X))dA_k(t)^*) =$$

$$Tr((\rho_s(0) \otimes (|\phi(u) >< \phi(u)|dA_k(t)^*).j_t(\theta_{2k}(X)))$$

$$= \bar{u}_k(t)dt.Tr((\rho_s(0) \otimes |\phi(u) >< \phi(u)|)j_t(\theta_{2k}(X)))$$

$$= \bar{u}_k(t).dt.Tr(\rho(0)j_t(\theta_{2k}(X))) = \bar{u}_k(t)dt.Tr(\rho_s(t)\theta_{2k}(X)),$$

$$Tr(\rho(0)j_t(\theta_{3k}(X))d\Lambda_k(t)) =$$

$$Tr(\rho_s(0) \otimes d\Lambda_k(t)|\phi(u) < \phi(u)|.j_t(\theta_{3k}(X)))$$

$$= dt^{-1}Tr(\rho_s(0) \otimes dA_k(t)^*dA_k(t)|\phi(u) >< \phi(u)|.j_t(\theta_{3k}(X)))$$

$$= u_k(t).Tr(\rho_s(0) \otimes |\phi(u) >< \phi(u)|dA_k(t)^*.j_t(\theta_{3k}(X)))$$

$$= |u_k(t)|^2 dtTr(\rho(0)j_t(\theta_{3k}(X)) = |u_k(t)|^2 dtTr(\rho_s(0)\theta_{3k}(X))$$

If now θ maps $\mathcal{B}(\mathfrak{h})$ into itself, then we have for each state ρ_s in $\mathfrak{h}$, a unique operator $\theta^*(\rho_s)$ in $\mathfrak{h}$ with the property

$$Tr(\theta^*(\rho_s)X) = Tr(\rho_s\theta(X))$$

for all $X \in \mathcal{B}(\mathfrak{h})$. The operator θ^* that maps states in $\mathfrak{h}$ into $\mathcal{B}(\mathfrak{h})$ is called the dual map of θ. For example, if

$$\theta(X) = \sum_k L_k X M_k$$

then,

$$\theta^*(\rho_s) = \sum_k M_k \rho_s L_k$$

Combining all these relations and using the arbitrariness of the system operator X results finally in the master equation/GKSL equation in its most generalized form:

$$\rho_s'(t) = \theta_0^*(\rho_s(t)) + \sum_k [u_k(t)\theta_{1k}^*(\rho_s(t)) + \bar{u}_k(t)\theta_{2k}^*(\rho_s(t)) + |u_k(t)|^2\theta_{3k}(\rho_s(t))]$$

It is easily shown that all the commonly used master equations in the theory of open quantum systems are special cases of this. That such a general master equation arises from a dilation to a unitary evolution using fundamental quantum noise processes is perhaps one of the pinnacle achievements of the Hudson-Parthasarathy theory.

[g1] The need for non-demolition measurements from the standpoint of Heisenberg's uncertainty principle for constructing quantum conditional expectations.

9.3.1 Non-demolition measurements on the bath space passed through the system

For obtaining a filtering theory for quantum noisy systems, in which the state at time t is $j_t(X)$, we need to calculate its conditional expectation given measurements upto time t. This means that if $Y(s), s \leq t$ is the set of measurements upto time t, then we must be in a position to define the conditional expectation

$$\pi_t(X) = \mathbb{E}(j_t(X)|Y(s), s \leq t)$$

Now in the quantum theory just as the system state at time t $j_t(X)$ is an operator in the system$\otimes$ bath space, the measurements $Y(.)$ must also be operators in this space. To define the above conditional expectation when the system $\otimes$ bath is in a given state and observables evolve with time, (ie, the Heisenberg picture of quantum mechanics), we must give meaning to the joint probability distribution of $(j_t(X), Y(s), s \leq t)$ and this is possible in view of the Heisenberg uncertainty principle iff all the observables $j_t(X), Y(s), s \leq t$ commute with each other for each $t \geq 0$. In other words, the measurement algebra $\eta_t = \sigma(Y(s) : s \leq t)$ must be Abelian for each t and further $[Y(s), j_t(X)] = 0, t \geq s$, ie, the measurements must commute with the future state so that on making a measurement, the future values of the state do not get disturbed. Whenever such measurements exist, we say that they follow the non-demolition property. An example of such measurements was first constructed by V.P.Belavkin [] in a series of path-breaking papers and refined in a mathematically rigorous way by John Gough and Koestler []. The idea of constructing non-demolition measurements proposed by Belavkin was to first define an input measurement process $Y_i(t)$ as a linear combination of the fundamental noise processes $A_i(t), A_i(t)^*, \Lambda_i(t), i = 1, 2, ..., d$, ie,

$$Y_i(t) = \sum_{i=1}^{d} (c[i] A_i(t) + \bar{c}[i] A_i(t)^* + d[i] \Lambda_i(t))$$

and then pass this input process through the HP noisy system to get the output measurements as

$$Y_o(t) = U(t)^* Y_i(t) U(t) = U(t)^* (I \otimes Y_i(t)) U(t)$$

Belavkin observed that for such an output process, one has

$$Y_o(t) = U(T)^* Y_i(t) U(T), T \geq t$$

The trick in proving this is to use the quantum Ito formulas combined with the fact that $U(T)$ is unitary and that the unitarity of $U(T)$ is solely dependent on the system operators appearing in the HP equation which therefore commute with the input noise process values $Y_i(t)$. Specifically, the differential of $U(T)^* Y_i(t) U(T)$ w.r.t T for $T > t$ is given by

$$d_T(U(T)^* Y_i(t) U(T)) = dU(T)^* Y_i(t) U(T) + U(T)^* Y_i(t) dU(T) + dU(T)^* Y_i(t) dU(T) = 0$$

which follows by explicitly substituting for $dU(T)$ and $dU(T)^*$ from the HP equation. This can equivalently be seen by noting that the HP equation can be expressed as

$$dU(T) = \sum_j E_j dM_j(T) U(T)$$

where the processes $M_j(t)'s$ consist of $t, A_k(t), A_k(t)^*$ and $\Lambda_k(t)$ while the $E_j's$ are system operators. It is clear from the quantum independent increment property of the fundamental noise processes that the $dM_j(T)$'s commute with all system operators as well as with $Y_i(t)$ since $T > t$. Thus, the above equation can also be expressed in the form

$$d_T(U(T)^* Y_i(t) U(T)) = U(T)^* (\sum_j (E_j^* dM_j(T)^*$$

$$+ E_j dM_j(T)) + \sum_{j,k} E_j^* E_k dM_j(T)^* dM_k(T)) Y_i(t) U(T)$$

and by the unitarity of $U(T)$ and the quantum Ito formula applied to $d(U(T)^* U(T)) = 0$, we have already the result

$$\sum_j (E_j^* dM_j(T)^* + E_j dM_j(T)) + \sum_{j,k} E_j^* E_k dM_j(T)^* dM_k(T) = 0$$

thus establishing $d_T(U(T)^*Y_i(t)U(T)) = 0, T > t$ and hence

$$U(T)^*Y_i(t)U(T) = U(t)^*Y_i(t)U(t), T > t$$

It is clear that since $[dY_i(t), dY_i(s)] = 0, t \neq s$ (the independent increment property of the fundamental processes) we have that $[Y_i(t), Y_i(s)] = 0, t \neq s$ and hence $Y_i(t), t \geq 0$ forms an Abelian family of operators. Hence from the unitarity of $U(T)$,

$$U(T)^*Y_i(t)U(T), t \leq T$$

also forms an Abelian family for each $T > 0$. Further since the operators $Y_i(t), t \geq 0$ act in the bath space, they all commute with any system operator X and hence again by the unitarity of $U(T)$, the family of operators $U(T)^*Y_i(t)U(T), t \leq T$ commute with the operator $U(T)^*XU(T)$ for any $T > 0$. Combined with the above observation, we obtain the result that the family of operators $Y_o(t), t \leq T$ commutes with $U(T)^*XU(T) = j_T(X)$ and this completes the proof of the non-demolition property. This remarkable fact was first noted by Belavkin and he used these measurements to obtain a real time filter that generalizes the Kushner-Kallianpur filter.

9.3.2 Derivation of the general Belavkin filter both in observable and in state for quadrature and photon counting measurements using Gough's reference probability approach

The HP equation has the general form

$$dU(t) = (-(iH + P)dt + \sum_{i=1}^{d} L_i dA_i + M_i dA_i^* + S_i d\Lambda_i)U(t)$$

and using the quantum Ito formulae, the condition for $U(t)$ to be unitary at all times is that

$$P = \sum_i M_i M_i^*/2, S_i^* + S_i + S_i^* S_i = 0, M_i^* + L_i + M_i^* S_i = 0,$$

$$L_i^* + +M_i + S_i^* M_i = 0$$

the last two of which are equivalent. Here, L_i, M_i, S_i, H are all system space observables.

9.3.3 Why the Belavkin filter is the non-commutative generalization of the Kushner-Kallianpur filter ?

The Belavkin filter for quadrature measurements in a coherent state has the form

$$d\pi_t(X) = \pi_t(\mathcal{L}_t X)dt + (\pi_t(M_t X + X M_t^*) - \pi_t(M_t + M_t^*)\pi_t(X))(dY_o(t)$$

$$-\pi_t(M_t + M_t^*)) - - - (1)$$

where $M_t, t \geq 0$ is a family of system operators. Taking $\mathfrak{h} = L^2(\mathbb{R})$, X as the operator of multiplication by a function $\phi(x)$ in $L^2(\mathbb{R})$ and $j_t(\phi) = \phi(\xi(t))$ where $\xi(t)$ is a classical diffusion process

$$d\xi(t) = \mu(\xi(t))dt + \sigma(\xi(t))dB(t)$$

we find that by the classical Ito formula,

$$dj_t(\phi) = \phi'(\xi(t))\sigma(\xi(t))dB(t) + L\phi(\xi(t))dt$$

$$= j_t(L\phi)dt + j_t(\theta(\phi))dB(t)$$

where

$$L = \mu(x)d/dx + (\sigma^2(x)/2)d^2/dx^2$$

is the generator of the diffusion and

$$\theta = \sigma(x)d/dx$$

We now briefly take a look at the Kushner-Kallianpur filter in classical non-linear filtering theory. The state process $x(t)$ satisfies the above sde and the measurement process has the form

$$dy(t) = h(x(t))dt + \sigma_v dv(t)$$

where the measurement noise process $v(.)$ is a Brownian motion independent of the state process noise $B(.)$ which is another Brownian motion. The measurement σ-field upto time t is given by

$$\eta_t = \sigma(y(s) : s \leq t)$$

but now everything in the state and measurement system commute with each other since all processes are defined on a fixed classical probability space $(\Omega, \mathcal{F}, P)$. The aim is to obtain a stochastic pde for $p_t(x|\eta_t)$ defined as the probability density of the state $x(t)$ at time t given measurement upto time t. We first do not make any restrictions about the state process $x(t)$ except that it is a Markov process with generator K_t (In the above special diffusion process case, $K_t = \mu(x)d/dx + (\sigma^2(x)/2)d^2/dx^2$ but if, for example, $x(t)$ is driven by a Poisson field $N(t, d\xi)$ having a rate function $\lambda dF(\xi)dt$, with the sde given by

$$dx(t) = \mu(t, x(t))dt + \int_{\xi \in E} g(t, x(t), \xi)N(dt, d\xi)$$

where $(E, \mathcal{E})$ is a measure space on which the Poisson field measure is defined, then $x(t)$ is a Markov process with generator given by

$$K_t\phi(x) = lim_{dt \to 0} dt^{-1}\mathbb{E}(\phi(x(t) + dx(t)) - \phi(x(t))|x(t) = x)$$

$$= \mu(t, x)d\phi(x)/dx + \int_{\xi \in E} (\phi(x + g(t, x, \xi)) - \phi(x))\lambda dF(\xi)$$

The filtering equation is obtained by applying the Bayes rule:

$$p(x(t+dt)|\eta_{t+dt}) = p(x(t+dt),\eta_t,dy(t))/p(\eta_t,dy(t))$$

$$= \int p(dy(t)|x(t+dt),x(t))p(x(t+dt)|x(t))p(x(t)|\eta_t)dx(t)/\int numeratordx(t+dt)$$

$$= \int p(dy(t)|x(t))p(x(t+dt)|x(t))p(x(t)|\eta_t)dx(t)/\int numeratordx(t+dt)$$

where we've made use of the fact that the difference between $p(dy(t)|x(t))$ and $p(dy(t)|x(t+dt),x(t))$ is $o(dt)$ and hence can be neglected. Continuing further, we get on multiplying both sides by $\phi(x(t+dt))$ and integrating w.r.t. $x(t+dt)$,

$$\pi_{t+dt}(\phi) = \int \phi(x(t+dt))p(x(t+dt)|\eta_{t+dt})dx(t+dt)$$

$$= \mathbb{E}(\phi(x(t+dt))|\eta_{t+dt}) =$$

$$\int exp(-(dy(t)-h(x(t))dt)^2/2\sigma_v dt^2)(\phi(x(t))+K_t\phi(x(t))dt)p(x(t)|\eta_t)dx(t)/num(\phi=1)$$

$$= \int exp(h(x(t))dy(t)/\sigma_v^2-h(x(t))^2dt/2\sigma_v^2)(\phi(x(t))+K_t\phi(x(t))dt)p(x(t)|\eta_t)dx(t)/num(\phi=1)$$

which on application of Ito's formula for Brownian motion in the form $(dy(t))^2 = \sigma_v^2 dt$, becomes

$$\int (1+h(x(t))dy(t))(\phi(x(t))+K_t\phi(x(t))dt)p(x(t)|\eta_t)dx(t)/num(\phi=1)$$

$$= (\pi_t(\phi)+\pi_t(h\phi)dy(t)+\pi_t(K_t\phi)dt)/(1+\pi_t(h)dy(t))$$

from which another application of Ito's formula after using the expansion

$$1/(1+\pi_t(h)dy) = 1-\pi_t(h)dy+\pi_t(h)^2(dy)^2/2+o(dt)$$

yields the Kushner-Kallianpur filter:

$$d\pi_t(\phi) = \pi_t(K_t\phi)dt+(\pi_t(h\phi)-\pi_t(h)\pi_t(\phi))(dy(t)-\pi_t(h)dt)$$

This is easily seen to be a special commutative case of the Belavkin filter (1) once we identify K_t with L_t and h with $M_t+M_t^*$ assuming now that M_t is a multiplication operator. The conditional expectation operation $\pi_t(X)$ in the case of the Belavkin filter is the same as $Tr(\rho_B(t)X)$ where $\rho_B(t)$ is the Belavkin filtered state which can be regarded as a random density matrix in the System Hilbert space that is measurable w.r.t the Abelian Von-Neumann algebra $\eta_t = \sigma(Y_o(s):s\leq t)$. In our classical probabilistic scenario, $\rho_B(t)$ is identified with a random diagonal matrix, ie, a multiplication operator $p_t(x|\eta_t), x \in \mathbb{R}$. The system observable $j_t(X)$ in the Belavkin filter is identified with the function $\phi(x(t))$. More precisely, if $x(0) = x$, we write $x(t,x)$ for $x(t)$. Then X stands for the multiplication operator by the function $\phi(x)$ while $j_t(X)$ stands for the multiplication operator by the function $\phi(x(t,x))$.

9.3.4 Quantum control for Lindblad noise reduction based on Luc-Bouten's approach

There is another interesting viewpoint that one can take in quantum filtering theory. It is related to quantum communication. Here, we wish to transmit the system state $\rho_s(0)$ over a noisy channel. The dynamics of the channel is dictated by a noisy Schrodinger equation. So the system state at time t after transmitting it through the noisy channel is

$$\rho_s(t) = Tr_2(U(t)(\rho_s(0) \otimes |\phi(u) > \phi(u)|)U(t)^*)$$

where $U(t)$ satisfies the noisy HP-Schrodinger equation. To recover $\rho_s(0)$ from $\rho_s(t)$, we can use the recovery operators of the Knill-Laflamme theorem if appropriate conditions are satisfied. This is because, one can express

$$\rho_s(t) = \sum_{k=1}^{p} E_k(t)\rho_s(0)E_k(t)^*$$

where the system operators $E_k(t)$ are obtained by solving the GKSL equation. However, when the required conditions on the system and noise subspaces for the Knill-Laflamme theorem are not satisfied, then we must adopt a different approach. This involves taking non-demolition measurements and applying the Belavkin filter to obtain an estimate $\rho_B(t)$ of the dynamically evolving system state. However, this estimate will contain the GKSL noise operators, which along with the Hamiltonian appearing in the HP-Schrodinger equation, dictate the true noisy state evolution. In order to obtain the initial system state, we must therefore apply some sort of control operation that would remove the GKSL noise. Such an algorithm was proposed by Belavkin and developed to perfection by Luc-Bouten in his Ph.D thesis []. We briefly summarize this idea here: At time $t = 0$, the system state is assume to be $\rho_c(0)$ which is assumed to be the Belavkin filtered state followed by application of control operations. We note that the Belavkin filter

$$d\pi_t(X) = \pi_t(\mathcal{L}_t X)dt + (\pi_t(M_t X + X M_t^*)$$
$$-\pi_t(M_t + M_t^*)\pi_t(X))(dY_o(t) - \pi_t(M_t + M_t^*)) ---(1)$$

can be expressed as a stochastic Schrodinger equation for $\rho_B(t)$ by replacing $\pi_t(Z)$ with $Tr(\rho_B(t)Z)$ for any system operator Z and using the arbitrariness of the system observable X as:

$$d\rho_B(t) = L_t^*(\rho_B(t))dt +$$

$$(\rho_B(t)M_t + M_t^*\rho_B(t) - Tr(\rho_B(t)(M_t + M_t^*))\rho_B(t))(dY_o(t) - Tr(\rho_B(t)(M_t + M_t^*))$$

It should be noted that this is a stochastic Schrodinger equation and not a quantum stochastic equation since the driving noise process $Y_o(.)$ is commutative. Further, the innovations process $W(t) = Y_o(t) - \int_0^t Tr(\rho_B(s)(M_s + M_s^*))ds$ is a scalar multiple of the Wiener process when the measurement is quadrature, ie,

$$Y_o(t) = U(t)^*(\sum_{k=1}^{d} \alpha_k A_k(t) + \bar{\alpha}_k A_k(t)^*)U(t)$$

$\rho_c(0)$ evolves under the Belavkin dynamics to $\rho_B(dt)$ at time dt given by

$$\rho_B(dt) = \rho_c(0) + (L_t^*(\rho_c(0))dt + (\rho_c(0)M_0 + M_0^*\rho_c(0) - Tr(\rho_c(0)(M_0 + M_0^*))\rho_c(0)dW(t))$$

and then we apply an infinitesimal control unitary

$$U_c(dt) = exp(iZdY_o(t))$$

to the Belavkin filtered state where Z is an appropriately chosen system space Hermitian operator. This gives us the filtered and controlled state at time dt as:

$$\rho_c(dt) = U_c(dt)\rho_B(dt)U_c(dt)^*$$

and it is easy to show Bouten) as we shall in chapter [] using the quantum Ito formula, that Z can be chosen so that $\rho_c(dt)$ is $\rho_c(0)$ plus terms which have one lesser Lindblad noise component. This was already proved by Bouten but in our thesis, we do something more, ie we also include quantum Poisson noise in our HP dynamics, and define a new control objective, namely to design Z so that the controlled state at time dt, namely $\rho_c(dt)$ is as close as possible in norm to a given state ρ_d, in other words, we achieve state tracking.

9.3.5 The Von-Neumann entropy of a state and its significance and properties

If ρ is a state in a separable Hilbert space with spectral resolution

$$\rho = \sum_{k=1}^{\infty} |e(k) > p(k) < e(k)|$$

then

$$S(\rho) = -Tr(\rho.log(\rho)) = -\sum_k p(k)log(p(k))$$

ie, $S(\rho)$ can be regarded as the classical entropy of ρ relative to its eigenbasis. More generally, if $M = \{M_\alpha\}$ is a measurement system, ie, POVM, then the classical entropy of ρ relative to M is given by

$$S_M(\rho) = -\sum_{\alpha} Tr(\rho M_\alpha)log(\rho M_\alpha)$$

and the maximum of $S_M(\rho)$ over all measurement systems M is $S(\rho)$ [Mark Wilde, Hayashi, Quantum information].

9.3.6 Maassen's Guichardet kernel approach to solving quantum stochastic differential equations

Let $(X, \mathcal{F}, \mu)$ be a measurable space and assume the measure μ to be non-atomic, ie, $\mu(\{x\}) = 0, x \in X$. Denote by Γ the set of all subsets of X. Denote by Γ_n the set of all subsets of X having n elements. let $Gamma_0 = \phi$, the empty set. Thus,

$$\Gamma = \bigcup_{n \geq 0} \Gamma_n$$

Define a measure μ_Γ on Γ by the prescription that if $f : \Gamma \to \mathbb{C}$ is measurable, then

$$\int_\Gamma f d\mu_\Gamma = 1 + \sum_{n \geq 1} (1/n!) \int_{\Gamma_n} f|_{\Gamma_n}(\sigma) d\mu^n(\sigma)$$

where μ^n is the product measure on Γ_n, ie, for $\sigma = (x_1, ..., x_n)$,

$$d\mu^n(\sigma) = \Pi_{k=1}^n d\mu(x_k)$$

For $f : X \to \mathbb{C}$, or more precisely, $f \in L^2(X, \mathcal{F}, \mu)$, define $e(f) \in L^2(\Gamma, \mu_\Gamma)$ by

$$e(f)(\sigma) = \Pi_{x \in \sigma} f(x), \sigma \in \Gamma_n, n \geq 1$$

and $e(f)(\phi) = 1$. We easily verify that

$$< e(f), e(g) >= \int \bar{e}(f)e(g)d\mu_\Gamma = exp(< f, g >) = exp(\int_X \bar{f}(x)g(x)d\mu(x))$$

For $f \in L^2(X, \mathcal{F}, \mu)$ define the operator

$$a(f) : L^2(\Gamma, \mu_\Gamma) \to L^2(\Gamma, \mu_\Gamma)$$

by

$$a(f)\psi(\sigma) = \int_X \bar{f}(x)\psi(\sigma \cup x)d\mu(x), \sigma \in \Gamma_n$$

Then, for $\sigma \in \Gamma_n$,

$$a(f)e(u)(\sigma) = \int_X \bar{f}(x)e(u)(\sigma \cup x)d\mu(x)$$

$$= \int_X \bar{f}(x)(\Pi_{y \in \sigma \cup \sigma \cup x} u(y))d\mu(x)$$

$$= (\int_X \bar{f}(x)u(x)d\mu(x))e(u)(\sigma) =< f, u > e(u)(\sigma)$$

Equivalently,

$$a(f)e(u) =< f, u > e(u)$$

We calculate the adjoint $a(f)^*$ of $a(f)$: $a(f)^*$ must satisfy

$$< a(f)^*e(v), e(u) >=< e(v), a(f)e(u) >=< f, u >< e(v), e(u) >$$

or equivalently,

$$\sum_n n!^{-1} \int_{\Gamma_n} \bar{(a(f)^*e(v))}(\sigma)e(u)(\sigma)d\mu^n(\sigma)$$

$$=< f, u > \sum_n n!^{-1} \int_{\Gamma_n} \bar{e}(v)(\sigma)e(u)(\sigma)d\mu^n(\sigma)$$

Suppose we try for $\psi \in L^2(\Gamma, \mu_\Gamma)$,

$$(a(f)^*\psi)(\sigma) = \sum_{x \in \sigma} f(x)\psi(\sigma - \{x\}), \sigma \in \Gamma_n$$

Then, we get

$$\int_{\Gamma_n} \overline{(a(f)^*e(v))(\sigma)}e(u)(\sigma)d\mu^n(\sigma) =$$

$$\int_{\Gamma_n} \sum_{x \in \sigma} \bar{f}(x)\Pi_{y \in \sigma, y \neq x}\bar{v}(y)\Pi_{z \in \sigma}u(z)d\mu^n(\sigma)$$

$$= n\int_{\Gamma_{n-1}} (\int_X \bar{f}(x)u(x)dx)(\Pi_{y \in \sigma}\bar{v}(y)u(y))d\mu^{n-1}(\sigma)$$

This gives the correct result since $n/n! = 1/(n-1)!$. Now we can explain Maasen's method for solving Hudson-Parthasarathy like qsde's. Consider first the annihilation process

$$A_t(f) = a(f\chi_{[0,t]})$$

where $f \in L^2(X, \mu)$ with $X = \mathbb{R}_+, \mu = $ Lebesgue measure on $\mathbb{R}_+$. Then, we have

$$A_t(f)\psi(\sigma) = \int_{\mathbb{R}_+} \bar{f}(s)\chi_{[0,t]}(s)\psi(\sigma \cup \{s\})ds$$

$$= \int_0^t \bar{f}(s)\psi(\sigma \cup \{s\})ds$$

Likewise,

$$A_t(f)^*\psi(\sigma) = \sum_{s \in \sigma cap[0,t]} f(s)\psi(\sigma - \{s\})$$

It follows that

$$dA_t(f)\psi(\sigma) = \bar{f}(t)\psi(\sigma \cup \{t\})dt$$

and

$$dA_t(f)^*\psi(\sigma) = \chi_\sigma(t)f(t)\psi(\sigma - \{t\})$$

We then find that

$$dA_t(f)dA_t(f)^*\psi(\sigma) = f(t)\bar{f}(t)dt\chi_{\sigma \cup \{t\}}(t)\psi(\sigma)$$

$$= |f(t)|^2\psi(\sigma)dt$$

Thus, we deduce the quantum Ito formula

$$dA_t(f)dA_t(f)^* = |f(t)|^2dt$$

from Maasen's kernel theory. Now writing

$$\psi_t = U(t)\psi_0$$

ie

$$\psi(\sigma) = (U(t)\psi_0)(\sigma), \sigma \in \Gamma_n, n \geq 1$$

where $U(t)$ satisfies the Hudson-Parthasarathy equation

$$dU(t) = (-(iH + P)dt + L_1 dA_t(f) + L_2 dA_t(f)^*)U(t)$$

or equivalently,

$$d\psi_t(\sigma) = (-(iH + P)dt + L_1 dA_t(f) + L_2 dA_t(f)^*)\psi_t)(\sigma)$$

$$= -(iH + P)\psi_t(\sigma)dt + L_1 \bar{f}(t)\psi(\sigma \cup t)dt + L_2 \chi_\sigma(t)f(t)\psi(\sigma - \{t\})$$

It is not difficult to obtain an explicit solution for ψ in this form. Note that we assume that for any $\sigma \in \Gamma_n$, $\psi_t(\sigma) \in \mathfrak{h}$ where $\mathfrak{h}$ is the system Hilbert space in which the operators H, P, L_1, L_2 act.

9.3.7 Unsolved problems in quantum filtering and control

[a] Develop quantum filtering theory when the measurements are obtained as arbitrary independent increment processes.

[b] Develop quantum filtering theory in non-coherent states.

[c] Develop quantum control theory when the objective cost function to be minimized is the expected value of the integral of any function of the evolving quantum observable in a coherent state.

[d] Develop quantum filtering theory for arbitrary Evans-Hudson flows not necessarily obtained via the unitary dynamics of the Hudson-Parthasarathy-Schrodinger qsde.

9.3.8 Simulating the HP and Belavkin filter using finite matrix algorithms

The HP equation

$$dU(t) = [-(iH + P)dt + \sum_{m=1}^{p}(L_m dA_m(t) + M_m dA_m(t)^* + S_m d\Lambda_m(t))]U(t)$$

is simulated as an ordinary matrix differential equation and hence after discretizing time, as an ordinary matrix difference equation (no random variables/random processes are involved). This equation is given by (after approximating it by a truncation of orthonormal bases)

$$d < \eta_k \otimes \xi_r | U(t) | \eta_l \otimes \xi_s > \approx$$

$$-\sum_{l',s'} < \eta_k | iH + P | \eta_{l'} > \delta_{r,s'} < \eta_{l'} \otimes \xi_{s'} | U(t) | \eta_l \otimes \xi_s > dt$$

$$+ \sum_{m,l',s',l'',s''} <\eta_k|L_m|\eta_{l'}> \delta_{rs'} <\eta_{l'}\otimes\xi_{s'}|U(t)|\eta_{l''}\otimes\xi_{s''}><\eta_{l''}\otimes\xi_{s''}|dA_m(t)|\eta_l\otimes\xi_s>$$

$$+ \sum_{m,l',s',l'',s''} <\eta_k|L_m|\eta_{l'}> \delta_{rs'} <\eta_{l'}\otimes\xi_{s'}|U(t)|\eta_{l''}\otimes\xi_{s''}><\eta_{l''}\otimes\xi_{s''}|dA_m(t)^*|\eta_l\otimes\xi_s>$$

$$+ \sum_{m,l',s',l'',s''} <\eta_k|L_m|\eta_{l'}> \delta_{rs'} <\eta_{l'}\otimes\xi_{s'}|U(t)|\eta_{l''}\otimes\xi_{s''}><\eta_{l''}\otimes\xi_{s''}|d\Lambda_m(t)|\eta_l\otimes\xi_s>$$

where we keep in mind the fact that the $\eta'_k s$ form an onb for the system Hilbert space and the $\xi'_r s$ for an approximate onb for the bath space, ie the Boson Fock space. It should be noted that this approximation is in fact very crude since the Boson Fock space is not a separable Hilbert space, however it works well during practical simulation just as the Feynman path integral does not converge on formally replacing real variances in Gaussian distributions by purely imaginary variances and yet it yields the correct physical results like amplitudes for Compton scattering, vacuum polarization and the anomalous magnetic moment. In the above equations, we substitute

$$< \eta_k \otimes \xi_r |dA_m(t)|\eta_l \otimes \xi_s >= \delta_{kl} < \xi_r|dA_m(t)|\xi_s >=$$

$$= \delta_{kl} \sum_{r',s'} \bar{c}(r,r')c(s,s') < e(u_{r'}|dA_m(t)|e(u_{s'}) >=$$

$$= \delta_{kl} \sum_{r',s'} \bar{c}(r,r')c(s,s')u_{s'm}(t) < e(u_{r'}|e(u_{s'} > dt$$

$$< \eta_k \otimes \xi_r |dA_m(t)^*|\eta_l \otimes \xi_s >= \delta_{kl} < \xi_r|dA_m(t)^*|\xi_s >=$$

$$= \delta_{kl} < dA_m(t)\xi_r|\xi_s >=$$

$$= \delta_{kl} \sum_{r',s'} \bar{c}(r,r')c(s,s') < dA_m(t)e(u_{r'}|e(u_{s'}) >=$$

$$= \delta_{kl} \sum_{r',s'} \bar{c}(r,r')c(s,s')\bar{u}_{r'm}(t) < e(u_{r'}|e(u_{s'} > dt,$$

and finally,

$$< \eta_k \otimes \xi_r |d\Lambda_m(t)|\eta_l \otimes \xi_s >= \delta_{kl} < \xi_r|d\Lambda_m(t)|\xi_s >=$$

$$= \delta_{kl} \sum_{r',s'} \bar{c}(r,r')c(s,s') < e(u_{r'}|d\Lambda_m(t)|e(u_{s'}) >=$$

$$= \delta_{kl} \sum_{r',s'} \bar{c}(r,r')c(s,s')\bar{u}_{r'm}(t)u_{s'm}(t) < e(u_{r'}|e(u_{s'} > dt$$

The details of how this simulation of the HP equation is actually carried out by choosing the functions $u_s(t)$ as normalized orthogonal sinusoids over a finite time interval $[0,T]$ are discussed in Chapter 2 of the thesis. It should be noted that as regards the quantum Poisson/conservation processes $\Lambda_m(t)$, we can more generally consider processes $\Lambda_m^k(t)$ defined as

$$\Lambda_m^k(t) = \lambda(\chi_{[0,t]}|f_k><f_m|), 1 \le k, m \le d$$

defined by its matrix elements

$$< e(u)|\Lambda_m^k(t)|e(v) >=< \chi_{[0,t]}u|f_k><f_m|\chi_{[0,t]}v <=$$

$$\int_0^t \bar{u}_k(s)v_m(s)ds$$

We can equivalently represent its differential as

$$d\Lambda_m^k(t) = dA_k(t)^*dA_m(t)/dt$$

as is readily verified by forming the matrix elements

$$< e(u)|dA_k(t)^*dA_m(t)/dt|e(v) >= \bar{u}_k(t)v_m(t)dt$$

We then have the generalized Ito rules

$$dA_k(t)d\Lambda_s^r(t) = dA_k(t)dA_r(t)^*dA_s(t)/dt = \delta_{kr}dA_s(t),$$

$$d\Lambda_s^r(t)dA_k(t)^* = dA_r(t)^*dA_s(t)dA_k(t)^*/dt = \delta_{sk}dA_r(t)^*$$

Then the term $\sum_{k=1}^d S_k d\Lambda_k(t)$ can be replaced by the more general term $\sum_{k,m=1}^d S_m^k d\Lambda_k^m(t)$. However, for simulation purposes, it is easier to consider the above special case of this since the general case is a straightforward extension of this special case.

The Belavkin filter may be simulated both the observable and the state domain either as ordinary non-random matrix differential equations or equivalently as classical stochastic differential equations. Specifically, the Belavkin filter for quadrature measurements

$$d\pi_t(X) = \pi_t(L_tX)dt+(\pi_t(M_tX+XM_t^*)-\pi_t(M_t+M_t^*)\pi_t(X))(dY_o(t)-\pi_t(M_t+M_t^*)dt)$$

where X ranges over all system observables is equivalent to the following in view of the linearity of the equation w.r.t. X: First choose a basis $\{X_1, ..., X_K\}$ of Hermitian matrices for the vector space of all Hermitian matrices in the system Hilbert space. Then, we can write

$$L_tX_k = \sum_{m=1}^K a_{km}(t)X_m, M_tX_k + X_kM_t^* = \sum_{m=1}^K b_{km}(t)X_m,$$

$$M_t + M_t^* = \sum_{k=1}^K e_k(t)X_k$$

where $a_{km}(t), b_{km}(t), e_k(t)$ are real valued functions. Now define the commutative processes

$$\xi_k(t) = \pi_t(X_k), 1 \le k \le K$$

Then recalling the fact that for quadrature measurements, the process

$$W(t) = Y_o(t) - \int_0^t \pi_s(M_s + M_s^*)ds$$

is a multiple of Brownian motion, the Belavkin filter reduces to the following system of K coupled stochastic differential equations driven by the classical Brownian motion process $W(t)$:

$$d\xi_k(t) = \sum_{m=1}^{K} a_{km}(t)\xi_m(t)dt$$

$$+(\sum_{m=1}^{K} b_{km}(t)\xi_m(t) - \sum_{m=1}^{K} e_m(t)\xi_m(t)\xi_k(t))dW(t), k = 1, 2, ..., K$$

We may simulate this evolution as a system of classical stochastic differential equations and obtain the correct statistics for the estimate of any observable when the the bath is the coherent state $|\phi(u)>$, however such a simulation would not tell us how the observable estimate depended on the output measurements $Y_o(.)$. To obtain this information, we must regard $\xi_k(t)$ as a $pN \times pN$ matrix and also $Y_o(t)$ as the $pN \times pN$ matrix whose $(N(r-1)+k, N(s-1)+l)^{th}$ entry is given by

$$< \eta_k \otimes \xi_r | Y_o(t) | \eta_k \otimes \xi_s >$$

The Belavkin filter can also be simulated in the state domain in the form of a classical sde by regarding the filtered state $\rho_B(t)$ as a random density matrix in the system Hilbert space that is measurable w.r.t the commutative family of output measurements $Y_o(s), s \leq t$ which in turn, owing to their commutativity, can be regarded as a classical random process when the noisy bath is in a coherent state. This involves writing

$$\pi_t(Z) = Tr(\rho_B(t)Z)$$

where $\rho_B(t)$ is regarded as a classical random process with values in the space of density matrices in the system Hilbert space and Z is any system operator. From the above observable form of Belavkin's equation, by applying duality in the Banach space of system Hilbert-Schmidt operators, we deduce the state form of Belavkin's equation as a stochastic non-linear Schrodinger equation:

$$d\rho_B(t) = L_t^*(\rho_B(t))dt+$$

$$(\rho_B(t)M_t + M_t^*\rho_B(t) - Tr(\rho_B(t)(M_t + M_t^*)))(dY_o(t) - Tr(\rho_B(t)(M_t + M_t^*)))$$

The simulation of this equation is by direct time discretization. However, this simulation will not, in general, lead to a positive definite matrix of unit trace for $\rho_B(t), t \geq 0$. So, after each iteration, we extract out the part of $\rho_B(t)$

that satisfies these properties required of a density matrix by performing the transformation

$$\rho_B(t) \rightarrow \frac{\sqrt{\rho_B(t)^*\rho_B(t)}}{Tr(\sqrt{\rho_B(t)^*\rho_B(t)})}$$

and prove numerical stability using the Dunford-Taylo integral for the square root of an operator (T.Kato[]):

$$\sqrt{T} = (2\pi i)^{-1} \int_\Gamma \sqrt{z}(zI - T)^{-1}dz$$

where the contour Γ is chosen appropriately within the resolvent set of T.

9.3.9 Generalizing the Belavkin filter when the measurements are mixture of creation, annihilation and conservation processes, ie superpositions of quantum Brownian motions and quantum Poisson processes or equivalently quadrature plus photon counting

The Belavkin filter has been constructed in the existing literature [J.Gough and Kostler) for the special cases when the measurement noise is either quadrature (ie, classical Brownian motion $A(t) + A(t)^*$ passed through the HP system and also in the case when the measurement noise is a photon counting process, ie, the quantum Poisson process $\Lambda(t)$ passed through the HP system. In both the cases, the construction is simple owing to the fact that the output measurement process differential can be expressed in terms of either the creation and annihilation processes differentials whose cubic and higher powers vanish by quantum Ito's formula, or in terms of quantum Poisson process differentials all of whose powers are proportional to the process differential: $(d\Lambda)^n = d\Lambda, n = 1, 2, ...$ When however, the measurement is a mixture of quadrature and photon counting processes, ie,

$$Y_o(t) = U(t)^*(\sum_{k=1}^{d} c[k]A_k(t) + \bar{c}[k]A_k(t)^* + d[k]\Lambda_k(t))U(t)$$

then application of quantum Ito's formula shows that $dY_o(t)$ can be expressed in the form

$$dY_o(t) = j_t(N_0)dt + \sum_{k=1}^{d}(j_t(N_{1k})dA_k(t) + j_t(N_{2k})dA_k(t)^* + j_t(N_{3k})d\Lambda_k(t))$$

and it is clear that all the powers of $dY_o(t)$ will not be expressible in an elementary way in terms of the fundamental processes. The trick is to assume

$$(dY_o(t))^n = j_t(N_0[n])dt + \sum_{k=1}^{d}(j_t(N_{1k}[n])dA_k(t) + j_t(N_{2k}[n])dA_k(t)^*$$
$$+ j_t(N_{3k}[n])d\Lambda_k(t)), n \geq 1 ---(2)$$

and calculate the system operators $N_0[n], N_{mk}[n], m = 1, 2, 3$ for different $n's$ by applying quantum Ito's formula to obtain a recursion for these system matrices. Having done so, we assume that the Belavkin filter has the form

$$d\pi_t(X) = F_t(X)dt + \sum_{k \geq 1} G_{kt}(X)(dY_o(t)^k$$

where $F_t(X), G_{kt}(X)$ are functions of the Abelian family $\eta_t = \sigma(Y_o(s), s \leq t)$ (because $\pi_t(X) = \mathbb{E}(j_t(X)|\eta_t)$ and then calculate $F_t(X), G_{kt}(X), k \geq 1$ by applying the reference probability method of Gough and Kostler. This method involves choosing arbitrary real valued functions $f_k(t), k \geq 1$, considering the process $C(t)$ that is η_t measurable because it is assumed to satisfy the sde

$$dC(t) = \sum_{k \geq 1} C(t)f_k(t)(dY_o(t))^k, t \geq 0, C(0) = I$$

and then applying quantum Ito's formula to the orthogonality relation (satisfied by the conditional expectation):

$$\mathbb{E}[(j_t(X) - \pi_t(X))C(t)] = 0$$

and then using the arbitrariness of the functions $f_k(t), k \geq 1$ to deduce that

$$\mathbb{E}[dj_t(X) - d\pi_t(X)|\eta_t] = 0,$$

$$\mathbb{E}[(j_t(X) - \pi_t(X))(dY_o(t))^k|\eta_t] + \mathbb{E}[(dj_t(X) - d\pi_t(X))(dY_o(t))^k|\eta_t] = 0$$

Further simplifications of these to calculate $F_t(X), G_{kt}(X)$ are based on the homomorphism property of the map j_t, the definition of π_t as a conditional expectation, expressions for $(dY_o(t))^k$ in terms of the fundamental noise processes in the form (2) and further applications of the quantum Ito formula. These aspects as well as the simulation of the Belavkin filter upto second degree terms in the noise differential are discussed in chapter 3.

9.3.10   Quantum control with more generalized objectives: Removing some components of Lindblad noise, state trajectory tracking

We've already discussed this aspect above.

9.3.11   Use of the differential of the exponential map and the Baker-Campbell-Hausdorff formula in Lie group theory for evaluating the rate of Von-Neumann entropy increase in the HP, Belavkin and controlled Belavkin state

The HP system state satisfies the generalized GKSL equation obtained after tracing out over the bath state. It is given by

$$\rho'_s(t) = -i[H, \rho_s(t)] + \theta_t(\rho_s(t))$$

where θ_t is a time dependent generalized Linear Lindblad operator depending on the vector $u(t)$ which defines the coherent state vector $|\phi(u) >= exp(-\parallel u \parallel^2 /2)|e(u) >$ of the bath and also on the system space operators $L_k, M_k, S_k, k = 1, 2, ..., d$ appearing in the HP equation. We can use this equation combined with the standard formula for the differential of the exponential map [V.S.Varadarajan, Lie groups, Lie algebras and their representations] to calculate the rate of change of the Von-Neumann entropy of the system. That would tell us at what rate the environmental bath is pumping in entropy into the system and if we are able to select our system operators so that this rate is positive then we can say that our system dynamics viewed as our complete system is in agreement with the second law of thermodynamics. We can also by regarding the Belavkin equation as a classical stochastic nonlinear Schrodinger equation, compute the average rate of entropy of increase of the Belavkin filtered state when the bath is in a coherent state. This computation would tell us whether conditioning the HP state on the non-demolition measurements increases or decreases the entropy. Ideally, we know that conditioning decreases the entropy: $H(X|Y) \leq H(X)$. Our simulations confirm this in the quantum case for a majority of experiments but we are yet to discover a conclusive proof that this is always the case. Finally, after applying quantum control using control unitaries as in the PhD thesis of Luc Bouten, we compute the approximate increase in system entropy after control. We expect that the entropy of the state will decrease after control since control involves Lindblad noise removal.

9.3.12 Performance analysis of the filtering and control algorithms

9.3.13 Simulations of the Hudson-Parthasarathy QSDE and Belavkin filter equations

Study projects

[a] Simulating the Hudson-Parthasarathy equation.

[b] MATLAB simulations of observable and state evolution in noise.

[c] Simulating the Belavkin filtering equation for quadrature noise measurements

[d] Plots of the nsr

$$nsr(t) = \mathbb{E}(j_t(X) - \pi_t(X))^2 / \mathbb{E}(j_t(X)^2)$$

where expectations are taken in the coherent state.

9.3.14 Simulating the Belavkin filter for mixture of quadrature and photon counting noise

The qsde has all powers of the measurement noise differential.

[a] Simulation of the nsr.

9.3.15 Study project:Quantum control of the Belavkin filtered state for (a) Lindblad noise removal and (b) state tracking

9.3.16 Study project:Von-Neumann entropy rate for HP,Belavkin and controlled Belavkin filtered states

[a] Theoretical derivations using Lie algebra theory.
  [b] Simulation of the entropy evolution.

9.4 Filter design for physical applications

Ref:Mridul, Ph.D thesis.

An ideal integrator in discrete time has the impulse response $u[n]$. Thus, when a signal $x[n]$ is passed through an integrator, the output is

$$y[n] = u * x[n] = \sum_{k=0}^{n} x[k]$$

The transfer function of an integrator is

$$\sum_{n} u[n]z^{-n} = \frac{1}{1 - z^{-1}}$$

with the ROC of $|z| > 1$. An ideal integrator is an unstable system since its impulse response is not summable. We therefore seek a stable approximation to an ideal integrator. This can be obtained either by truncating the impulse response, ie, by approximating $1/(1 - z^{-1})$ by

$$1 + z^{-1} + ... + z^{-N}$$

for some sufficiently large N, or by approximating $1/(1 - z^{-1})$ by the ratio of two polynomials $B(z)/A(z)$ with the zeros of $A(z)$ constrained to fall inside the unit circle. One way to design such an approximate integrator would be to fix $A(z)$ so that its zeroes are inside the unit circle and to calculate $B(z)$ so that

$$\int_{-\pi}^{\pi} W(\omega)|(1 - exp(-j\omega))B(e^{j\omega}) - A(e^{j\omega})|^2 d\omega$$

is a minimum. Since B is a polynomial in z, minimization of this error energy w.r.t the coefficients of $B(z)$ would amount to solving a linear least squares problem. After designing $B(z)$, we can approximate $B(z)/A(z)$ with a function $1/H(z)$, where $H(z)$ is obtained by truncating the infinite series for $A(z)/B(z)$. This process is explained below.

[1] The aim is to express the ratio $B(z)/A(z)$ of two polynomials in z^{-1}, namely

$$A(z) = \sum_{k=0}^{p} a[k]z^{-k}, B(z) = 1 + \sum_{k=1}^{p} b[k]z^{-k}$$

as $1/H(z)$ where

$$H(z) = \sum_{n=0}^{\infty} h[n]z^{-n}$$

with z varying over an appropriate region in the complex plane. Suppose we impose the condition

$$\left|\sum_{k=1}^{p} b[k]z^{-k}\right| < 1$$

This can be satisfied if for example,

$$\sum_{k=1}^{p} |b[k]||z|^{-k} < 1$$

which can in turn be guaranteed provided that we choose z so that $|z| > 1$ and simultaneously

$$|z|^{-1} \sum_{k=1}^{p} |b[k]| < 1$$

ie

$$|z| > max(1, \sum_{k=1}^{p} |b[k]|)$$

Then, we have the convergent geometric series

$$B(z)^{-1} = (1 + \sum_{k=1}^{p} b[k]z^{-k})^{-1} = 1 + \sum_{n=0}^{\infty} (-1)^n (\sum_{k=1}^{p} b[k]z^{-k})^n$$

By the multinomial theorem in the form

$$\left(\sum_{k=1}^{p} x_k\right)^n = \sum_{n_1,\dots,n_p \geq 0, n_1+\dots+n_p=n} \frac{n!}{n_1!\dots n_p!} x_1^{n_1}\dots x_p^{n_p}$$

we get

$$\left(\sum_{k=1}^{p} b[k]z^{-k}\right)^n = \sum_{n_1+\dots+n_p=n} \frac{n!}{n_1!\dots n_p!} b[1]^{n_1}\dots b[p]^{n_p} z^{-(n_1+2n_2+\dots+pn_p)}$$

it follows that

$$(1 + \sum_{k=1}^{p} b[k]z^{-k})^{-1} = 1 + \sum_{l \geq 0} c[l]z^{-l}$$

where for $l \geq 1$.

$$c[l] = \sum_{n_1,..,n_p \geq 0, n_1 + 2n_2 + ... + pn_p = l} (-1)^{n_1 + ... + n_p} \frac{(n_1 + ... + n_p)!}{n_1! ... n_p!} b[1]^{n_1} ... b[p]^{n_p}$$

Finally, with $c[0] = 1$, we have

$$H(z) = \sum_{n \geq 0} h[n] z^{-n} = (\sum_{k=0}^{p} a[k] z^{-k}).(\sum_{l \geq 0} c[l] z^{-l})$$

so that

$$h[n] = \sum_{k=0}^{min(p,n)} a[k] c[n-k], n \geq 0$$

Example: $p = 2$. Then

$$c[l] = \sum_{n_1 + 2n_2 = l} (-1)^{n_1 + n_2} \frac{(n_1 + n_2)!}{n_1! n_2!} b[1]^{n_1} b[2]^{n_2}$$

$$= \sum_{n=0}^{[l/2]} (-1)^{l-n} \frac{(l-n)!}{(l-2n)! n!} b[1]^{l-2n} b[2]^n$$

In particular,

$$c[0] = 1,$$

$$c[1] = -b[1],$$

$$c[2] = \sum_{n=0,1} (-1)^{2-n} \frac{(2-n)!}{(2-2n)! n!} b[1]^{2-2n} b[2]^n = b[1]^2 - b[2],$$

Design of filters using transmission line elements. The T matrix of the mth section of a cascade of M finite transmission line elements has the general form

$$T_m(z) = A_m + B_m z^{-1}, m = 1, 2, ..., M$$

where A_m, B_m are constant 2×2 matrices:

$$A_m = \begin{pmatrix} A_m(1,1) & A_m(1,2) \\ A_m(2,1) & A_m(2,2) \end{pmatrix},$$

$$B_m = \begin{pmatrix} B_m(1,1) & B_m(1,2) \\ B_m(2,1) & B_m(2,2) \end{pmatrix},$$

The aim is to express the matrix product

$$T(z) = T_M(z) T_{M-1}(z) ... T_2(z) T_1(z) = \Pi_{m=1}^{M} T_m(z)$$

as a matrix polynomial in z^{-1} of degree M, ie, in the form

$$T(z) = \sum_{m=0}^{M} S_m z^{-m}$$

where $S_0, S_1, ..., S_M$ are constant 2×2 matrices. We write

$$K_m(z) = T_m(z)T_{m-1}(z)...T_1(z), 1 \leq m \leq M$$

Then we have the obvious recursion

$$K_{m+1}(z) = T_{m+1}(z)K_m(z)$$

Writing

$$K_m(z) = \sum_{r=0}^{m} K_m[r] z^{-r}$$

where $K_m[r]$ is a constant 2×2 matrix, we have

$$\sum_{r=0}^{m+1} K_{m+1}[r] z^{-r} = (A_{m+1} + B_{m+1} z^{-1}). \sum_{r=0}^{m} K_m[r] z^{-r}$$

which gives on equating coefficients of the same powers of z^{-1},

$$K_{m+1}[r] = A_{m+1} K_m[r] + B_{m+1} K_m[r-1], r = 0, 1, 2, ..., M-1, K_m[-1] = 0$$

Define the matrix

$$P_m = \begin{pmatrix} K_m[0] \\ K_m[1] \\ K_m[2] \\ ... \\ K_m[M] \end{pmatrix} \in \mathbb{C}^{2M+2 \times 2}$$

where

$$K_m[r] = 0, r > m$$

Then, we can write the above recursion as

$$P_{m+1} = C_{m+1} P_m, 1 \leq m \leq M-1$$

where C_{m+1} is the $2M+2 \times 2M+2$ block structured matrix

$$C_{m+1} = \begin{pmatrix} A_{m+1} & 0 & 0 & 0.. & 0 \\ B_{m+1} & A_{m+1} & 0 & 0.. & 0 \\ 0 & B_{m+1} & A_{m+1} & 0... & 0 \\ .. & .. & .. & .. & .. \\ 0 & 0 & 0 & B_{m+1} & A_{m+1} \end{pmatrix}$$

We can express this as

$$C_{m+1} = I_M \otimes A_{m+1} + Z_M \otimes B_{m+1}$$

where I_M is the $M \times M$ identity matrix and Z_M is the $M \times M$ unit delay matrix:

$$Z_M = C_{m+1} = \begin{pmatrix} 0 & 0 & 0 & 0.. & 0 \\ 1 & 0 & 0 & 0.. & 0 \\ 0 & 1 & 0 & 0... & 0 \\ .. & .. & .. & .. & .. \\ 0 & 0 & 0 & 1 & 0 \end{pmatrix}$$

This matrix difference equation has the solution

$$P_m = C_m C_{m-1}...C_2 P_1, 2 \leq m \leq M$$

Note that

$$P_1 = \begin{pmatrix} K_1[0] \\ K_1[1] \\ 0 \\ 0 \\ ... \\ 0 \end{pmatrix}$$

where

$$K_1[0] = A_1, K_1[1] = B_1$$

Finally,

$$T(z) = K_M(z) = \sum_{r=0}^{M} z^{-r} K_M[r] = [I_2, z^{-1}I_2, ..., z^{-M}I_2] P_M$$

$$= ([1, z^{-1}, z^{-2}, ..., z^{-M}] \otimes I_2) P_M$$

To implement this algorithm for computing $T(z)$, first we must write a program for computing P_M as a product of matrices.

Chapter 10

Gravity interacting with waveguide quantum fields with filtering and control

10.1 Waveguides placed in the vicinity of a strong gravitational field

In the absence of the gravitational field, let the electromagnetic field tensor within the guide be given by $F_{\mu\nu}^{(0)}$. The components of this tensor are easily determined using the standard expressions in the frequency domain:

$$E_\perp = \sum_n (-\gamma_E[n]/h_E[n]^2)\nabla_\perp E_{z,n} exp(-\gamma_E[n]z)$$

$$-(j\omega\mu/h_H[n]^2)\nabla_\perp H_{z,n}\times\hat{z}.exp(-\gamma_H[n]z))$$

$$H_\perp = \sum_n (-\gamma_H[n]/h_H[n]^2)\nabla_\perp H_{z,n} exp(-\gamma_H[n]z)+(j\omega\epsilon/h_E[n]^2)\nabla_\perp E_{z,n}\times\hat{z})exp(-\gamma_E[n]z))$$

with

$$(\nabla_\perp^2 + h_E[n]^2)E_{z,n} = 0, E_{z,n}|_{\partial D} = 0,$$

$$(\nabla_\perp^2 + h_E[n]^2)H_{z,n} = 0, \frac{\partial H_{z,n}}{\partial\hat{n}}|_{\partial D} = 0$$

In the presence of a gravitational field, let the perturbation to the em field tensor be denote by $F_{\mu\nu}^{(1)}$. In other words,

$$A_\mu = A_\mu^{(0)} + A_\mu^{(1)},$$

$$F_{\mu\nu} = F_{\mu\nu}^{(0)} + F_{\mu\nu}^{(1)},$$

$$F_{\mu\nu}^{(0)} = A_{\nu,\mu}^{(0)} - A_{\mu,\nu}^{(0)},$$

$$F_{\mu\nu}^{(1)} = A_{\nu,\mu}^{(1)} - A_{\mu,\nu}^{(1)}$$

165

Write

$$g_{\mu\nu} = \eta_{\mu\nu} + h_{\mu\nu}(x)$$

so that

$$g \approx -(1+h), h = \eta_{\mu\nu}h_{\mu\nu} = h_\mu^\mu$$

and hence,

$$\sqrt{-g} \approx 1 + h/2$$

This gives us using the exact Maxwell equations

$$(F^{\mu\nu}\sqrt{-g})_{,\nu} = 0$$

the first order perturbed term

$$(F_{\alpha\beta}^{(0)}\delta(g^{\mu\alpha}g^{\nu\beta}\sqrt{-g}) + \eta_{\mu\alpha}\eta_{\nu\beta}F_{\alpha\beta}^{(1)})_{,\nu} = 0$$

or equivalently,

$$(\eta_{\mu\alpha}\eta_{\nu\beta}F_{\alpha\beta}^{(1)})_{,\nu} =$$

$$-(F_{\alpha\beta}^{(0)}\delta(g^{\mu\alpha}g^{\nu\beta}\sqrt{-g})_{,\nu}$$

Now,

$$\delta(g^{\mu\alpha}g^{\nu\beta}\sqrt{-g}) =$$

$$(\delta g^{\mu\alpha})(g^{\nu\beta}\sqrt{-g}) + \delta(g^{\nu\beta})g^{\mu\alpha}\sqrt{-g} + g^{\mu\alpha}g^{\nu\beta}h/2$$

where we substitute

$$\delta g^{\mu\alpha} = -g^{\mu\rho}g^{\nu\sigma}h_{\rho\sigma}$$

$$\delta g^{\nu\beta} = -g^{\nu\rho}g^{\beta\sigma}h_{\rho\sigma}$$

Exercise: Using the above perturbed Maxwell equations along with the gauge condition

$$(A^\mu\sqrt{-g})_{,\mu} = 0,$$

or equivalently,

$$(A_\nu g^{\mu\nu}\sqrt{-g})_{,\mu} = 0$$

whose first order perturbed term gives

$$(A_\nu^{(1)}g^{\mu\nu}\sqrt{-g})_{,\mu} + (A_\nu^{(0)}\delta(g^{\mu\nu}\sqrt{-g}))_{,\nu} = 0,$$

derive the modified wave equation with source for $A_\mu^{(1)}$.

10.2 Some study projects regarding waveguides and cavity resonators in a gravitational field

[1] Rectangular waveguides near a gravitational field of a Schwarzchild blackhole

[2] Cylindrical waveguides with cladding near a gravitational field.

Inner radius is a and outer radius is b. In the region $\rho < a$, the permittivity and permeability are (ϵ_1, μ_1) and in the region $a < \rho < b$, these parameters are (ϵ_2, μ_2). The transverse components of the field are

$$E_\perp^{(k)} = (-\gamma/h_k^2)\nabla_\perp E_z^{(k)} - (j\omega\mu_k/h_k^2)\nabla_\perp H_z^{(k)} \times \hat{z}$$

$$H_\perp^{(k)} = (-\gamma/h_k^2)\nabla_\perp H_z^{(k)} + (j\omega\epsilon_k/h_k^2)\nabla_\perp E_z^{(k)} \times \hat{z}$$

where $k = 1, 2$. Note that the propagation constant γ for the fields along the z direction must be the same in both the regions in view of the continuity of E_z and H_z at $\rho = a$. Note that there is no surface current density at the dielectric interface $\rho = a$ and hence the tangential components of H are also continuous. We have

$$h_k^2 = \omega^2 \epsilon_k \mu_k + \gamma^2, k = 1, 2$$

Also the z components of the field satisfy the Helmholtz equation

$$(\nabla_\perp^2 + h_k^2)(E_z^{(k)}, H_z^{(k)}) = 0$$

Keeping in mind the fact that the second Bessel function $Y_m(x)$ is singular at $\rho = 0$, the solutions are

$$E_z^{(1)} = J_m(h_1\rho)(C_1.cos(m\phi) + C_2.sin(m\phi)),$$

$$H_z^{(1)} = J_m(h_1\rho)(C_1'.cos(m\phi) + C_2'.sin(m\phi))$$

$$E_z^{(2)} = (A_1 J_m(h_2\rho) + A_2 Y_m(h_2\rho)).(C_1.cos(m\phi) + C_2.sin(m\phi))$$

$$H_z^{(2)} = (A_1' J_m(h_2\rho) + A_2' Y_m(h_2\rho)).(C_1'.cos(m\phi) + C_2'.sin(m\phi))$$

These are consistent with the continuity of E_z and H_z at $\rho = a$ provided that

$$J_m(h_1a) = A_1 J_m(h_2a) + A_2 Y_m(h_2a) = (A_1' J_m(h_2a) + A_2' Y_m(h_2a)) - - - (1)$$

Assuming that γ is known, it follows that h_1, h_2 are also known and hence these furnish us with two equations for the four constants A_1, A_2, A_1', A_2'. The outer surface $\rho = b$ is a perfect conductor. Hence, $E_z^{(2)} = 0$ at $\rho = b$ and this gives us

$$A_1 J_m(h_2b) + A_2 Y_m(h_2b) = 0 - - - (2a)$$

which is yet another equation for the above four constants. We also have that H_ρ vanishes at the perfectly conducting surface $\rho = b$. This gives

$$(-\gamma/h_2^2)\partial H_z^{(2)}/\partial\rho + (j\omega\epsilon_2/bh_2^2)\partial E_z^{(2)}/\partial\phi = 0 - - - (2b)$$

at $\rho = b$. This yields us

$$(-\gamma/h_2)(A_1' J_m(h_2 b) + A_2' Y_m(h_2 b))(C_1' cos(m\phi) + C_2' sin(m\phi))$$

$$+(jw\epsilon_2/bh_2^2)(A_1 J_m(h_2 b) + A_2 Y_m(h_2 b))(-mC_1 sin(m\phi) + mC_2 cos(m\phi)) = 0 ----(2c)$$

which gives us two equations on equating separately the coefficients of $cos(m\phi)$ and $sin(m\phi)$. Continuity of ϵE_ρ at $\rho = a$ gives us two more equations:

$$(-\gamma\epsilon_1/h_1^2)\partial E_z^{(1)}/\partial\rho - (jw\epsilon_1\mu_1/ah_1^2)\partial H_z^{(1)}/\partial\phi$$

$$= (-\gamma\epsilon_2/h_2^2)\partial E_z^{(2)}/\partial\rho - (jw\epsilon_2\mu_2/ah_2^2)\partial H_z^{(2)}/\partial\phi ---(3a)$$

evaluated at $\rho = a$, or equivalently,

$$(-\gamma\epsilon_1/h_1)J_m'(h_1 a)(C_1 cos(m\phi) + C_2 sin(m\phi))$$

$$-(jw\epsilon_1\mu_1/ah_1^2)J_m(h_1 a)(-mC_1'.sin(m\phi) + m.C_2'.cos(m\phi))$$

$$= (-\gamma\epsilon_2/h_2)(A_1 J_m'(h_2 a) + A_2 Y_m'(h_2 a))(C_1 cos(m\phi) + C_2 sin(m\phi))$$

$$+(jw\epsilon_2\mu_2/ah_2^2)(A_1' J_m(h_2 a) + A_2' Y_m(h_2 a))(-mC_1'.sin(m\phi) + mC_2'.cos(m\phi)) ----(3b)$$

Equating the coefficients of $cos(m\phi)$ and $sin(m\phi)$ in these equations gives us two equations for the four constants C_1, C_2, C_1', C_2'. Likewise equating μH_ρ at $\rho = a$ gives us two more equations relating these constants. These are obtained from

$$(-\gamma\mu_1/h_1^2)\partial H_z^{(1)}/\partial\rho + (jw\mu_1\epsilon_1/ah_1^2)\partial E_z^{(1)}/\partial\phi$$

$$= (-\gamma\mu_2/h_2^2)\partial H_z^{(2)}/\partial\rho + (jw\mu_2\epsilon_2/ah_2^2)\partial E_z^{(2)}/\partial\phi ---(4a)$$

evaluated at $\rho = a$, or equivalently,

$$(-\mu_1\gamma/h_1)J_m'(h_1 a)(C_1'.cos(m\phi) + C_2'.sin(m\phi))$$

$$+(jw\epsilon_1\mu_1/ah_1^2)J_m(h_1 a)(-mC_1.sin(m\phi) + mC_2.cos(m\phi))$$

$$= (-\gamma\mu_2/h_2)(A_1' J_m'(h_2 a) + A_2' Y_m'(h_2 a))(C_1' cos(m\phi) + C_2' sin(m\phi))$$

$$-(jw\epsilon_2\mu_2/ah_2^2)(A_1 J_m(h_2 a) + A_2 Y_m(h_2 a))(-mC_1.sin(m\phi) + mC_2.cos(m\phi)) ----(4b)$$

Equating E_ϕ and H_ϕ at $\rho = a$ gives us another set of four equations. Specifically, these are

$$(-\gamma/ah_1^2)\partial E_z^{(1)}/\partial\phi + jw\mu_1\partial H_z^{(1)}/\partial\rho =$$

$$(-\gamma/ah_2^2)\partial E_z^{(2)}/\partial\phi + jw\mu_2\partial H_z^{(2)}/\partial\rho ---(5)$$

evaluated at $\rho = a$ and

$$(-\gamma/h_1^2 a)\partial H_z^{(1)}/\partial\phi - (jw\epsilon_1/h_1^2)\partial E_z^{(1)}/\partial\rho$$

$$= (-\gamma/ah_2^2)\partial H_z^{(2)}/\partial\phi - (jw\epsilon_2/h_2^2)\partial E_z^{(2)}/\partial\rho ---(6)$$

evaluated at $\rho = a$. Now, the continuity of E_z and H_z at $\rho = a$ implies the continuity of $\partial E_z/\partial\phi$ and $\partial H_z/\partial\phi$ at $\rho = a$. In view of this fact, (6) is equivalent to the equation

$$(1 - \gamma^2/h_1^2)a^{-1}\partial H_z^{(1)}/\partial\phi - (jw\gamma\epsilon_1/h_1^2)\partial E_z^{(1)}/\partial\rho$$

$$= (1 - \gamma^2/h_2^2)a^{-1}\partial H_z^{(2)}/\partial\phi - (j\omega\gamma\epsilon_2/h_2^2)\partial E_z^{(2)}/\partial\rho$$

at $\rho = a$, which in view of the fact that $h_k^2 - \gamma^2 = \omega^2\mu_k\epsilon_k, k = 1, 2$ becomes

$$(\omega^2\mu_1\epsilon_1/ah_1^2)\partial H_z^{(1)}/\partial\phi - (j\omega\gamma\epsilon_1/h_1^2)\partial E_z^{(1)}/\partial\rho$$

$$= (\omega^2\mu_2\epsilon_2/ah_2^2)\partial H_z^{(2)}/\partial\phi - (j\omega\gamma\epsilon_2/h_2^2)\partial E_z^{(2)}/\partial\rho$$

at $\rho = a$, or after making cancellations, is the same as $(3a)$, namely the equation of continuity of ϵE_ρ at $\rho = a$. Likewise, (5) is equivalent to $(4a)$, namely, the equation of continuity of μH_ρ at $\rho = a$. Thus, in all, our only independent equations are $(1), (2a), (2c), (3b), (4b)$ which yield $2+1+2+2+2 = 9$ equations for the nine variables $\gamma, A_1, A_2, A_1', A_2', C_1, C_2, C_1', C_2'$. These yield a characteristic equation for γ which determines discrete set of possible propagation constants.

A neater way to obtain the required characteristic function for γ is to denote the constants $A_1C_1, A_1C_2, A_2C_1, A_2C_2$ by B_1, B_2, B_3, B_4 and likewise the constants $A_1'C_1', A_1'C_2', A_2'C_1', A_2'C_2'$ by B_1', B_2', B_3', B_4' so that the continuity of E_z and H_z at $\rho = a$ gives us four equations obtained by equating the coefficients of $cos(m\phi)$ and $sin(m\phi)$ in place of (1). Then the constants to be determined are $C_1, C_2, C_1', C_2', B_1, B_2, B_3, B_4, B_1', B_2', B_3', B_4'$, namely twelve in number and the number of linear equations for these constants is 4 corresponding to continuity of E_z and H_z at $\rho = a$ plus 4 corresponding to continuity of E_z and H_ρ at $\rho = b$ plus four corresponding to continuity of ϵE_ρ and μH_ρ at $\rho = a$, ie, in all twelve homogeneous linear equations. Setting the determinant of the corresponding 12×12 matrix to zero then yield the possible discrete values of γ.

[3] cavity resonators in a gravitational field.

[4] Waveguides with inhomogeneity and anisotropicity placed close to a strong gravitational field.

Lie groups, Lie algebras and differential equations

10.3 A comparison between the EKF and Wavelet based block processing algorithms for estimating transistor parameters in an amplifier drived by the Ornstein-Uhlenbeck process

A problem in group theory related to fluid velocity pattern recognition

$v(t,r) = (v_x(t,r), v_y(t,r)), r = (x,y)$ is a 2-D fluid velocity field. It satisfies the Navier-Stokes equation

$$(v, \nabla)v + v_{,t} = -\nabla p/\rho + \nu\nabla^2 v + f$$

where $f(t,r) = (f_x(t,r), f_y(t,r))$ is a random driving force field assume to be white Gaussian w.r.t the time variable. We apply a Galilean transformation to this field comprising of a rotation $R(\phi)$ in $SO(2)$ followed by a translation $a \in \mathbb{R}^2$. The resulting velocity field is $w(t,r)$ and after noise corruption, this rotated and translated velocity field is assumed to satisfy a Navier-Stokes equation with a different driving noise $g(t,r)$ again white Gaussian w.r.t. the time variable:

$$(w, \nabla)w + w_{,t} = -\nabla p_1/\rho + \nu\nabla^2 w + g$$

The fluid is incompressible, ie,

$$div\,v = 0, div\,w = 0$$

Thus there exist stream functions $\psi_1(t,r)$ and $\psi_2(t,r)$ such that

$$v = \nabla\psi_1 \times \hat{z}, w = \nabla\psi_2 \times \hat{z}$$

10.4 Computing the Haar measure on a Lie group using left invariant vector fields and left invariant one forms

G is a Lie group and $(X_1, ..., X_n)$ is a basis of left invariant vector fields on G. L_g denotes left translation on G, ie, $L_g h = gh$. By left invariance of the vector fields X_k, we have that

$$dL_g(X_k) = X_k$$

ie

$$X_k(f o L_g)(x) = X_k(gx), g, x \in G$$

or equivalently,

$$L_{g*}X_k(x) = X_k(L_g x)$$

or equivalently,

$$L_{g*}X_k = X_k L_g, g \in G$$

where L_{g*} denotes the push-forward map, ie if $T : \mathcal{M} \to \mathcal{N}$ is a differentiable map from a differentiable manifold $\mathcal{M}$ into another differentiable manifold $\mathcal{N}$ and if $X(x) \in T\mathcal{M}_x$, then

$$T_* X(x) = Y(T(x)) \in T\mathcal{N}_{T(x)}$$

where
$$Y(T(x))(f) = X(foT)(x)$$

Equivalently, in terms of local coordinates,
$$Y^a(T(x)) = \frac{\partial T^a(x)}{\partial x^b} X^b(x)$$

with the Einstein summation convention adopted. It follows that
$$X_k^a(gx) = \frac{\partial L_g^a(x)}{\partial x^b} X_k^b(x)$$

which gives
$$X_k^a(g) = \frac{\partial L_g^a(e)}{\partial x^b} X_k^b(e)$$

from which we get on taking determinants,
$$det(((X_k^a(g))) = (detL_g'(e))det((X_k^a(e)))$$

so that the unnormalized left invariant Haar measure density $(detL_g'(e))^{-1}$ is proportional to $1/det((X_k^a(g)))$. Equivalently, by duality of bases, if $\omega_k(g), k = 1, 2, ..., n$ is a basis of left invariant one forms on G, then the left invariant Haar measure density is proportional to $det((\omega_k^a(g)))$.

10.5 How background em radiation affects the expansion of the universe

Terms that are square in the electromagnetic field components determine the first order perturbations to the metric tensor of the expansion universe. The unperturbed metric of the universe is the Robertson-Walker metric in the coordinates $t = x^0, r = x^1, \theta = x^2, \phi = x^3$:

$$g_{00} = 1, g_{11} = -f(r)S^2(t), g_{22} = -r^2 S^2(t), g_{33} = -r^2 sin^2(\theta)S^2(t)$$

The unperturbed Maxwell equations determine the electromagnetic field which in turn drives the metric perturbations via the Einstein field equations.

$$R_{\mu\nu} = -8\pi G S_{\mu\nu}^M,$$

$$\delta R_{\mu\nu} = -8\pi G(\delta S_{\mu\nu}^M + S_{\mu\nu}^M)$$

where
$$T_{\mu\nu}^M = (\rho(t) + p(t))v^\mu v^\nu - p(t)g^{\mu\nu}$$

with $(v^\mu) = (1, 0, 0, 0)$ being the comoving four-velocity field.

$$S_{\mu\nu}^M = T_{\mu\nu}^M - T^M g_{\mu\nu}/2$$

$$T^M = g^{\mu\nu} T^M_{\mu\nu} = \rho(t) - 3p(t)$$

$$\delta S^M_{\mu\nu} = \delta T^M_{\mu\nu} - \delta T^M . g_{\mu\nu}/2 - T^M \delta g_{\mu\nu}/2$$

$$\delta T^M_{\mu\nu} = (\delta\rho(x) + \delta p(x))v^\mu v^\nu - \delta p(x) g_{\mu\nu} - p(t)\delta g_{\mu\nu}(x)$$

$$+(\rho(t) + p(t))(v^\mu \delta v^\nu(x) + v^\nu \delta v^\mu(x))$$

$$\delta T^M(x) = \delta\rho(x) - 3\delta p(x)$$

$$S^{EM}_{\mu\nu} = (-1/4)F_{\alpha\beta}F^{\alpha\beta}g_{\mu\nu} + F_{\mu\alpha}F^\alpha_\nu$$

Check of the above formula in special relativity: We use the Maxwell equations

$$F^{\mu\nu}_{,\nu} = -\mu_0 J^\mu$$

$$S^{EM\mu\nu} = [(-1/4)F_{\alpha\beta}F^{\alpha\beta}g^{\mu\nu} + F^{\mu\alpha}F^\nu_\alpha]_{,\nu}$$

$$S^{EM\mu\nu}_{,\nu} = (-1/2)g^{\mu\nu}F^{\alpha\beta}F_{\alpha\beta,\nu} + F^{\mu\alpha}_{,\nu}F^\nu_\alpha + F^{\mu\alpha}F^\nu_{\alpha,\nu}$$

$$= (1/2)g^{\mu\nu}F^{\alpha\beta}(F_{\beta\nu,\alpha} + F_{\nu\alpha,\beta}) + F^{\mu\alpha}_{,\nu}F^\nu_\alpha + \mu_0 F^{\mu\alpha}J_\alpha$$

where we have made use of the Maxwell equations

$$F_{\mu\nu,\alpha} + F_{\nu\alpha,\mu} + F_{\alpha\mu,\nu} = 0$$

which may also be seen as a consequence of the definition

$$F_{\mu\nu} = A_{\nu,\mu} - A_{\mu,\nu}$$

Continuing further, the above equals

$$(1/2)F^{\alpha\beta}(F^\mu_{\beta,\alpha} + F^\mu_{\alpha,\beta}) + F^{\mu\alpha}_{,\nu}F^\nu_\alpha + \mu_0 F^{\mu\alpha}J_\alpha$$

$$= (1/2)F^{\alpha\beta}(F^\mu_{\beta,\alpha} + F^\mu_{\alpha,\beta}) + F^\beta_\alpha F^{\mu\alpha}_{,\beta} + \mu_0 F^{\mu\alpha}J_\alpha$$

$$= (1/2)F^{\alpha\beta}(F^\mu_{\beta,\alpha} + F^\mu_{\alpha,\beta}) + F^\beta_\alpha F^{\mu\alpha}_{,\beta} + \mu_0 F^{\mu\alpha}J_\alpha$$

$$= (1/2)F^{\alpha\beta}(F^\mu_{\beta,\alpha} + F^\mu_{\alpha,\beta} - F^\mu_{\alpha,\beta} + \mu_0 F^{\mu\alpha}J_\alpha$$

$$= (1/2)F^{\alpha\beta}g^{\mu\nu}(-F_{\nu\beta,\alpha} + F_{\nu\alpha,\beta} - 2F_{\nu\alpha,\beta}) + \mu_0 F^{\mu\alpha}J_\alpha$$

$$= (-1/2)F^{\alpha\beta}g^{\mu\nu}(F_{\nu\beta,\alpha} + F_{\nu\alpha,\beta}) + \mu_0 F^{\mu\alpha}J_\alpha$$

$$= \mu_0 F^{\mu\alpha}J_\alpha$$

as required by the four-divergence of the energy-momentum tensor of the electromagnetic field.

10.6 Stochastic BHJ equations in discrete and continuous time for stochastic optimal control based on instantaneous feedback

. First we consider a general example of a Markov process satisfying a stochastic differential equation driven by Brownian motion and a Poisson field:

$$dX(t) = \mu(t, X(t))dt + \sigma(t, X(t))dB(t) + \int_{\xi \in E} \psi(t, X(t), \xi)dN(t, \xi)$$

where $N(.,.)$ is a Poisson field with intensity $dF(t, \xi)$, ie,

$$\mathbb{E}[N(dt, d\xi)] = dF(t, \xi)$$

Here $(E, \mathcal{E})$ and the space-time Poisson random field $N(.,.)$ is defined on the measure space $(\mathbb{R} \times E, \mathcal{B}(\mathbb{R}) \otimes \mathcal{E})$. F is a measure on this space. Ito's formula for $\phi(X(t))$ gives

$$d\phi(X(t)) = L_t\phi(X(t))dt + dB(t)^T \sigma(t, X(t))^T \nabla \phi(X(t))$$

$$+ \int_{\xi \in E} (\phi(X(t) + \psi(t, X(t), \xi)) - \phi(X(t)))N(dt, d\xi)$$

from which we deduce that the generator of $X(t)$ defined by

$$K_t\phi(x)dt = \mathbb{E}(d\phi(X(t))|X(t) = x)$$

is given by

$$K_t\phi(x) = L_t\phi(x) + \int_{\xi \in E} (\phi(x + \psi(t, x, \xi)) - \phi(x))f(t, d\xi)$$

where

$$f(t, d\xi) = F(dt, d\xi)/dt = \frac{\partial F(t, d\xi)}{\partial t}$$

and L_t is the generator of the diffusion part of the Markov process $X(t)$, ie,

$$L_t\phi(x) = \mu(t, x)^T \nabla \phi(x) + (1/2)Tr(\sigma(t, x)\sigma(t, x)^T \nabla \nabla^T \phi(x))$$

ie,

$$L_t = \mu(t, x)^T \nabla + (1/2)Tr(\sigma(t, x)\sigma(t, x)^T \nabla \nabla^T)$$

Equivalently, if we interpret the operator $exp(x^T\nabla)$ by the familiar Taylor expansion formula

$$exp(x^T\nabla)f(y) = f(y + x)$$

then we can write

$$K_t = L_t + \int_E (exp(\psi(t, x, \xi)^T\nabla) - 1)f(t, d\xi)$$

Now consider the optimal control problem in which the generator K_t is a function of the instantaneous input $u(t)$, ie, $K_t = K_t(u(t))$. We are allowed to take the control input $u(t)$ only as a function of the instantaneous state $X(t)$, ie, $u(t) = \chi_t(X(t))$, where χ_t maps the state space to the control input space. Only then the Markovianity of $X(t)$ is not destroyed. In fact the generator of the controlled Markov process $X(t)$ is then given by

$$\tilde{K}_t\phi(x) = (K_t(\chi_t(x))\phi)(x)$$

We wish to select the feedback control functions $\chi_t(.)$ so that

$$C_T(x) = \mathbb{E}[\int_0^T \mathcal{L}(X(t), u(t))dt | X(0) = x]$$

$$= \mathbb{E}[\int_0^T \mathcal{L}(X(t), \chi_t(X(t)))dt | X(0) = x]$$

is minimized. To this end, we define

$$C(t, T, x) = min_{u(s), t \le s \le T}\mathbb{E}[\int_t^T \mathcal{L}(X(s), u(s))ds | X(t) = x]$$

$$= min_{\chi_s, t \le s \le T}\mathbb{E}[\int_t^T \mathcal{L}(X(s), \chi_s(X(s)))ds | X(t) = x]$$

Then, by applying the Markov property, we easily deduce that

$$C(t, T, x) = min_{\chi_t(.)}(\mathcal{L}(x, \chi_t(x))dt + \mathbb{E}(C(t + dt, T, X(t + dt)) | X(t) = x))$$

$$= min_{\chi_t(.)}(\mathcal{L}(x, \chi_t(x))dt + C(t, T, x) + dt.\frac{\partial C(t, T, x)}{\partial t} + dt.K_t(\chi_t(x))(C(t, T, x)))$$

or equivalently,

$$\frac{\partial C(t, T, x)}{\partial t} + min_{\chi_t(.)}(\mathcal{L}(x, \chi_t(x)) + K_t(\chi_t(x))(C(t, T, x))) = 0$$

This is the stochastic BHJ (Bellman-Hamilton-Jacobi) equation. As a by product, it yields the optimal feedback control maps $\chi_t(.), 0 \le t \le T$. The final point condition while solving this pde is given by

$$lim_{t \to T}C(t, T, x) = 0$$

The discrete time case: Here $X(n), n \ge 0$ is a Markov process in discrete time with one step transition probability generator $K_n(u(n))$ where $u(n) = \chi_n(X(n))$. Thus,

$$\mathbb{E}(\phi(X(n + 1)) | X(n) = x) = (K_n(\chi_n(x))\phi)(x)$$

The control maps $\chi_n(.), n \geq 0$ are to be chosen so that

$$C_N(x_0) = \mathbb{E}[\sum_{n=0}^{N} \mathcal{L}(X(n), u(n))|X(0) = x_0]$$

$$= \mathbb{E}[\sum_{n=0}^{N} \mathcal{L}(X(n), \chi_n(X(n)))|X(0) = x_0]$$

is a minimum. We define

$$C(n, N, x) = min_{\chi_k(.), n \leq k \leq N} \mathbb{E}[\sum_{k=n}^{N} \mathcal{L}(X(k), \chi_k(X(k)))|X(n) = x]$$

Using the Markov property, we find that

$$C(n, N, x) = min_{\chi_n(.)}(\mathcal{L}(x, \chi_n(x)) + \mathbb{E}[C(n + 1, N, X(n + 1))|X(n) = x])$$

$$= min_{\chi_n(.)}(\mathcal{L}(x, \chi_n(x)), K_n(\chi_n(x))(C(n + 1, N, x)))$$

This is the stochastic BHJ equation in discrete time. This is to be solved with the final point condition

$$\chi_N(.) = argmin_\chi \mathcal{L}(x_0, \chi(x_0))$$

10.7 Quantum stochastic optimal control of the HP-Schrodinger equation

The controlled HP equation is

$$dU(t) = ((-iH+P)dt+LdA(t)+MdA(t)^*+Sd\Lambda(t)-iK(t)(X_d(t)-\pi_t(X))dt)U(t)$$

where
$$j_t(X) = U(t)^*XU(t), \pi_t(X) = \mathbb{E}(j_t(X)|\eta_t)$$

$X_d(t)$ is the desired state trajectory to be tracked and $K(t)$ is the controller coefficient. η_t is the Von-Neumann algebra generated by output non-demolition measurements upto time t. Equivalently, we can express the controlled HP equation as

$$dU(t) = [(-i(H + K(t)(X_d(t) - \pi_t(X)) + P)dt + LdA(t) + MdA(t)^* + Sd\Lambda(t)]dt$$

Study project: Explain how you can solve the above controlled HP equation coupled with the Belavkin filter equation for $\pi_t(X)$ using perturbation theory.

10.8   Bath in a superposition of coherent states interacting with a system

The system dynamics corresponds to that of a fan motor subject to quantum noise:

$$\theta''(t) + a\theta'(t) + f(t, \theta(t)) = w(t)$$

where $w(t)$ is white noise, a is the damping coefficient and $f(t, \theta) = -I(t)BLsin(\theta(t))$. To give a quantum mechanical description of this motor, we introduce a Lagrangian

$$L(t, \theta, \theta') = \theta'^2/2 - \int_0^\theta f(t, \theta)d\theta + w(t)\theta$$

which yields using the Euler-Lagrange equations

$$\theta''(t) + f(t, \theta) - w(t) = 0$$

namely, the same as that of the fan but with the damping term excluded. To include the damping term using quantum mechanics, we consider the canonical momentum

$$p = \partial L/\partial\theta'$$

and apply the Legendre transformation to obtain the Hamiltonian

$$H(t, \theta, p) = p\theta' - L = p^2/2 + V(t, \theta) - w(t)\theta$$

where

$$V(t, \theta) = \int_0^\theta f(t, \theta)d\theta$$

and then introduce the Lindblad operators to obtain damping. The resulting master equation is of the form

$$\rho'(t) = -i[H(t, \theta, p), \rho(t)] - \theta(\rho(t))$$

where

$$\theta(\rho) = (1/2)(L^*L\rho + \rho L^*L - 2L\rho L^*)$$

where

$$L = \alpha\theta + \beta p$$

Chapter 11

Basic triangle geometry required for understanding Riemannian geometry in Einstein's theory of gravity

11.1 Problems in mathematics and physics for school students

[1] A triangle has sides a, b, c and corresponding angles A, B, C. Draw the triangle and mark all the sides and angles. Express b in terms of a, c, C by solving a quadratic equation. Likewise, express c in terms of a, b, B and a in terms of b, c, C.

[2] Let ABC be a triangle with side lengths a, b, c and opposite angles A, B, C. Drop the altitude from vertex A onto the side $a = BC$. Let D denote the intersection point. Draw the diagram marking all the points. Let $h = AD$ denote this altitude. Calculate h in terms of a, B, C. Also calculate h in terms of c and B and finally, h in terms of b and C.

[3] Two straight lines having equations $y = mx + c$ and $y = m'x + c'$ are given. Determine their intersection point coordinates (x, y) by solving the simultaneous equation. For $m = tan(60^o)$ and $m' = tan(30^o)$ and $c = -5, c' = -4$, draw these lines using ruler and protractor and verify your result.

[4] Factorize:
[a]

$$(x + y)^2 - x^2 - 2zy$$

[b]

$$(x^2 - y^2) + a(x + y) + c(x - y + a)$$

[4] Calculate the incircle radius of a triangle ABC with side-lengths a, b, c in terms of $B/2, C/2, a$. Also calculate the radius of the excircle that touches the side $a = BC$ in terms of B, C, a.

11.2   Geometry on a curved surface, study problems

[1] Define a straight line on a curved surface as the path of shortest Euclidean distance on the surface between two points. This presupposes that the curved surface of dimension p is immersed in an $N > p$ dimensional Euclidean space.

[2] Define parallel displacement of a vector on a curved surface by infinitesimally translating the vector parallely in the Euclidean sense and then projecting the resulting vector onto the tangent plane at the neighbouring point.

[3] Define a geodesic triangle on a curved two dimensional surface and prove Gauss'm theorem that the sum of the angles of such a triangle equals the integral of the Gauss curvature over the triangle.

[4] Show that an alternate equivalent definition of a straight line on a curved surface is given by that path such that when the tangent vector to this curve at any point is parallely displaced along the curve to another point, it continues to be a tangent to the curve at the displaced point.

Chapter 12

Design of gates using Abelian and non-Abelian gauge quantum field theories with performance analysis using the Hudson-Parthasarathy quantum stochastic calculus

12.1 Design of quantum gates using Feynman diagrams

If $1/A$ denotes the electron propagator, then corrections to it coming from terms like vacuum polarization, external field effects etc. will result in the corrected propagator $1/(A + B)$. Denoting the parameters of the external fields by θ, we have the corrected propagator

$$(A + B(\theta))^{-1} = A^{-1} - A^{-1}BA^{-1} + A^{-1}BA^{-1}BA^{-1} + \ldots$$

Let X denote the desired propagator that will yield the desired scattering matrix. Then we must design the parameters θ so that

$$\| X - (A + B(\theta))^{-1} \|$$

is minimized. Writing $Z = X - A^{-1}$, the approximate minimization problem is

$$\hat{\theta} = argmin_\theta \| Z + A^{-1}B(\theta)A^{-1} - A^{-1}B(\theta)A^{-1}B(\theta)A^{-1} \|^2$$

If the minimization is over all B, then we expand the above propagator error energy upto quadratic terms in B and then the optimal equations will be linear in B which are easily inverted. We leave it as an exercise to show that with this second order approximation,

$$\hat{B} = argmin_B(Tr(A^{-1}BA^{-2}BA^{-1}) + 2Re(Tr(ZA^{-1}BA^{-1})$$
$$-2Re(Tr(ZA^{-1}BA^{-1}BA^{-1}))$$

An electron with a four momentum of p_1, a spin of σ_1 interacts with a positron with a four momentum of p_2 and a spin of σ_2, annihilate each other to produce a γ-ray photon of four momentum $k = p_1 + p_2$ which again polarizes into an electron-positron pair in the form of a loop (vacuum polarization) and again a pair annihilation takes place from this loop to result in a γ-ray photon which propagates and again polarizes into an electron positron pair with four momenta p_3, σ_3 and p_4, σ_4 respectively. To calculate the amplitude for this process, we must assume four momentum conservation, ie, $p_1 + p_2 = p_3 + p_4$ which can be ensured by introducing a momentum conserving δ-function $\delta^4(p_1 + p_2 - p_3 - p_4)$. Denoting the electron wave function by $u(p, \sigma)$ and the positron wave function by $\bar{v}(p, \sigma) = v(p, \sigma)^* \gamma^0$, the photon propagator by $D_{\mu\nu}(k)$ and the electron propagator matrix by $S(p) = (\gamma.p - m)^{-1} = (\gamma.m + p)/(p^2 - m^2)$, with

$$p^2 = (p^0)^2 - (p^1)^2 - (p^2)^2 - (p^3)^2 = p^{02} - P^2$$

so that

$$p^2 = m^2 = p^{02} - E(P), E(P) = m^2 + P^2$$

we obtain the amplitude for the above process as

$$S(p_3, \sigma_3, p_4, \sigma_4 | p_1, \sigma_1, p_2, \sigma_2) =$$

$$(\bar{v}(p_2, \sigma_2)\gamma^\mu u(p_1, \sigma_1))(\bar{v}(p_4, \sigma_4)\gamma^\nu u(p_3, \sigma_3)\int D_{\mu\rho}(p_1 + p_2)D_{\nu\alpha}(p_3 + p_4)$$
$$Tr(S(p_1 + p_2 - q)\gamma^\alpha S(q)\gamma^\rho)d^4q$$

$$= (p_1 + p_2)^{-4}(v(p_2, \sigma_2)\gamma_\rho u(p_1, \sigma_1))(\bar{v}(p_4, \sigma_4)\gamma_\alpha u(p_3, \sigma_3)Tr(S(p_1 + p_2 - q)\gamma^\alpha S(q)\gamma^\rho)d^4q$$

Note that this amplitude is to be multiplied by $\delta^4(p_1 + p_2 - p_3 - p_4)$.

Remark 1: To make the integrals converge we insert factors like $-\lambda^2/(q^2 - \lambda^2)$. In the limit as $\lambda \to \infty$, this factor becomes unity. The meaning of this factor is that it corresponds to a pseudo-photon propagator with pseudo-photon mass of λ.

Remark 2: Consider introducing an external em field in the vacuum polarization loop of a photon. The process is described in the following way. An electron of four momentum p_1 and spin σ_1 interacts with a positron of four momentum p_2 and spin σ_2, annihilating each other to give a photon which propagates and then again polarizes into an electron-positron pair appearing in the form of a loop in the Feynman diagram. An external field $A_\mu(x)$ is connected to this loop, ie, it interacts with the electron-positron pair after polarization of the photon.

The electron-positron pair again, after this interaction annihilate to produce a photon which propagates and finally polarize into an electron-positron pair of four momenta and spins (p_3, σ_3) and (p_4, σ_4) respectively. Using the Feynman rules, the amplitude for this overall correction to the scattering matrix is given by assuming k to be the four momentum of the external photon line

$$S(p_3, \sigma, p_4, \sigma_4 | p_1, \sigma_1, p_2, \sigma_2) =$$

$$\delta^4(p_1 + p_2 - p_3 - p_4).(\bar{v}(p_2, \sigma_2)\gamma^\mu u(p_1, \sigma_1)) D_{\mu\nu}(p_1 + p_2) A_\beta(k)$$

$$.\left(\int Tr[S(p_1 + p_2 - q + k)\gamma^\beta S(p_1 + p_2 - q)\gamma^\nu S(q)\gamma^\rho] d^4 q \right)$$

$$D_{\rho\alpha}(p_3 + p_4)\bar{v}(p_4, \sigma_4)\gamma^\alpha u(p_3, \sigma_3)$$

where $S(p) = (\gamma.p - m)^{-1}$ is the electron propagator while $D_{\mu\nu}(q) = \eta_{\mu\nu}/q^2$ is the photon propagator. This scattering amplitude correction can alternately be viewed as coming from a correction to the electron propagator $1/A$ coming from the external photon line. We can express the contribution of such multiple loops with external field lines by the corrected propagator $1/(A + B)$ where B depends on the external photon line $A_\mu(k)$. Thus, the corrected propagator can be expressed as

$$(A + B)^{-1} = A^{-1} - A^{-1}BA^{-1} + A^{-1}BA^{-1}BA^{-1} + ...$$

$$= A^{-1} + \sum_{n \geq 1} (-1)^n (BA^{-1})^n$$

For a desired scattering matrix, let X denote the desired propagator, then the external photon field $A_\mu(k)$ must be "controlled" so that

$$\| X - (A + B)^{-1} \|$$

is a minimum.

12.2 An optimization problem in electromagnetism

We consider here the problem of calculating the multipole em radiation fields E, H expressing them in terms of components tangential to the radial direction $\hat{r}$ and parallel to the radial direction. They satisfy the free space Maxwell equations

$$div E = 0, div H = 0, (\nabla^2 + k^2)E = 0, (\nabla^2 + k^2)H = 0$$

From these equations, it easily follows that

$$(\nabla^2 + k^2)(r.E) = 0, (\nabla^2 + k^2)(r.H) = 0$$

This suggests to us the possibility of decomposing em fields into radial an tangential components with each of these components satisfying the Helmholtz equation. To this end we denote by

$$L = r \times p = -ir \times \nabla$$

the angular momentum vector operator. Let us define

$$E_{lm} = f_l(r)LY_{lm}(\hat{r})$$

Then clearly since $r.L = 0$ or equivalently, $\hat{r}.L = 0, \hat{r} = r/|r|$, we have

$$r.E_{lm} = 0$$

ie E_{lm} is a purely tangential solution to the Helmholtz equation provided that $(\nabla^2 + k^2)E_{lm} = 0$ and $(\nabla^2 + k^2)(r.E_{lm}) = 0$. Note that these two equations guarantee that $divE_{lm} = 0$. Also since $r.E_{lm} = 0$, the latter equation is already guaranteed to be satisfied. Since L commutes with L^2, we have using

$$\nabla^2 = r^{-2}\frac{\partial}{\partial r}r^2\partial/\partial r - L^2/r^2$$

and

$$L^2Y_{lm}(\hat{r}) = l(l+1)Y_{lm}(\hat{r})$$

that

$$(\nabla^2 + k^2)E_{lm} = f_l''(r) + (2/r)f_l'(r) + (-l(l+1)/r^2 + k^2)f_l(r)$$

and for this to vanish, we must have

$$r^2 f_l''(r) + 2r f_l'(r) + (k^2r^2 - l(l+1))f_l(r) = 0$$

which has two linearly independent solutions $j_l(kr), h_l(kr)$ so that

$$f_l(r) = c(l,m)j_l(kr) + d(l,m)h_l(kr)$$

These linearly independent functions are called the modified Bessel functions and they can be expressed in terms of the usual Bessel functions by formulae of the form

$$j_l(x) = x^{-1/2}J_{l+1/2}(x)$$

Assuming these, we have that E_{lm} is a valid electric field. The corresponding magnetic field $\tilde{H}_{lm}$ is given by

$$\nabla \times E_{lm} = -j\omega\mu\tilde{H}_{lm}$$

or equivalently,

$$\tilde{H}_{lm} = (j/\omega\mu)\nabla \times (f_l(r)LY_{lm}(\hat{r}))$$

since the operator $\nabla\times$ commutes with ∇^2, it follows that since $(\nabla^2+k^2)E_{lm}=0$ that

$$(\nabla^2 + k^2)\tilde{H}_{lm} = 0$$

and it is clear that

$$r.\tilde{H}_{lm} \neq 0$$

In fact,

$$r.\tilde{H}_{lm} = (j/\omega\mu)r.(\nabla \times f_l(r)LY_{lm}(\hat{r}))$$
$$= (j/\omega\mu)(r \times \nabla).(f_l(r)LY_{lm}(\hat{r}))$$
$$= (-1/\omega\mu)L.(f_l(r)LY_{lm}(\hat{r})) = (-1/\omega\mu)f_l(r)L^2Y_{lm}(\hat{r})$$
$$= (-1/\omega\mu)l(l+1)f_l(r)Y_{lm}(\hat{r}) \neq 0$$

thus proving that $\tilde{H}_{lm}$ is a non-tangential solution to for the magnetic field. Likewise, we can start with another solution $g_l(r)$ to the equation

$$r^2g_l''(r) + 2rg_l'(r) + (k^2r^2 - l(l+1))g_l(r) = 0$$

and construct a tangential solution to the magnetic field as

$$H_{lm} = g_l(r)LY_{lm}(\hat{r})$$

with the corresponding non-tangential solution to the electric field given by

$$\nabla \times H_{lm} = j\omega\epsilon\tilde{E}_{lm}$$

or equivalently,

$$\tilde{E}_{lm} = (-j/\omega\epsilon)\nabla \times (g_l(r)LY_{lm}(\hat{r}))$$

Note that since $\nabla.\nabla \times F = 0$ for any vector field F, we must necessarily have

$$div\tilde{E}_{lm} = 0, div\tilde{H}_{lm} = 0$$

which means that the non-tangential components of the electric and magnetic field constructed above satisfy all the requirements for an electric and magnetic field. Superposing both the tangential and non-tangential components, we get the general radiation fields at a given frequency ω as

$$E(r) = \sum_{l,m}[f_{lm}(r)LY_{lm}(\hat{r}) - (j/\omega\epsilon)\nabla \times (g_{lm}(r)LY_{lm}(\hat{r}))]$$

$$H(r) = \sum_{l,m}[g_{lm}(r)LY_{lm}(\hat{r}) + (j/\omega\mu)\nabla \times (f_{lm}(r)Y_{lm}(\hat{r}))]$$

where

$$f_{lm}(r) = c_E(l,m)j_l(kr) + d_E(l,m)h_l(kr),$$
$$g_{lm}(r) = c_H(l,m)j_l(kr) + d_H(l,m)h_l(kr)$$

and the coefficients c_E, d_E, c_H, d_H are obtained by measuring the fields on the surface of a sphere and using orthogonality properties of the vector valued complex functions

$$LY_{lm}(\hat{r}), \nabla \times LY_{lm}(\hat{r})$$

on the unit sphere.

12.3 Design of quantum gates using non-Abelian gauge theories

The Yang-Mills equations for the $4N \times 1$ wave function $\psi(x)$ in the presence of mass terms, external electromagnetic fields $A_\mu(x)$ and non-Abelian gauge field terms $B_\mu^\alpha(x)\tau_\alpha$ are given by

$$[\gamma^\mu(i\partial_\mu + eA_\mu(x) + B_\mu^\alpha(x)\tau_\alpha - m]\psi(x) = 0$$

or more precisely

$$(i\gamma^\mu \otimes I_N)\partial_\mu\psi(x) + eA_\mu(x)(\gamma^\mu \otimes I_N)\psi(x)$$

$$+B_\mu^\alpha(x)(\gamma^\mu \otimes \tau_\alpha)\psi(x) - m\psi(x) = 0$$

Expressing this equation in Hamiltonian form, we get

$$i\partial_0\psi(x) = H(x)\psi(x)$$

where

$$H(x) = -i\gamma^0\gamma^r\partial_r - e(\gamma^0\gamma^\mu \otimes I_N)A_\mu(x)$$

$$-(\gamma^0\gamma^\mu \otimes \tau_\alpha)B_\mu^\alpha(x) + m\gamma^0 \otimes I_N$$

By regarding the potentials $A_\mu(x), B_\mu^\alpha(x)$ as control fields and replacing the wave function $\psi(x)$ by a unitary operator kernel $U_t(r, r') \in \mathbb{C}^{4N \times 4N}$ so that

$$\int U_t(r, r')^* U(r'', r')d^3r' = I_{4N}\delta^3(r - r'')$$

our aim is to get as close as possible after time T to a given unitary kernel $U_g(r, r')$ in the sense that

$$\int \| U_g(r, r') - U_T(r, r') \|^2 d^3r d^3r'$$

is a minimum. One can also try to introduce quantum noise processes in the sense of Hudson and Parthasarathy into the control fields to model the effects of noise on the designed gate. Specifically, this would involve replacing $A_\mu(t, r)dt$ by a classical field plus $c_{\mu\beta}^\alpha(r)d\Lambda_\alpha^\beta(t)$ and likewise $B_\mu^\alpha(t, r)dt$ by a classical field plus $d_{\alpha\sigma}^{\mu\rho}(r)d\Lambda_\rho^\sigma(t)$ where the fundamental noise processes $\Lambda_\beta^\alpha(t)$ satisfy the quantum Ito formula

$$d\Lambda_\beta^\alpha d\Lambda_\nu^\mu = \epsilon_\nu^\alpha d\Lambda_\beta^\mu$$

with ϵ_ν^μ assuming the value zero if either μ or ν is zero and otherwise assuming the value δ_ν^μ $(\mu, \nu = 0, 1, 2, ...)$.

12.4 Design of quantum gates using the Hudson-Parthasarathy quantum stochastic Schrodinger equation

$U(t)$ satisfies the HP equation

$$dU(t) = (-(iH + e^2 P)dt + eL_1 dA(t) + eL_2 dA(t)^* + e^2 Sd\Lambda(t))U(t)$$

e is a perturbation parameter introduced to show that the creation and annihilation process noise are small ie of $O(e)$ while the conservation process term is of order e^2 since $d\Lambda = dA^* dA/dt$ and finally the coefficient of the quantum Ito correction term Pdt is $O(e^2)$ since $P = (eL_2)^*(eL_2)$.

12.5 gravitational waves in a background curved metric

$$g_{\mu\nu}(x) = g^{(0)}_{\mu\nu}(x) + h_{\mu\nu}(x)$$

$$h_{\mu\nu} = \delta g_{\mu\nu}$$

is the metric of space-time.

$$\delta(\Gamma^\alpha_{\mu\nu}) = \delta(g^{\alpha\beta}\Gamma_{\beta\mu\nu}) =$$

$$(\delta g^{\alpha\beta})\Gamma^{(0)}_{\beta\mu\nu} + g^{(0)\alpha\beta}\delta\Gamma_{\beta\mu\nu}$$

$$= -g^{(0)\alpha\rho}g^{(0)\beta\sigma}h_{\rho\sigma}\Gamma^{(0)}_{\beta\mu\nu}$$

$$+(1/2)g^{(0)\alpha\beta}(h_{\beta\mu,\nu} + h_{\beta\nu,\mu} - h_{\mu\nu,\beta})$$

$$= g^{(0)\alpha\beta}(1/2(h_{\beta\mu,\nu} + h_{\beta\nu,\mu} - h_{\mu\nu,\beta}) - h_{\beta\sigma}\Gamma^{(0)\sigma}_{\mu\nu})$$

$$= (1/2)g^{(0)\alpha\beta}(h_{\beta\mu:\nu} + h_{\beta\nu:\mu} - h_{\mu\nu:\beta})$$

where the covariant derivative is taken w.r.t the unperturbed metric $g^{(0)}_{\mu\nu}$. Since the covariant derivative of the unperturbed metric is zero, assuming that raising and lowering of metric perturbations and their covariant derivatives are taken w.r.t. the unperturbed metric, we can also write this equation as

$$\delta(\Gamma^\alpha_{\mu\nu}) = (1/2)(h^\alpha_{\mu:\nu} + h^\alpha_{\nu:\mu} - h^{:\alpha}_{\mu\nu})$$

Now, a straightforward computation shows that the perturbation in the Ricci tensor is

$$\delta R_{\mu\nu} = \delta\Gamma^\alpha_{\mu\alpha:\nu} - \delta\Gamma^\alpha_{\mu\nu:\alpha}$$

which implies on substituting the above equation,

$$2\delta R_{\mu\nu} = (h^\alpha_{\mu:\alpha:\nu} + h^\alpha_{\alpha:\mu:\nu} - h^\alpha_{\mu:\alpha:\nu})$$

$$-(h^\alpha_{\mu:\nu:\alpha} + h^\alpha_{\nu:\mu:\alpha} - h^{:\alpha}_{\mu\nu:\alpha})$$

which simplifies to

$$2\delta R_{\mu\nu} = h^\alpha_{\alpha:\mu:\nu} - (h^\alpha_{\mu:\nu:\alpha} + h^\alpha_{\nu:\mu:\alpha} - h^{:\alpha}_{\mu\nu:\alpha})$$

or writing $h = h^\alpha_\alpha$, and $\Box h_{\mu\nu} = h^{:\alpha}_{\mu\nu:\alpha}$, we get

$$2\delta R_{\mu\nu} = \Box h_{\mu\nu} + h_{,\mu:\nu} - h^\alpha_{\mu:\nu:\alpha} - h^\alpha_{\nu:\mu:\alpha}$$

Suppose we use harmonic coordinates, ie $g^{\mu\nu}\Gamma^\alpha_{\mu\nu} = 0$. The perturbed form of this is

$$\delta(g^{\mu\nu}\Gamma^\alpha_{\mu\nu}) = 0$$

or equivalently,

$$g^{(0)\mu\nu}(h^\alpha_{\mu:\nu} + h^\alpha_{\nu:\mu} - h^{:\alpha}_{\mu\nu}) + \delta g^{\mu\nu}\Gamma^{(0)\alpha}_{\mu\nu} = 0$$

Instead, we modifiy our coordinate condition by removing the last term to get a modified version of perturbed harmonic coordinates:

$$2h^\mu_{\alpha:\mu} - h_{,\alpha} = 0$$

Then noting that

$$h_{,\mu:\nu} = h_{,\nu:\mu}$$

we get

$$h^\alpha_{\mu:\alpha:\nu} = h^\alpha_{\nu:\alpha:\mu}$$

our perturbed field equations

$$\delta R_{\mu\nu} = 0$$

in this perturbed coordinate system assume the form

$$0 = \Box h_{\mu\nu} + h_{,\mu:\nu} - h^\alpha_{\mu:\nu:\alpha} - h^\alpha_{\nu:\mu:\alpha} =$$

$$= \Box h_{\mu\nu} + (h^\alpha_{\mu:\alpha:\nu} - h^\alpha_{\mu:\nu:\alpha})$$

$$+(h^\alpha_{\nu:\alpha:\mu} - h^\alpha_{\nu:\mu:\alpha}) = 0$$

The last two brackets can be expressed in terms of the Riemann curvature tensor of the unperturbed metric and the perturbed metric coefficients $h_{\mu\nu}$ not involving their partial derivatives.

12.6 Topics for a short course on electromagnetic field propagation at high frequencies

To transmit and detect high frequency electromagnetic waves, our transmitter and receiver antennae must be of very small size, ie of the Angstrom scale where quantum mechanical effects become dominant. This is because, the wavelength of an em wave is inversely proportional to frequency ($\lambda = c/\nu$ or equivalently, $k = \omega/c, \lambda = 2\pi/k$). It is therefore impossible to discuss a theory of electromagnetic wave propagation at high frequencies without introducing fundamental quantum mechanical principles such as second quantization, Feynman path integrals for fields, interaction between the electromagnetic field, matter and the gravitational field at the quantum level. Keeping this in mind, we discuss the following topics for a short course on high frequency communication.

[1] Quantization of the em field in terms of creation and annihilation operator fields.

[2] Canonical commutation relations.

[3] Quantization with constraints–The Dirac bracket.

[4] Creation, annihilation and conservation processes in the sense of Hudson and Parthasarathy. Quantum Ito's formula, Derivation of the GKSL equation when the bath is in a coherent state or in a superposition of coherent states.

Remark: In the quantum theory of fields, we introduce creation and annihilation operator fields in three momentum space. However these operator fields are time independent. To represent quantum noise, however, we need to make these creation and annihilation fields time dependent in such a way that these time dependent processes behave like classical stochastic processes in certain states of the bath. The Hudson-Parthasarathy quantum stochastic calculus precisely achieves this. Whenever we have an operator field like $a(u), a(u)^*, \lambda(H)$ acting in the Boson Fock space of $\mathcal{H}$ where u is a vector in the Hilbert space $\mathcal{H}$ and H is an operator in the Hilbert space $\mathcal{H}$, we can introduce time dependence by replacing u with $\chi_{[0,t]}u$ and H with $\chi_{[0,t]}H$ provided that $\mathcal{H}$ has the form $L^2(\mathbb{R}_+) \otimes \mathcal{H}_0$ and $\chi_{[0,t]}$ denotes multiplication by the indicator function of $[0,t]$ in $L^2(\mathbb{R}_+)$. We must necessarily assume that $\chi_{[0,t]}$ commutes with H. This will happen when for example, H acts in $\mathcal{H}_0$. When time dependence of the creation, annihilation and conservation fields is thus introduced, we obtain quantum stochastic processes that satisfy quantum Ito's formula owing to the non-commutativity of these operator fields. Classical Ito's formula for Brownian motion and Poisson processes follow as special cases. Since non-commutativity of observables implies Heisenberg uncertainty, ie, impossibility of simultaneously measuring these observables, it follows that Ito's formula can be traced to the Heisenberg uncertainty principle.

[5] Dirac's equation for the electron in a noisy electromagnetic field.

[6] Approximate solution of the GKSL equation using time dependent perturbation theory on system space.

[7] Approximate solution of the Hudson-Parthasarathy noisy Schrodinger equation using time dependent perturbation theory on system$\otimes$ bath space.

[8] Quantum entropy pumped by a noisy em field into an atomic system—approximate expressions.

[9] Filtering in quantum mechanics using V.P.Belavkin's theory. The notion of non-demolition measurements associated to a given Hudson-Parthasarathy noisy Schrodinger evolution. Examples of non-demolition measurements using quadrature processes, photon counting processes and mixture of quadrature and photon counting processes.

[10] Quantum control after filtering with the objective of (a) reducing GKSL noise and (b) state tracking.

[11] Comparison of quantum filtering and control with classical filtering and control.

[12] Interaction of the gravitational field with the em field–the classical theory based on the Einstein-Maxwell equations.

[13] Interaction of the gravitational field with the em field–the quantum theory based on approximate linearization of the Einstein field equations.

[14] Gravitational waves in a flat and curved background metric.

[15] Proof that gravitons are spin 2 particles. Proof based on choosing a harmonic coordinate system and determining transformation properties of the tensor components of the gravitational wave amplitudes under rotations of the coordinate system around an axis.

[16] Energy-momentum tensor of the gravitational field.

[17] Noether's theorem on conserved charges for a classical field theory when the Lagrangian density is invariant under an infinitesimal Lie algebra of field transformations.

[18] How to draw Feynman diagrams for scattering, absorption and emission processes involving electrons, positrons, photons, mesons and gravitons. Derivation of the Feynman rules using operator theory, ie using canonical commutation rules for Bosons and canonical anticommutation rules for Fermions.

[19] Path integrals for fields with application to Yang-Mills quantization. Derivation of the invariance of the path integral under different gauge fixing conditions whenever the action and path measure are gauge invariant.

[20] The Galilean group and its projective unitary representations. Derivation of the energy, position, momentum, angular momentum and velocity operators from the multipliers of the Galilean group.

[21] Application of the theory of induced representations for estimating the rotation, translation and uniform velocity of motion of an antenna from noisy em pattern measurements.

The initial current density field is $J(r)$ at frequency ω. Let

$$G(|r|) = (\mu/4\pi|r|)exp(-j\omega|r|/c)$$

Then the initial magnetic vector potential is

$$A(r) = \int G(|r - r'|)J(r')d^3r'$$

After rotation and translation, the current density is

$$\tilde{J}(r) = RJ(R^{-1}(r - a)), R \in SO(3), a \in \mathbb{R}^3$$

The corresponding magnetic vector potential is

$$\tilde{A}(r) = \int G(|r - r'|)\tilde{J}(r')d^3r' + w(r)$$

where $w(r)$ is a noise field. This evaluates to

$$\tilde{A}(r) = \int G(|r - Rr' - a|)RJ(r')d^3r' + w(r)$$

$$= R\int G(|R^{-1}(r - a) - r'|)J(r')d^3r' + w(r)$$

$$= RA(R^{-1}(r - a)) + w(r)$$

since $detR = 1$. The initial and final magnetic fields are respectively

$$B(r) = \nabla \times A(r),$$

$$\tilde{B}(r) = \nabla \times \tilde{A}(r) = ((R^{-1}\nabla) \times (RA))(R^{-1}(r - a)) + \nabla \times w(r)$$

Likewise, the electric field can be transformed: The initial electric field is

$$E(r) = -\nabla V(r) - j\omega A(r)$$

where

$$V(r) = (jc^2/\omega)divA(r)$$

so that

$$E(r) = (-jc^2/\omega)\nabla(divA(r)) - j\omega A(r)$$

This can equivalently be expressed using the Maxwell equation

$$\nabla \times B/\mu = J + j\omega\epsilon E, B = \nabla \times A$$

as

$$E(r) = (-j/\omega\epsilon)(\nabla \times (\nabla \times A(r))/\mu - J(r))$$

The equivalence of the two expressions follows from the wave equation for $A(r)$:

$$(\nabla^2 + \omega^2/c^2)A(r) = -\mu J(r), c^2 = 1/\epsilon\mu$$

It is however more convenient to work using special relativistic tensors:

$$F_{\mu\nu} = A_{\nu,\mu} - A_{\mu,\nu}$$

where we use time domain expressions. We get on applying the Lorentz gauge condition

$$A^{\mu}_{,\mu} = 0$$

to the Maxwell equations

$$F^{\mu\nu}_{,\nu} = -\mu_0 J^{\mu}$$

that

$$\Box A^\mu(x) = \mu_0 J^\mu(x), \Box = \partial_\alpha \partial^\alpha$$

with solution

$$A^\mu(x) = \int J^\mu(x') G(x - x') d^4 x'$$

where

$$G(x - x') = \mu_0 \delta((x - x')^2) = (\mu_0/2\pi)\delta((t - t')^2 - |r - r'|^2)$$

$$= \mu_0 \delta(t - t' - |r - r'|)/4\pi|r - r'|$$

assuming $t > t'$. Now suppose, we apply a Poincare transformation, ie, a Lorentz transformation L along with a space-time translation a to the four current density $J^\mu = J$. The transformed four current density is then

$$\tilde{J}(x) = LJ(L^{-1}(x - a))$$

where

$$(LJ)^\mu = L^\mu_\nu J^\nu$$

The problem is to estimate the Poincare group element (L, a) where $a = (a^\mu)$. We get for the initial andtransformed em four potentials

$$A(x) = \int G(x - x') J(x') d^4 x'$$

$$\tilde{A}(x) = \int G(x - x') LJ(L^{-1}(x' - a)) d^4 x'$$

$$= L \int G(L^{-1}(x - a) - x') J(x') d^4 x' = LA(L^{-1}(x - a))$$

where we have used the fact that L preserves the space-time Minkowski metric $(x - x')^2$, ie, $(L(x - x'))^2 = (x - x')^2$.

Chapter 13

Quantum gravity with photon interactions, cavity resonators with inhomogeneities, classical and quantum optimal control of fields

13.1 Quantum control of the HP-Schrodinger equation by state feedback

$$dU(t) = (-(iH + P)dt + L_1 dA(t) + L_2 dA(t)^* + Sd\Lambda(t))U(t)$$

$$j_t(X) = U(t)^* XU(t), X \in \mathcal{L}(\mathfrak{h})$$

$$dj_t(X) = j_t(\theta_0(X))dt + j_t(\theta_1(X))dA(t) + j_t(\theta_2(X))dA(t)^* + j_t(\theta_3(X))d\Lambda(t)$$

We wish $j_t(X)$ to track the noiseless trajectory $X_d(t) \in \mathcal{L}(\mathfrak{h})$. Assume non-demoltion measurements $Y_o(t) = U(t)^* Y_i(t)U(t)$ are made and the Belavkin filter for

$$\pi_t(X) = \mathbb{E}(j_t(X)|\eta_t), \eta_t = \sigma(Y_o(s) : s \leq t)$$

has been constructed as

$$d\pi_t(X) = F_t(X)dt + G_t(X)dY_o(t)$$

trajectory estimation error $X_d(t) - \pi_t(X)$ is given as feedback to the state equations in the form

$$dj_t(X) = j_t(\theta_0(X))dt + j_t(\theta_1(X))dA(t) + j_t(\theta_2(X))dA(t)^* + j_t(\theta_3(X))d\Lambda(t)$$

$$+K(t)(X_d(t) - \pi_t(X))dt$$

Alternately, if we assume that X_d follows a noiseless trajectory, ie, it evolves according to the equation

$$X_d(t) = j_t^{(0)}(X_d) = U_0(t)^* X_d U_0(t), U_0(t) = exp(-itH_0)$$

then we can give as feedback the estimation error

$$E(t) = X_d(t) - Tr_2(\pi_t(X)(I \otimes |\phi(u) > < \phi(u)|))$$

which is a system observable. This error feedback is given to the above state equations, ie, to the Evans-Hudson flow resulting in the dynamics

$$dj_t(X) = j_t(\theta_0(X))dt + j_t(\theta_1(X))dA(t) + j_t(\theta_2(X))dA(t)^* + j_t(\theta_3(X))d\Lambda(t)$$

$$+K(t)E(t)$$

Alternately, we can incorporate this error feedback in the original HP equation as

$$dU(t) = (-(iH + P + K(t)E(t))dt + L_1 dA(t) + L_2 dA(t)^* + Sd\Lambda(t))U(t)$$

resulting in the Evans-Hudson flow the same as above but with $\theta_0(X)$ replaced by $\theta_0(X) + iK(t)[E(t), X]$. Luc-Bouten's method of control is slightly different. Here, we choose a system observable Z and give an infinitesimal control unitary

$$U_c(t, t + dt) = U(t + dt)^* exp(-iZdY_i(t))U(t + dt) = exp(iZ(t + dt)dY_o(t))$$

Note that

$$Y_o(t) = U(t + dt)^* Y_i(t)U(t + dt), Y_o(t + dt) = U(t + dt)^* Y_i(t + dt)U(t + dt)$$

and hence, forming the difference, we get

$$dY_o(t) = U(t + dt)^* dY_i(t)U(t + dt)$$

Also, Z commutes with $dY_i(t)$ and hence $Z(t + dt) = U(t + dt)^* ZU(t + dt)$ commutes with $U(t + dt)^* dY_o(t)U(t + dt)$. Let $\rho_c(t)$ denote the controlled state at time t, ie after applying the Belavkin filter and control upto time t to the HP evolved state. Then, we apply the Belavkin filter from t to $t + dt$, via

$$\rho_B(t + dt) = \rho_c(t) + \delta\rho_B(t)$$

where

$$\delta\rho_B(t) = L_t^*(\rho_c(t))dt + (M_t\rho_c(t) + \rho_c(t)M_t^* - Tr(M_t + M_t^*)\rho_c(t))\rho_c(t))(dY_o(t)$$
$$-Tr(\rho_c(t)(M_t + M_t^*))dt)$$

Finally, the filtered and controlled state at time $t + dt$ is given by

$$\rho_c(t + dt) = \rho_c(t) + \delta\rho_c(t) =$$

$$U_c(t, t+dt)(\rho_c(t) + \delta\rho_B(t)).U_c(t, t+dt)^*$$
$$= exp(iZ(t+dt)dY_o(t))(\rho_c(t) + \delta\rho_B(t)).exp(-iZ(t+dt)dY_o(t))$$

Comparison with classical filter/state observer and controller:

$$X'(t) = \psi(t, X(t)) + G(t, X(t))(\tau_c(t) + W(t))$$

$$\tau_c(t) = G(t, \hat{X}(t))^{-1}(K(t)(X_d(t) - \hat{X}(t)) + X_d'(t) - \psi(t, \hat{X}(t)))$$

$$X_d'(t) = \psi(t, X_d(t)) + G(t, X_d(t))\tau_d(t)$$

$$\hat{X}'(t) = \psi(t, \hat{X}(t)) + L(t)(dZ(t) - h(t, \hat{X}(t))dt)$$

$$dZ(t) = h(t, X(t)dt + \sigma_V dV(t)$$

$$e(t) = X_d(t) - X(t), f(t) = X(t) - \hat{X}(t)$$

Then,

$$e'(t) = \psi(t, X_d(t)) - \psi(t, X(t)) + G(t, X_d(t))\tau_d(t) - G(t, X(t))\tau_c(t) - G(t, X(t))W(t)$$

$$= \psi(t, X_d(t)) - \psi(t, X(t)) + G(t, X_d(t))\tau_d(t) - G(t, X(t))\tau_c(t) - G(t, X(t))W(t)$$

$$= \psi(t, X_d(t)) - \psi(t, X(t)) + G(t, X_d(t))\tau_d(t) - G(t, X(t))G(t, \hat{X}(t))^{-1}K(t)(e(t) + f(t))$$

$$-G(t, X(t))G(t, \hat{X}(t))^{-1}(X_d(t) - \psi(t, \hat{X}(t))) - G(t, X(t))W(t)$$

On linearizing this equation around $\hat{X}(t)$, this equation appears in the form

$$e'(t) = A_1(t)e(t) + A_2(t)f(t) + A_3(t)W(t)$$

where $A_1(t), A_2(t), A_3(t)$ are functions of $t, \hat{X}(t)$ only.

13.2 Some applications of Poisson processes

Change of measure theorem of Girsanov for Poisson Martingales
Consider the process

$$X(t) = \sum_{a=1}^{p} c(a)N_a(t)$$

where $N_1, ..., N_p$ are p independent Poisson processes with rates $\lambda_1, ..., \lambda_p$ respectively. The process

$$Y(t) = X(t) - \sum_{a=1}^{p} c(a)\lambda_a t = \sum_{a=1}^{p} c(a)(N_a(t) - \lambda_a t)$$

is a Martingale. Let $\eta(t)$ be a finite variation adapted process so that $exp(Y(t) + \eta(t))$ is a Martingale. Then, we require to compute $\eta(t)$. We have with

$$Z(t) = exp(Y(t) + \eta(t)),$$

$$dZ(t) = d(exp(Y(t) + \eta(t))) = d(exp(X(t) + \eta(t) - \sum_a c(a)\lambda_a t)) =$$

$$Z(t)[\sum_a (exp(c(a)) - 1)dN_a(t) + d\eta(t) - \sum_a c(a)\lambda_a dt]$$

$$= Z(t)[\sum_a (exp(c(a)) - 1)(dN_a(t) - \lambda_a dt) + d\eta(t) + \sum_a \lambda_a (exp(c(a)) - 1 - c(a))dt)$$

So for $Z(t)$ to be a Martingale, we require that

$$\eta(t) = -\sum_a \lambda_a (exp(c(a)) - 1 - c(a))t$$

Thus the exponential Martingale associated with $X(t)$ is given by

$$Z(t) = exp(X(t) - \sum_a \lambda_a (exp(c(a)) - 1 - c(a))t)$$

and, in fact, we have

$$dZ(t) = Z(t). \sum_a (exp(c(a)) - 1)(dN_a(t) - \lambda_a dt)$$

We next define a measure Q so that

$$dQ_t/dP_t = Z(t), t \geq 0$$

where Q_t, P_t respectively are restrictions of Q and P to $\mathcal{F}_t$. This is a consistent definition since if $t > s$ and $B \in \mathcal{F}_s$, then

$$Q_t(B) = \int_B Z(t)dP_t = \int_B Z(t)dP = \int_B Z(s)dP = \int_B dQ_s = Q_s(B)$$

where the martingale property of Z w.r.t P has been used. Now, we wish to determine an adapted process $f(t)$ of finite variation such that $U(t) = Y(t) + f(t)$ is a Q-martingale. Note that $Y(t)$ is a P-martingale. For this, we must have that

$$\mathbb{E}_P[d(Z(t)U(t))|\mathcal{F}_t] = 0$$

ie, the process ZU is a P-Martingale, for then, this would imply that for any $\mathcal{F}_t$-measurable; r.v. V, we have

$$\mathbb{E}_P[d(Z(t)U(t)).V] = 0$$

which would in turn imply that

$$\mathbb{E}_P[(Z(t + dt)U(t + dt) - Z(t)U(t))V] = 0$$

or equivalently,

$$\mathbb{E}_Q[(U(t + dt) - U(t))V] = 0$$

ie $U(t)$ is a Q-martingale. Now application of Ito's formula for Poisson processes gives

$$d(ZU)) = ZdU + UdZ + dU.dZ = Z(dY + df) + UdZ + dYdZ + df.dZ$$

Since $\int UdZ$ and $\int ZdY$ are martingales, we therefore require that $\int (Zdf + dYdZ + df dZ)$ be a Martingale. But,

$$Zdf + dYdZ = Zdf + Z.\sum_a exp(c(a)) - 1)(dN_a - \lambda_a dt).\sum_b c(b)(dN_b - \lambda_b dt)$$

$$= Z[df + \sum_a (exp(c(a)) - 1)c(a)dN_a]$$

Writing

$$f(t) = \sum_a d(a)N_a(t)$$

we get

$$df.dZ = Z.\sum_a (exp(c(a)) - 1)d(a)dN_a(t)$$

and hence, we deduce that U is a Q-martingale provided that

$$f(t) = -\sum_a (exp(c(a)) - 1)(c(a) + d(a))dN_a(t)$$

This means that we should have

$$d(a) + (exp(c(a)) - 1)(c(a) + d(a)) = 0$$

or equivalently,

$$d(a) = -c(a)(1 - exp(-c(a)))$$

In other words,

$$U(t) = Y(t) - \sum_a c(a)(1 - exp(-c(a)))N_a(t)$$

is a Q-martingale.

A Feynman-Kac formula for Poisson processes.
Then define
Let as before $X(t) = \sum_a c(a)N_a(t)$.

$$u(t,x) = \mathbb{E}[exp(\int_0^t V(X(s))ds)\phi(X(t))|X(0) = x]$$

We get using the Markov property of $X(.)$ that

$$u(t + dt, x) = (1 + V(x)dt)\mathbb{E}(u(t, X(dt))|X(0) = x)$$

$$= (1 + V(x)dt)(u(t,x) + \sum_a (u(t, x + c(a)) - u(t,x))\lambda_a dt)$$

so that u satisfies the equation

$$u_{,t}(t,x) = V(x)u(t,x) + \sum_a \lambda_a(u(t,x+c(a)) - u(t,x)), u(0,x) = \phi(x)$$

Replacing t by $-it$ by defining

$$\psi(t,x) = u(-it,x)$$

we get

$$i\psi_{,t}(t,x) = V(x)\psi(t,x) + \sum_a \lambda_a(\psi(t,x+c(a)) - \psi(t,x)), \psi(0,x) = \phi(x)$$

This equation has the following physical interpretation. At time $t = 0$, $\phi(x)$ is the wave function of a quantum particle. The amplitude of the particle to go from $x + c(a)$ to x in time dt is given by $-i\lambda_a dt, a = 1, 2, ..., p$ in the absence of an external potential. In the presence of an external potential $V(x)$, the amplitude of the particle to go from $x + c(a)$ to x in time dt is given by $-i\lambda_a dt$ and the amplitude to stay at x in time dt is given by $1 + i(\sum_a \lambda_a dt - V(x))dt$. In other words, this version of the Feynman path integral assumes that quantum transitions take place by discrete jumps rather than continuous motion.

Suppose that an electron moves in a one dimensional crystal. At any time t, there is an amplitude $c_t(n)$ for the electron to be at the site $n\Delta$ at time t. Transitions can take place only between neighbouring sites. If there is an amplitude $-iadt$ for the electron to make a transition from $(n+1)\Delta$ to $n\Delta$ and the same for the transition from $(n-1)\Delta$ to $n\Delta$, then quantum mechanics gives us the equation

$$c_{t+dt}(n) = c_t(n)(1 - i\lambda dt) - c_t(n-1)iadt - c_t(n+1)iadt$$

or

$$idc_t(n)/dt = \lambda c_t(n) + a(c_t(n+1) + c_t(n-1))$$

where $1 - i\lambda dt$ is the amplitude for no transition in time dt. We make the approximation,

$$c_t(n+1) + c_t(n-1) - 2c_t(n) \approx \Delta^2 c_t''(x), x = n\Delta$$

and then we get Schrodinger's equation

$$idc_t(x)/dt = a\Delta^2 c_t''(x) + (\lambda + 2a)c_t(x)$$

a should be chosen to be negative. Then a has the interpretation of being $-h^2/8\pi^2 m$ while $\lambda + 2a$ has the interpretation of being $V(x)$ the external potential field in which the electron moves. We can obtain the solution to the above equation using the Feynman path integral based on Poisson processes or equivalently, birth-death processes.

13.3   A problem in optimal control

The state equations are

$$dX(t) = AX(t)dt + Cu(t)dt + GdW(t)$$

where

$$X(t) \in \mathbb{R}^n, A \in \mathbb{R}^{n \times n}, u(t) \in \mathbb{R}^p, C \in \mathbb{R}^{n \times p}, G \in \mathbb{R}^{n \times d}, W(t) \in \mathbb{R}^d$$

with $W(.)$ being vector valued standard Brownian motion. The control input $u(t)$ is restricted to be of instantaneous feedback type, ie, of the form $u(t) = \chi_t(X(t))$ where $\chi_t : \mathbb{R}^n \to \mathbb{R}^p$ is a non-random function. The aim is to determine this control input over the time range $[0, T]$ so that

$$(1/2)\mathbb{E} \int_0^T (X(t)^T Q_1 X(t) + u(t)^T Q_2 u(t))dt$$

is a minimum. We already know from the stochastic Bellman-Hamilton-Jacobi dynamic (SBHJ) programming theory that if we define

$$V(t, X(t)) = min_{u(s), s \in [t,T]} \mathbb{E}[\int_t^T (X(s)^T Q_1 X(s) + u(s)^T Q_2 u(s))ds | X(t)]$$

then $V(t, x)$ satisfies the SBHJ equation

$$V_{,t}(t, x) + min_u(K_t(u)V(t, x) + (1/2)(x^T Q_1 x + u^T Q_2 u))$$

where $K_t(u)$ is the generator of the Markov process $X(t)$. It is given by

$$K_t(u) = (Ax + Cu)^T \nabla_x + (1/2)Tr(GG^T \nabla_x \nabla_x^T)$$

Thus, our SBHJ equation is

$$V_{,t}(t, x) + min_u((x^T A^T \nabla_x V(t, x) + u^T C^T \nabla_x V(t, x)$$

$$+(1/2)(x^T Q_1 x + u^T Q_2 u) + (1/2)Tr(GG^T \nabla_x \nabla_x^T V(t, x)))$$

The minimization is easily carried out and it gives the optimal value of $u = \chi_t(x)$ as

$$u = -Q_2^{-1} C^T \nabla_x V(t, x) = \chi(x)$$

Substituting, we get the SBHJ equation in the form

$$V_{,t}(t, x) + x^T A^T \nabla_x V(t, x) + (1/2)(-\nabla_x V(t, x))^T C Q_2^{-1} C^T \nabla_x V(t, x) + x^T Q_1 x$$

$$+Tr(GG^T \nabla_x \nabla_x^T V(t, x))) = 0$$

We rearrange this equation so that it comprises of three parts. The first part is linear in V and does not involve noise terms, the second part is nonlinear in V and again does not involve noise terms and finally, the third part is linear in V but involves noise terms. The nonlinear part is assumed to be of $O(\delta)$ and the noise part is assumed to be of of $O(\delta^2)$, where δ is a small perturbation parameter:

$$(V_{,t} + x^T A^T \nabla_x V(t,x) + x^T Q_1 x/2) - (\delta/2)(\nabla_x V(t,x))^T C Q_2^{-1} C^T \nabla_x V(t,x)$$

$$+ (\delta^2/2) Tr(GG^T \nabla_x \nabla_x^T V(t,x)) = 0$$

We solve this equation approximately upto $O(\delta^2)$ using perturbation theory:

$$V(t,x) = V_0(t,x) + \delta.V_1(t,x) + \delta^2.V_2(t,x) + O(\delta^3)$$

Substituting this and equating coefficients of $\delta^m, m = 0, 1, 2$ successively gives us

$$V_{0,t}(t,x) + x^T A^T \nabla_x V_0(t,x) + x^T Q_1 x/2 = 0$$

$$V_{1,t}(t,x) + x^T A^T \nabla_x V_1(t,x) = (\nabla_x V_0(t,x))^T C Q_2^{-1} C^T \nabla_x V_0(t,x),$$

$$V_{2,t}(t,x) + x^T A^T \nabla_x V_2(t,x) = \nabla_x V_1(t,x)^T C Q_2^{-1} C^T \nabla_x V_0(t,x)$$

$$- Tr(GG^T \nabla_x \nabla_x^T V_0(t,x))$$

It is clear that the boundary condition $V(T,x) = 0$ gives us

$$V_0(T,x) = V_1(T,x) = V_2(T,x) = 0$$

and hence,

$$V_0(t,x) = - \int_t^T exp((s-t)x^T A^T \nabla_x) x^T Q_1 x ds/2$$

$$V_1(t,x) = - \int_t^T exp((s-t)x^T A^T \nabla_x)(\nabla_x V_0(s,x)^T C Q_2^{-1} C^T \nabla_x V_0(s,x) ds$$

$$V_2(t,x) = - \int_t^T exp((s-t)x^T A^T \nabla_x)(\nabla_x V_1(s,x)^T C Q_2^{-1} C^T \nabla_x V_0(s,x)$$

$$- Tr(GG^T \nabla_x \nabla_x^T V_0(s,x))) ds$$

13.4   Interaction between photons and gravitons

The Einstein tensor is

$$G^{\mu\nu} = R^{\mu\nu} - (1/2)Rg^{\mu\nu}$$

We write

$$G^{\mu\nu} = G^{(1)\mu\nu} + G^{(2)\mu\nu}$$

where $G^{(1)\mu\nu}$ is linear in the metric perturbations $h_{\mu\nu}(x)$ and $G^{(2)\mu\nu}$ consists of quadratic and higher order terms in the metric perturbations. It is easy to see that

$$G^{(1)\mu\nu}_{,\nu} = 0$$

and therefore, the Einstein field equations

$$G^{\mu\nu} = -8\pi G T^{\mu\nu}$$

which can also be expressed as

$$G^{(1)\mu\nu)} = -8\pi G(T^{\mu\nu} - G^{(2)\mu\nu}/8\pi G)$$

imply that

$$(T^{\mu\nu} - G^{(2)\mu\nu}/8\pi G)_{,\nu} = 0$$

and this is a conservation law. Since $T^{\mu\nu}$ is the energy-momentum tensor of the matter and radiation field, we can thus interpret $\tau^{\mu\nu} = -G^{(2)\mu\nu}/8\pi G$ as the energy-momentum pseudo-tensor of the gravitational field. In what follows we first show that

$$G^{(1)\mu\nu}_{,\nu} = 0,$$

then calculate $\tau^{\mu\nu} = -G^{(2)\mu\nu}/8\pi G$ upto quadratic orders in the $h_{\mu\nu}$ and their partial derivatives, then we evaluate the energy of the gravitational field, namely,

$$H_G = \int \tau^{00} d^3 r$$

in terms of the gravitational field creation and annihilation operators $d(K,\sigma)^*, d(K,\sigma)$ upto second order where in view of the plane wave expansion

$$h_{\mu\nu}(x) = \int [e_{\mu\nu}(K,\sigma)d(K,\sigma)exp(-ik.x) + \bar{e}_{\mu\nu}(K,\sigma)d(K,\sigma)^*exp(ik.x)]d^3 K$$

The form of H_G is given by

$$H_G = \int C(K,\sigma,\sigma')d(K,\sigma)d^*(K,\sigma')d^3 K$$

We then evaluate the interaction Hamiltonian $H_I(t)$ between the electromangnetic field and the gravitational field using the energy-momentum tensor of the electromagnetic field

$$S^{\mu\nu} = (-1/4)F_{\alpha\beta}F^{\alpha\beta}g^{\mu\nu} + F^{\mu\alpha}F^{\nu}_{\alpha}$$

Then Interaction energy of the gravitational field with the em field is therefore given by the spatial integral of the $(00)^{th}$ component of $S^{\mu\nu}$ that contains terms linear in $h_{\mu\nu}$. It is given by

$$\int [(-1/4)F_{\mu\nu}F_{\alpha\beta}\delta(g^{mu\alpha}g^{\nu\beta}\sqrt{-g}g^{00})$$

$$+\delta(g^{0\mu}g^{\alpha\beta}g^{0\rho}\sqrt{-g})F_{\mu\beta}F_{0\alpha}]d^3 r$$

In this expression we note that

$$g_{\mu\nu} = \eta_{\mu\nu} + \delta g_{\mu\nu}(x), \delta g_{\mu\nu}(x) = h_{\mu\nu}(x)$$

$$h^{\alpha}_{\mu} = \eta_{\alpha\beta}h_{\beta\mu}, h^{\alpha\beta} = \eta^{\alpha\mu}\eta^{\beta\nu}h_{\mu\nu},$$

$$g^{\mu\nu} = \eta_{\mu\nu} - \eta_{\mu\alpha}\eta_{\nu\beta}h_{\alpha\beta} + O(h^2)$$
$$= \eta_{\mu\nu} - h^{\mu\nu} + O(h^2),$$
$$g = -(1+h), \sqrt{-g} = 1 + h/2, h = h^{\mu}_{\mu} = \eta_{\mu\nu}h_{\mu\nu}$$

where $O(h^2)$ terms have been neglected. Thus,

$$\delta g^{\mu\nu} = -\eta^{\mu\alpha}\eta^{\nu\beta}h_{\alpha\beta} = -h^{\mu\nu}$$

The free gravitational field in the absence of gravitational interactions, satisfies the wave equation

$$\Box h_{\mu\nu}(x) = 0$$

provided that we retain only terms linear in the $h_{\mu\nu}$ in the Einstein field equations and further assume harmonic coordinates, ie,

$$h^{\mu}_{\nu,\mu} - h_{,\nu}/2 = 0$$

The solution for the free gravitational field is thus expandable as above as a superposition of plane waves with the coefficient functions $e_{\mu\nu}(K, \sigma)$ satisfying the coordinate condition

$$e^{\mu}_{\nu}k_{\mu} - e^{\alpha}_{\alpha}k_{\nu}/2 = 0$$

These are 4 constraints on the ten coefficients $e_{\mu\nu}$ and hence, we have just six degrees of freedom. Actually, these reduce further to five if we ignore an arbitrary scaling factor. That is why a graviton is a spin two particle ($l = 2$ implies $2l + 1 = 5$). Now the energy of the gravitational field is computed as above. Likewise, the interaction energy between the electromagnetic field and the gravitational field upto linear orders in the $h_{\mu\nu}$ can be obtained from the plane wave expansion for the free gravitational field discussed above and the plane wave expansion of the free electromagnetic field:

$$\Box A_{\mu}(x) = 0$$

gives

$$A_{\mu}(x) = \int (e_{\mu}(K, s)a(K, s)exp(-ik.x) + \bar{e}_{\mu}(K, s)a(K, s)^* exp(ik.x))d^3K$$

where $s = 1, 2$, ie, there are only two degrees of freedom for the polarization of the em field. The first comes from the Lorentz gauge condition

$$A^{\mu}_{,\mu} = 0$$

which results in

$$e^{\mu}(K, s)k_{\mu} = 0$$

and the second from the fact that some part of the electromagnetic field is a matter field. This can be seen more clearly from the Coulomb gauge $divA = A^r_{,r} = 0$ which results in $\nabla^2 A^0 = -\mu J^0$ implying that A^0 is a pure matter field.

Note that since both the weak gravitational field and the em field satisfy the wave equation, it follows that both photons and gravitons travel at the speed of light, ie $k^0 = |K|$ for both of them. The interaction Hamiltonian between the gravitational field and the electromagnetic field is therefore approximately given by an expression of the form

$$H_{GEM}(t) = \int [C_1(K, K', \sigma, \sigma', s))a(-K - K', s)d(K, \sigma)d(K', \sigma')+$$

$$C_2(K, K', \sigma, \sigma', s)a(K - K', s)d(K, \sigma)^*d(K', \sigma')$$

$$+C_3(K, K', \sigma\sigma', s)a(K + K', s)^*d(K, \sigma)d(K', \sigma')$$

$$+C_4(K, K', \sigma, \sigma', s)a(K - K', s)^*d(K, \sigma)d(K, \sigma)^*d(K', \sigma)]d^3K$$

$$+c.c.$$

where *c.c.* denotes the complex adjoint of the previous terms. The commutation relations are the usual Bosonic relations:

$$[a(K, s), a(K', s')^*] = \delta^3(K - K')\delta_{s,s'},$$

$$[a(K, s), a(K', s')] = 0,$$

$$[d(K, \sigma), d(K', \sigma')^*] = \delta^3(K - K')\delta_{\sigma,\sigma'}$$

$$[d(K, \sigma), d(K', \sigma')] = 0,$$

$$[a(K, s), d(K', \sigma')] = 0,$$

$$[a(K, s), d(K', \sigma')^*] = 0$$

Computation of $G^{(1)\mu\nu}$ and $G^{(2)\mu\nu}$:

$$R^{(1)\mu\nu} = (g^{\mu\alpha}g^{\nu\beta}R_{\alpha\beta})^{(1)}$$

$$= (\eta_{\mu\alpha} - h^{\mu\alpha})(\eta_{\nu\beta} - h^{\nu\beta})R^{(1)}_{\alpha\beta}$$

$$= \eta_{\mu\alpha}\eta_{\nu\beta}R^{(1)}_{\alpha\beta}$$

$$R^{(1)} = (g^{\mu\nu}R_{\mu\nu})^{(1)} = (\eta_{\mu\nu} - h^{\mu\nu})R_{\mu\nu})^{(1)}$$

$$= \eta_{\mu\nu}R^{(1)}_{\mu\nu},$$

Thus,

$$G^{(1)\mu\nu} = \eta_{\mu\alpha}\eta_{\nu\beta}R^{(1)}_{\alpha\beta} - (\eta_{\alpha\beta}R^{(1)}_{\alpha\beta})\eta_{\mu\nu}$$

$$G^{(2)\mu\nu} = R^{(2)\mu\nu} - (1/2)(Rg^{\mu\nu})^{(2)}$$

$$R^{(2)}_{\mu\nu} = [(\eta_{\mu\alpha} + h_{\mu\alpha})(\eta_{\nu\beta} + h_{\nu\beta})(R^{(1)}_{\alpha\beta} + R^{(2)}_{\alpha\beta})]^{(2)}$$

$$= \eta_{\mu\alpha}\eta_{\nu\beta}R^{(2)}_{\alpha\beta} + \eta_{\mu\alpha}h_{\nu\beta}R^{(1)}_{\alpha\beta}$$

$$+\eta_{\nu\beta}h_{\mu\alpha}R^{(1)}_{\alpha\beta}$$

$$(Rg^{\mu\nu})^{(2)} = -R^{(1)}h_{\mu\nu} + R^{(2)}\eta_{\mu\nu},$$

$$R^{(1)} = \eta_{\mu\nu}R^{(1)}_{\mu\nu},$$

$$R^{(2)} = (g^{\mu\nu}R_{\mu\nu})^{(2)} = \eta_{\mu\nu}R^{(2)}_{\mu\nu} - h^{\mu\nu}R^{(1)}_{\mu\nu}$$

We now evaluate $G^{(1)\mu\nu}$ and show that $G^{(1)\mu\nu}_{,\nu} = 0$. First observe that

$$R^{(1)}_{\mu\nu} = (\Gamma^{\alpha}_{\mu\alpha,\nu} - \Gamma^{\alpha}_{\mu\nu,\alpha})^{(1)} =$$

$$= (g^{\alpha\beta}\Gamma_{\beta\mu\alpha})^{(1)}_{,\nu} - (g^{\alpha\beta}\Gamma_{\beta\mu\nu})^{(1)}_{,\alpha}$$

$$= (1/2)\eta_{\alpha\beta}(h_{\beta\mu,\alpha\nu} + h_{\beta\alpha,\mu\nu} - h_{\mu\alpha,\beta\nu}$$

$$-h_{\beta\mu,\nu\alpha} - h_{\beta\nu,\mu\alpha} + h_{\mu\nu,\alpha\beta})$$

$$= (1/2)(h_{,\mu\nu} + \Box h_{\mu\nu} - h^{;\alpha}_{\mu\alpha,\nu} - h^{;\alpha}_{\nu\alpha,\mu})$$

It follows that

$$R^{(1)} = \eta_{\mu\nu}R^{(1)}_{\mu\nu} =$$

$$\Box h - h^{;\alpha\beta}_{\alpha\beta}$$

and hence,

$$G^{(1)}_{\mu\nu} = R^{(1)}_{\mu\nu} - (1/2)R^{(1)}\eta_{\mu\nu}$$

$$= (1/2)(h_{,\mu\nu} + \Box h_{\mu\nu} - h^{;\alpha}_{\mu\alpha,\nu} - h^{;\alpha}_{\nu\alpha,\mu}) - \eta_{\mu\nu}\Box h + h^{;\alpha\beta}_{\alpha\beta}\eta_{\mu\nu})$$

It follows that

$$G^{(1)\mu\nu} = (1/2)(h^{,\mu\nu} + \Box h^{\mu\nu} - h^{\mu\alpha,\nu}_{,\alpha} - h^{\nu\alpha,\mu}_{,\alpha}$$

$$-\eta^{\mu\nu}\Box h + h^{;\alpha\beta}_{\alpha\beta}\eta^{\mu\nu})$$

from which, we deduce that

$$G^{(1)\mu\nu}_{,\nu} =$$

$$(1/2)(\Box h^{,\mu} + \Box h^{\mu\nu}_{,\nu} - \Box h^{\mu\alpha}_{,\alpha} - h^{\nu\alpha,\mu}_{,\nu\alpha} - \Box h^{,\mu} + h^{\alpha\beta,\mu}_{,\alpha\beta}) = 0$$

13.5 A version of quantum optimal control

Abstract: The state equations for the evolution of a quantum system observable in the presence of bath noise is modeled using noisy Heisenberg dynamics defined by a unitary evolution operator in system⊗ bath which satisfies the Hudson-Parthasarathy noisy Schrodinger equation with the noise processes being families of observables in the bath Boson Fock space. These noise processes are the creation, annihilation and conservation operators that exhibit in some special cases statistical properties like classical Brownian motion and Poisson processes while in the general case, these processes are non-commutative and therefore have no analogue in classical stochastic processes. In fact, these processes do not in general have any joint probability distribution because their values at two different times generally do not commute. Furthermore, the Heisenberg uncertainty principle ensures that general non-commutative measurements are not only impossible but even commutative measurements in general may not commute with the future values of the states. Belavkin thereofore constructed a family of non-demolition measurements which form an Abelian family and which also commute with the future values of the state. Now, Belavkin constructed a quantum filter which provides real time estimates of either noisy Heisenberg states or equivalently of noisy Schrodinger states coming from the Hudson-Parthasarathy Schrodinger equation. based on non-demolition measurements. These estimates are functions of the output non-demolition measurements and hence are commutative. Owing to the commutativity of all the variables in the Belavkin filter, this filter is also known as a stochastic Schrodinger equation. The Belavkin filter is a non-commutative generalization of the classical Kushner filter in the sense that if the system Hilbert space is $L^2(\mathbb{R}^n)$, the system observable whose evolution is to be studied is multiplication by some function $f(x)$ in $L^2(\mathbb{R}^n)$ and $X(t)$ is a classical Markov process so that we can define the homomorphism $j_t(f) = f(X(t))$, then based on noisy measurements

$$dz(t) = h_t(X(t))dt + \sigma_v dV(t)$$

we define the conditional expectation

$$\pi_t(f) = \mathbb{E}(f(X(t))|\eta_o(t)), \eta_o(t) = \sigma(z(s), s \leq t)$$

then if K_t denotes the generator of $X(t)$, we obtain the classical Kushner-Kallianpur filter

$$d\pi_t(f) = \pi_t(K_t f)dt + \sigma_v^{-2}(\pi_t(h_t f) - \pi_t(h_t)\pi_t(f))(dz(t) - \pi_t(h_t)dt)$$

which can also be derived from the Belavkin filter by replacing the non-commutative operators appearing in it with multiplication operators by functions. Now we can formulate a control problem by including in the Hudson-Parthasarathy equation, polynomial functions of the input measurement process $Y_i(t)$ (which are superpositions of the fundamental noise processes) with coefficients being operators in the system Hilbert space. These functions may be chosen arbitrarily with the only constraint that the HP evolution operator should be unitary at all times. It should be noted that the system operators commute with the input measurements but not with the output measurements $Y_o(t) = j_t(Y_i(t)) = U(t)^* Y_i(t) U(t)$. Corresponding to this revised version of the HP equation, the unitary evolution gives rise to Heisenberg dynamics for the evolving state $j_t(X)$ similar to the Evans-Hudson flow but with functions of the output measurements appearing as coefficients. These functions are the control functions. When one formulates the Belavkin filter for such processes using the reference probability approach of Gough et.al, then one ends up with the Belavkin filter having the form

$$d\pi_t(X) = F_t(X, u(t))dt + \sum_{k \geq 1} G_{kt}(X, u(t))(dY_o(t))^k$$

This is a direct consequence of the revised Evans-Hudson flow assuming the form

$$dj_t(X) = \theta_b^a(u(t), j_t(F), j_t(X))d\Lambda_a^b(t)$$

where F denotes the set of all the system observables appearing in the Hudson-Parthasarathy equation and θ_b^a are the structure maps.

Here, $u(t)$ is of the form $\chi_t(Y_o(t)) \in \eta_o(t)$ and it is then an easy problem to derive the quantum stochastic Bellman-Hamilton-Jacobi equation for the optimal control $\chi_t(.)$ that would minimize a cost function of the form

$$\mathbb{E} \int_0^T \mathcal{L}(j_t(X), u(t))dt$$

with the expectation being taken in any initial state of the system$\otimes$ bath. The method for carrying out the minimization is based on the Hamilton-Jacobi function

$$V(t, \pi_t) = min_{u(s), t \leq s \leq T} \mathbb{E}[\int_t^T \mathcal{L}(j_s(X), u(s))ds | \eta_o(t)]$$

Problem formulation

The process to be controlled is j_t which satisfies the qsde

$$dj_t(X) = j_t(\theta_b^a(X))d\Lambda_a^b(t)$$

where X is in $\mathcal{B}(\mathfrak{h})$, $\mathfrak{h}$ being the system Hilbert space and $\theta_b^a : \mathcal{B}(\mathfrak{h}) \to \mathcal{B}(\mathfrak{h})$ are linear operators which are called structure maps. They satisfy certain relations that guarantee that j_t is a *-unital homomorphism. $\Lambda_b^a(t), a, b \geq 0$ are the fundamental processes of Hudson and Parthasarathy. They satisfy the quantum Ito's formula:

$$d\Lambda_b^a \Lambda_d^c = \epsilon_d^a d\Lambda_b^c$$

where ϵ_d^a is zero if either a or d is zero and δ_d^a otherwise. The structure maps θ_b^a are assumed to depend on a control input $u(t)$ which is restricted to be a function of j_t only. More precisely, we choose a basis $\{Z_a < a = 1, 2, ...\}$ for $\mathcal{B}(\mathfrak{h})$ and then any $j_t(X)$ is a complex linear combination of $j_t(Z_a), a = 1, 2, ...,$ so we can regard the homomorphism j_t as being equivalent to the family of operators $j_t(Z_a), a = 1, 2, ...$ choose a basis $\{\eta_k\}$ for the system Hilbert space $\mathfrak{h}$ and an approximate basis $\{\xi_r\}$ for the Boson Fock space such that $\xi_r = \sum_s c(r, s)|e(u_s) >$ where $|e(u_s) >, s = 1, 2, ...$ are exponential vectors in the Boson Fock space. Then we can represent the operator $j_t(Z_a)$ by the matrix elements $< \eta_k \otimes \xi_r | j_t(Z_a) | \eta_l \otimes \xi_s >= J_t(a, k, r, l, s)$. Now let ρ be a state in the system$\otimes$ bath space, ie, is $\mathfrak{h} \otimes \Gamma_s(\mathcal{H})$ where $\mathcal{H} = L^{(}\mathbb{R}_+) \otimes \mathbb{C}^d$. Then let $X_d(t)$ be process in this space to be tracked. The θ_b^a being dependent on $u(t)$ can be expressed as

$$\theta_b^a(X) = \theta_b^a(u(t), X)$$

Then after applying this control input

$$u(t) = F(t, j_t) = F(t, j_t(Z_a)), a = 1, 2, ...) - - - (1)$$

our qsde in the sense of Hudson-Parthasarathy and Evans-Hudson, can be expressed as

$$dj_t(X) = j_t(\theta_b^a(u(t), X))d\Lambda_a^b(t) - - - (2)$$

The quantum Markovianity is still preserved after applying such a state dependent control. The control input $u(t), 0 \le t \le T$ is to be selected so that it minimizes the cost function

$$C(u) \int_0^T \mathcal{L}(j_t(X), u(t))dt)$$

where for example we may take $\mathcal{L}$ as

$$\mathcal{L}(j_t(X), u(t)) = Tr(\rho.(X_d(t) - j_t(X))^2)$$

where ρ is a state in $\mathfrak{h} \otimes \Gamma_s(\mathcal{H})$ and $X_d(t)$ is an operator valued process in $\mathfrak{h} \otimes \Gamma_s(\mathcal{H})$ to be tracked. As in the classical BHJ theory, we introduce the energy function

$$V(t, j_t) = V(t, j_t(Z_a), a = 1, 2, ...) = V(t, J_t(a, k, r, l, s), a, k, r, l, s = 1, 2, ...)$$

$$= min_{t \le u(s) \le T} \int_s^T \mathcal{L}(j_s(X), u(s))ds$$

where $j_t(X)$ satisfies the above qsde (2) with $u(t)$ allowed to be of the form (1) only, ie, instantaneous state feedback. Then as in the classical BHJ theory of optimal control, we easily derive the equation

$$V_{,t}(t, j_t)) + min_{u(t)}(L(j_t(X), u(t)) + (V(t, j_t + dj_t) - V(t, j_t))/dt) = 0$$

Now, we can write

$$V(t, j_t + dj_t) - V(t, j_t) = \frac{\partial V(t, j_t)}{\partial j_t}.dj_t$$

$$= \frac{\partial V(t, J_t(a, k, r, l, s))}{\partial J_t(a, k, r, l, s)} dJ_t(a, k, r, l, s)$$

where the summation over repeated indices is assumed. We note that

$$dJ_t(a, k, r, l, s) = < \eta_k \otimes \xi_r | dj_t(Z_a) | \eta_l \otimes \xi_s > =$$

$$= < \eta_k \otimes \xi_r | j_t(\theta_q^p(u(t), Z_a)) d\Lambda_p^q(t) | \eta_l \otimes \xi_s > =$$

$$= < \eta_k \otimes \xi_r | j_t(\theta_q^p(u(t), Z_a)) | \eta_l \otimes \xi_s > \bar{c}(r, m) c(s, n) \bar{u}_{mq}(t) u_{np}(t) dt$$

We can write

$$\theta_q^p(u(t), Z_a) = \sum_b A(u(t), p, q, a, b) Z_b$$

where if $u(t)$ is a scalar function, then $A(u(t), p, q, b)$ are complex numbers and if $u(t)$ is a function of $j_t(Z_c), c = 1, 2, ...$, then $A(u(t), p, q, b)$ becomes an operator in $\mathfrak{h} \otimes \Gamma_s(\mathcal{H})$. We then have by the homomorphism property of j_t that

$$j_t(\theta_q^p(u(t), Z_a)) = \sum_b A(u(t), p, q, a, b) j_t(Z_b)$$

So we get

$$dJ_t(a, k, r, l, s)/dt = \bar{c}(r, m) c(s, n) \bar{u}_{mq}(t) u_{np}(t) A(u(t), p, q, a, b) J_t(b, k, r, l, s)$$

and so our quantum BHJ equation assumes the form

$$V_{,t}(t, J_t) +$$

$$min_{u(t)} (L(j_t(X), u(t)) + \frac{\partial V(t, J_t)}{\partial J_t(a, k, r, l, s)} \bar{c}(r, m) c(s, n) \bar{u}_{mq}(t) u_{np}(t) A(u(t), p, q, a, b) J_t(b, k, r, l, s)$$

$$= 0$$

In these expressions, J_t corresponds to the set of numbers $\{J_t(a, k, r, l, s)\}$ and writing

$$X = \sum_a d(a) Z_a$$

we can express this quantum stochastic BHJ equation as

$$V_{,t}(t, J_t) +$$

$$min_{u(t)} (L(\sum_a d(a) J_t(a, .), u(t)) + \frac{\partial V(t, J_t)}{\partial J_t(a, k, r, l, s)} \bar{c}(r, m) c(s, n) \bar{u}_{mq}(t) u_{np}(t) A(u(t), p, q, a, b) J_t(b, k, r, l, s))$$

$$= 0$$

In this expression, the solution for $u(t)$ on minimization comes out to be a function of j_t or equivalently, of the numbers $\{J_t(a, k, r, l, s)\}$

A more practically implementable approach to the quantum control problem is first to estimate the Belavkin filter estimate of the state $j_t(X)$ at time t based on non-demolition measurements $Y_o(s) = U(s)^*Y_i(s)U(s), s \leq t$ as

$$\pi_t(X) = \mathbb{E}(j_t(X)|\eta_o(t)], \eta_o(t) = \sigma(Y_o(s), s \leq t)$$

and then choose our control input $u(t)$ in the form

$$u(t) = \chi_t(\pi_t)$$

where by π_t, we mean the family of operators $\pi_t(Z_a), a = 1, 2, ..$ where $Z_a, a = 1, 2, ...$ forms a basis for $\mathcal{B}(\mathfrak{h})$, We choose the function χ_t so that

$$\mathbb{E}\int_0^T \mathcal{L}(t, j_t(X), u(t))dt = \mathbb{E}\int_0^T \mathcal{L}(t, j_t(X), \chi_t(\pi_t))dt$$

is a minimum. The above expectation may for example be taken when the system and bath are in the state $|f \otimes \phi(u) >$, where $|f >\in \mathfrak{h}, < f|f >= 1$ and $|\phi(u) >= exp(- \| u \|^2 /2)|e(u) >, u \in \mathcal{H}$ is a coherent state of the bath. As in the classical BHJ equation, we therefore seek to minimize

$$\mathbb{E}[\int_t^T \mathcal{L}(s, j_s(X), u(s))ds|\eta_o(t)]$$

w.r.t $u(s), t \leq s \leq T$. This minimum will be of the form

$$V(t, \pi_t) = V(t, \pi_t(Z_a), a = 1, 2, ...)$$

and as usual, we have

$$V(t, \pi_t) = min_{u(t)=\chi_t(\pi_t)}(\mathbb{E}[\mathcal{L}(t, j_t(X), u(t))|\eta_o(t)]dt + \mathbb{E}(V(t + dt, \pi_{t+dt})|\eta_o(t)))$$

The Belavkin filter when the structure map θ_b^a depend on the control input $u(t)$ is given by

$$d\pi_t(X) = F_t(X, u(t))dt + G_t(X, u(t))dY_o(t)$$

where the control input $u(t) = \chi_t(\pi_t)$. Everything is commutative in this Belavkin equation. Thus,

$$V(t + dt, \pi_{t+dt}) = V(t, \pi_t) + V_{,t}(t, \pi_t)dt + Tr(\frac{\partial V(t, \pi_t)}{\partial \pi_t(Z_a)}d\pi_t(Z_a))$$

$$+Tr(\frac{\partial^2 V(t, \pi_t)}{\partial \pi_t(Z_a)\partial \pi_t(Z_b)}d\pi_t(Z_a)d\pi_t(Z_b))$$

with summation over the repeated index a being implied assuming quadrature noise. This leads to the quantum stochastic BHJ equation

$$V_{,t}(t, \pi_t) + min_{u(t)=\chi_t(\pi_t)}(\pi_t(\mathcal{L}(t, X, u(t)) + dt^{-1}.Tr(\frac{\partial V(t, \pi_t)}{\partial \pi_t(Z_a)}\mathbb{E}[d\pi_t(Z_a)|\eta_o(t)])$$

$$+dt^{-1}.Tr(\frac{\partial^2 V(t,\pi_t)}{\partial\pi_t(Z_a)\partial\pi_t(Z_b)}\mathbb{E}[d\pi_t(Z_a)d\pi_t(Z_b)|\eta_o(t)]) = 0$$

Remark: We have used the following identities. Writing

$$\mathcal{L}(t,j_t(X),u(t)) = \mathcal{L}(t,j_t(X),\chi_t(\pi_t)) = \sum_{k\geq 0}\mathcal{L}_k(t,\chi_t(\pi_t))j_t(X)^k$$

(Note that $[j_t(X),\chi_t(\pi_t)] = 0 since [j_t(X),\pi_t(Z_a)] = 0 \forall a$)

$$= \sum_{k\geq 0}\mathcal{L}_k(t,\chi_t(\pi_t))\pi_t(X^k) = \pi_t(\sum_{k\geq 0}\mathcal{L}_k(t,\chi_t(\pi_t))X^k)$$

$$= \pi_t(\mathcal{L}(t,X,\chi_t(\pi_t))) = \pi_t(\mathcal{L}(t,X,u(t)))$$

Since all the operators $\pi_t(Z_a), a = 1, 2, \dots$ are commutative, whilst carrying out the above minimization, we may assume that these operators are all real numbers. We also note that in the case of quadrature noise,

$$dY_o(t) = dY_i(t) + dU(t)^* dY_i(t)U(t) + U(t)^* dY_i(t)dU(t) =$$

$$dY_i(t) + j_t(L_2 + L_2^*)dt$$

so that

$$dt^{-1}\mathbb{E}[d\pi_t(Z)|\eta_o(t)] = F_t(X) + G_t(X)(u(t) + \bar{u}(t) + \pi_t(L_2 + L_2^*))$$

Further,
$$dt^{-1}\mathbb{E}[d\pi_t(Z_a)d\pi_t(Z_b)|\eta_o(t)] = G_t(Z_a)G_t(Z_b)$$

13.6   A neater formulation of the quantum optimal control problem

Let L_1, L_2, S, H, P be functions of the input measurement $Y_i(t)$ at time t where we take $Y_i(t) = c_1 A(t) + \bar{c}_1 A(t)^* + c_2 \Lambda(t)$. We assume that these are polynomial functions of $Y_i(t)$ with coefficients being operators in the system Hilbert space $\mathfrak{h}$ and these functions have been chosen so that $U(t)$ is unitary for all $t \geq 0$ where $U(t)$ satisfies the qsde

$$dU(t) = (-(iH + P)dt + L_1 dA(t) + L_2 dA(t)^* + Sd\Lambda(t))U(t), U(0) = I$$

Note that the family of operators $Y_i(.)$ commutes with $\mathcal{B}(\mathfrak{h})$, so we can write

$$H = F_1(H_k, k = 1, 2, \dots, p, \chi_t(Y_i(t))), P = F_2(P_k, k = 1, 2, \dots, p, \chi_t(Y_i(t))),$$

$$L_1 = F_3(L_{1k}, k = 1, 2, \dots, p, \chi_t(Y_i(t))), L_2 = F_4(L_{2k}, k = 1, 2, \dots, p, \chi_t(Y_i(t))),$$

$$S = F_5(S_1, ..., S_p, \chi_t(Y_i(t)))$$

where $H_k, P_k, L_{1k}, L_{2k}, S_k$ are all system space operators, ie, operator in $\mathfrak{h}$, we may assume that they are in $\mathcal{B}(\mathfrak{h})$ and the control input at time t is $u(t) = \chi_t(Y_o(t))$ where χ_t is an ordinary function of a real variable. Here, $Y_o(t) = U(t)^* Y_i(t) U(t)$ is the output measurement process. Clearly,

$$Y_o(t) = U(t)^* Y_i(t) U(t) = U(T)^* Y_i(t) U(T), T \geq t$$

which follows by taking the differential of the rhs w.r.t. T and using the fact that the unitarity of $U(T)$ depends only on the operators $H_k, P_k, L_{1k}, L_{2k}, S_k, Y_i(T)$ all of which commute with $Y_i(t)$. Defining for any system$\otimes$ bath observable X,

$$j_t(X) = -U(t)^* X U(t)$$

We get that if X is a system observable,

$$dj_t(X) = dU(t)^* X U(t) + U(t)^* X dU(t) + dU(t)^* X dU(t) =$$

$$j_t(\theta_0(w(t), X))dt + j_t(\theta_1(w(t), X))dA(t) + j_t(\theta_2(w(t), X))dA(t)^* + j_t(\theta_3(w(t), X))d\Lambda(t)$$

where

$$w(t) = \chi_t(Y_i(t))$$

and $\theta_k(w(t), .)$ are linear maps that take system observables to system $\otimes$ bath observables. These maps are functions of $w(t)$ and the system operators $L_{1k}, L_{2k}, S_k, H_k, P_k,$ $k = 1, 2, ..., p$. We denote the set of system operators $H_k, P_k, S_k, L_{1k}, L_{2k}, k = 1, 2, ..., p$ by F. Then, to be precise, we must write the above qsde as

$$dj_t(X) = j_t(\theta_0(w(t), F, X))dt +$$

$$j_t(\theta_1(w(t), F, X))dA(t) + j_t(\theta_2(w(t), F, X))dA(t)^* + j_t(\theta_3(w(t), F, X))d\Lambda(t)$$

It is easy to see using the unitarity of $U(t)$ that

$$j_t(\theta_k(w(t), X)) = \theta_k(u(t), j_t(X)), u(t) = j_t(w(t)) = \chi_t(Y_o(t))$$

So, we get the qsde

$$dj_t(X) = \theta_0(u(t), j_t(F), j_t(X))dt + \theta_1(u(t), j_t(F), j_t(X))dA(t) + \theta_2(u(t), j_t(F), j_t(X))dA(t)^*$$

$$+ \theta_3(u(t), j_t(F), j_t(X))d\Lambda(t)$$

We also note the fact that $j_t(X)$ and $j_t * (F)$ both commute with $u(t)$ and that

$$\mathbb{E}[\theta_k(u(t), j_t(F), j_t(X))|\eta_o(t)] =$$

$$\pi_t(\theta_k(u(t), F, X))$$

which is easily seen by using the above stated commutativity and the fact that j_t is a homomorphism. It should be noted that if $f(u(t))$ is any function of $u(t) \in \eta_o(t)$ and X is a system operator, then by $\pi_t(f(u(t))X)$, we mean

$f(u(t))\pi_t(X) = f(u(t))\mathbb{E}[j_t(X)|\eta_o(t)]$ and not $\mathbb{E}[j_t(f(u(t))X)|\eta_o(t)]$. Now, we are in a position to formulate and solve the optimal control problem. We define

$$V(t,\pi_t) = min_{u(s),t\le s\le T}\mathbb{E}[\int_t^T \mathcal{L}(j_s(X),u(s))ds|\eta_o(t)]$$

where

$$u(s) = \chi_s(Y_o(s)) = j_s(\chi_s(Y_i(s))) = U(s)^*\chi_s(Y_i(s))U(s) = \chi_s(U(s)^*Y_i(s)U(s))$$

Then, we get

$$V_{,t}(t,\pi_t) + min_{u(t)}(\pi_t(L(X,u(t))))$$

$$+dt^{-1}\sum_{n\ge 1} Tr(\frac{\partial^n V(t,\pi_t)}{\partial\pi_t^{\otimes n}}\mathbb{E}[(d\pi_t)^{\otimes n}|\eta_o(t)])$$

$$= 0$$

where by $\pi_t^{\otimes n}$, we mean a lexicographically ordered set of the elements $\pi_t(Z_{a_1})...\pi_t(Z_{a_n})$ with $a_1,...,a_n = 1,2,....$ The above conditional expectation is easily computed using the Belavkin filtering equations

$$d\pi_t(X) = F_t(X,u(t))dt + \sum_{k\ge 1} G_{t,k}(X,u(t))(dY_o(t))^k$$

with the functions $F_t, G_{k,t}$ derived in the usual way using the reference probability method.

13.7 Calculating the approximate shift in the oscillation frequency of a cavity resonator having arbitrary cross section when the medium has a small inhomogeneity

$\epsilon(\omega,x,y,z), \mu(\omega,x,y,z)$ are the permittivity and permeability. They can be expressed as

$$\epsilon(\omega,r) = \epsilon_0(1 + \delta\chi_e(\omega,r)),$$

$$\mu(\omega,r) = \mu_0(1 + \delta\chi_m(\omega,r)), r = (x,y,z)$$

By virtue of the boundary conditions on a conducting surface and the Maxwell equations, we have the fact that H_z vanishes when $z = 0, d$, $E_{z,z}$ vanishes when $z = 0, d$, $E_\perp$ vanishes when $z = 0, d$ and hence, these fields can be expanded as

$$H_z(\omega,x,y,z) = \sum_p H_{zp}(\omega,x,y)sin(p\pi z/d),$$

$$E_z(\omega,x,y,z) = \sum_p E_{zp}(\omega,x,y)cos(p\pi z/d),$$

$$E_\perp(\omega, x, y, z) = \sum_p E_{\perp p}(\omega, x, y) sin(p\pi z/d)$$

Further, the Maxwell curl equations give us

$$E_{z,y} - E_{y,z} = -j\omega\mu H_x, E_{x,z} - E_{z,x} = -j\omega\mu H_y, E_{y,x} - E_{x,y} = -j\omega\mu H_z,$$

$$H_{z,y} - H_{y,z} = j\omega\epsilon E_x, H_{x,z} - H_{z,x} = j\omega\epsilon E_y, H_{y,x} - H_{x,y} = j\omega\epsilon E_z$$

Combining these with the above boundary conditions implies that $H_{\perp,z}$ vanishes when $z = 0, d$ and hence we have the expansion

$$H_\perp(\omega, x, y, z) = \sum_p H_{\perp,p}(\omega, x, y) cos(p\pi z/d)$$

Substituting these expansions into the Maxwell curl equations expressed in the form

$$\nabla_\perp E_z \times \hat{z} + \hat{z} \times E_{\perp,z} = -j\omega\mu H_\perp,$$

$$\nabla_\perp \times E_\perp = -j\omega\mu H_z\hat{z},$$
$$\nabla_\perp \times H_\perp = j\omega\epsilon E_z\hat{z}$$

we get

$$\sum_p \nabla_\perp E_{zp}(\omega, x, y) \times \hat{z}.cos(p\pi z/d) + \sum_p (\pi p/d)\hat{z} \times E_{\perp,p}(\omega, x, y)cos(p\pi z/d)$$

$$-j\omega\mu(\omega, r)\sum_p H_{\perp,p}(\omega, x, y)cos(p\pi z/d)$$

$$\sum_p \nabla_\perp H_{zp}(\omega, x, y) \times \hat{z}.sin(p\pi z/d) - \sum_p (\pi p/d)\hat{z} \times H_{\perp,p}(\omega, x, y)sin(p\pi z/d)$$

$$j\omega\epsilon(\omega, r)\sum_p E_{\perp,p}(\omega, x, y)sin(p\pi z/d)$$

Multiplying the first equation by $(2/d)cos(m\pi z/d)$, the second equation by $(2/d)sin(m\pi z/d)$ and integrating w.r.t. z over $[0, d]$ gives us

$$\nabla_\perp E_{zm}(\omega, x, y) \times \hat{z} + (\pi m/d)\hat{z} \times E_{\perp,m}(\omega, x, y)$$

$$= -j\omega\sum_p (\int_0^d \mu(\omega, x, y, z)(2/d)cos(p\pi z/d)cos(m\pi z/d)dz)H_{\perp,p}(\omega, x, y)$$

and likewise,

$$\nabla_\perp H_{zm}(\omega, x, y) \times \hat{z} - (\pi m/d)\hat{z} \times H_{\perp,m}(\omega, x, y)$$

$$= j\omega\sum_p (\int_0^d \epsilon(\omega, x, y, z)(2/d)sin(p\pi z/d)sin(m\pi z/d)dz)E_{\perp,p}(\omega, x, y)$$

Finally, the z component of the Maxwell curl equations give

$$\sum_p \nabla_\perp \times E_{\perp,p}(\omega, x, y) sin(p\pi z/d) = -j\omega\mu(\omega, r) \sum_p H_{z,p}(\omega, x, y) sin(p\pi z/d)\hat{z},$$

$$\sum_p \nabla_\perp \times H_{\perp,p}(\omega, x, y) cos(p\pi z/d) = j\omega\epsilon(\omega, r) \sum_p E_{z,p}(\omega, x, y) cos(p\pi z/d)\hat{z},$$

which given in the same manner,

$$\nabla_\perp \times E_{\perp,m}(\omega, x, y)$$

$$= -j\omega \sum_p (\int_0^d \mu(\omega, x, y, z)(2/d) sin(p\pi z/d) sin(m\pi z/d) dz) H_{z,p}(\omega, x, y)\hat{z},$$

$$\nabla_\perp \times H_{\perp,m}(\omega, x, y)$$

$$= j\omega \sum_p (\int_0^d \epsilon(\omega, x, y, z)(2/d) cos(p\pi z/d) cos(m\pi z/d) dz) E_{z,p}(\omega, x, y)]\hat{z},$$

So far, everything is exact. No approximations have been made. Writing these equations in perturbation theoretic form gives us

$$\nabla_\perp E_{zm}(\omega, x, y) \times \hat{z} + (\pi m/d)\hat{z} \times E_{\perp,m}(\omega, x, y)$$

$$+j\omega\mu_0 H_{\perp,m}(\omega, x, y)$$

$$= -j\omega\mu_0 \sum_p (\int_0^d \delta\chi_m(\omega, x, y, z)(2/d) cos(p\pi z/d) cos(m\pi z/d) dz) H_{\perp,p}(\omega, x, y),$$

$$\nabla_\perp H_{zm}(\omega, x, y) \times \hat{z} - (\pi m/d)\hat{z} \times H_{\perp,m}(\omega, x, y)$$

$$-j\omega\epsilon_0 E_{\perp,m}(\omega, x, y)$$

$$= j\omega\epsilon_0 \sum_p (\int_0^d \delta\chi_e(\omega, x, y, z)(2/d) sin(p\pi z/d) sin(m\pi z/d) dz) E_{\perp,p}(\omega, x, y),$$

$$\nabla_\perp \times E_{\perp,m}(\omega, x, y) + j\omega\mu_0 H_{z,m}(\omega, x, y) =$$

$$-j\omega\mu_0 \sum_p (\int_0^d \delta\chi_m(\omega, x, y, z)(2/d) sin(p\pi z/d) sin(m\pi z/d) dz) H_{z,p}(\omega, x, y)\hat{z},$$

$$\nabla_\perp \times H_{\perp,m}(\omega, x, y) - j\omega\epsilon_0 E_{z,m}(\omega, x, y)$$

$$= j\omega\epsilon_0 \sum_p (\int_0^d \chi_e(\omega, x, y, z)(2/d) cos(p\pi z/d) cos(m\pi z/d) dz) E_{z,p}(\omega, x, y)\hat{z},$$

These equations can be expressed as

$$\nabla_\perp E_{zm}(\omega, x, y) \times \hat{z} + (\pi m/d)\hat{z} \times E_{\perp,m}(\omega, x, y)$$

$$+j\omega\mu_0 H_{\perp,m}(\omega, x, y)$$

$$= \sum_p \delta F_1(\omega, x, y, m, p) H_{\perp,p}(\omega, x, y),$$

$$\nabla_\perp H_{zm}(\omega, x, y) \times \hat{z} - (\pi m/d)\hat{z} \times H_{\perp,m}(\omega, x, y)$$

$$-j\omega\epsilon_0 E_{\perp,m}(\omega, x, y)$$

$$= \sum_p \delta F_2(\omega, x, y, m, p) E_{\perp,p}(\omega, x, y),$$

$$\nabla_\perp \times E_{\perp,m}(\omega, x, y) + j\omega\mu_0 H_{z,m}(\omega, x, y) =$$

$$\sum_p \delta G_1(\omega, x, y, m, p) H_{z,p}(\omega, x, y)\hat{z},$$

$$\nabla_\perp \times H_{\perp,m}(\omega, x, y) - j\omega\epsilon_0 E_{z,m}(\omega, x, y)$$

$$= \sum_p \delta G_2(\omega, x, y, m, p) E_{z,p}(\omega, x, y)\hat{z},$$

Let us write the approximate solutions to these equations as

$$E_m = E_m^{(0)} + E_m^{(1)}, H_m = H_m^{(0)} + H_m^{(1)},$$

or equivalently,

$$E_{z,m} = E_{z,m}^{(0)} + E_{z,m}^{(1)}, H_{z,m} = H_{z,m}^{(0)} + H_{z,m}^{(1)},$$

$$E_{\perp,m} = E_{\perp,m}^{(0)} + E_{\perp,m}^{(1)}, H_{\perp,m} = H_{\perp,m}^{(0)} + H_{\perp,m}^{(1)}$$

and also assume that the characteristic frequency of oscillation gets perturbed from ω to $\omega + \delta\omega$. Then applying perturbation theory, we get the zeroth order perturbation equations as

$$\nabla_\perp E_{zm}^{(0)} \times \hat{z} + (\pi m/d)\hat{z} \times E_{\perp,m}^{(0)}$$

$$+j\omega\mu_0 H_{\perp,m}^{(0)} = 0,$$

$$\nabla_\perp H_{zm}^{(0)} \times \hat{z} - (\pi m/d)\hat{z} \times H_{\perp,m}^{(0)}$$

$$-j\omega\epsilon_0 E_{\perp,m}^{(0)} = 0$$

$$\nabla_\perp \times E_{\perp,m}^{(0)} + j\omega\mu_0 H_{z,m}^{(0)}\hat{z} = 0$$

$$\nabla_\perp \times H_{\perp,m}^{(0)} - j\omega\epsilon_0 E_{z,m}^{(0)}\hat{z}$$

and the first order perturbation equations as

$$\nabla_\perp E_{zm}^{(1)} \times \hat{z} + (\pi m/d)\hat{z} \times E_{\perp,m}^{(1)}$$

$$+j\omega\mu_0 H_{\perp,m}^{(1)} + j\mu_0 \delta\omega H_{\perp,m}^{(0)}$$

$$= \sum_p \delta F_1(\omega, x, y, m, p) H_{\perp,p}^{(0)},$$

$$\nabla_\perp H_{zm}^{(1)} \times \hat{z} - (\pi m/d)\hat{z} \times H_{\perp,m}^{(1)}$$

$$-j\omega\epsilon_0 E_{\perp,m}^{(1)} - j\epsilon_0\delta\omega E_{\perp,m}^{(0)}$$

$$= \sum_p \delta F_2(\omega,x,y,m,p)E_{\perp,p}^{(0)},$$

$$\nabla_\perp \times E_{\perp,m}^{(1)} + j\omega\mu_0 H_{z,m}^{(1)}\hat{z} + j\mu_0\delta\omega H_{z,m}^{(0)}\hat{z} =$$

$$\sum_p \delta G_1(\omega,x,y,m,p)H_{z,p}^{(0)}\hat{z}$$

$$\nabla_\perp \times H_{\perp,m}^{(1)} - j\omega\epsilon_0 E_{z,m}^{(1)}\hat{z} - j\epsilon_0\delta\omega E_{z,m}^{(0)}\hat{z}$$

$$= \sum_p \delta G_2(\omega,x,y,m,p)E_{z,p}^{(0)}\hat{z},$$

When the zeroth order equations are solved subject to the above mentioned boundary conditions, we get as in standard cavity resonator analysis, we get a discrete set of frequency values for ω, say $\omega(m,n)^{(0)}, n = 1, 2, ...$ for each value of m and correspondingly a normalized eigenvector for $E_m^{(0)}, H_m^{(0)}$, we denote these eigenvectors by $(E_{mn}^{(0)}, H_{mn}^{(0)}), n = 1, 2, ...$ Specifically, we can split these eigenvector into transverse and longitudinal components:

$$E_{mn}^{(0)} = E_{\perp,mn}^{(0)} + E_{zmn}^{(0)}\hat{z}, H_{mn}^{(0)} = H_{\perp,mn}^{(0)} + H_{zmn}^{(0)}\hat{z}$$

where all of these are functions of (x, y) only. There may also be a degeneracy of these eigenvectors. Specifically, for the unperturbed eigen-frequency $\omega(m,n)^{(0)}$, we may have $K(m,n)$ eigenvectors, say $\psi_{mnk}^{(0)} = (E_{mnk}^{(0)T}, H_{mnk}^{(0)T})^T, k = 1, 2, ..., K(m,n)$. We assume that these eigenvectors are all normalized in the standard sense, ie,

$$<\psi_{mnk}^{(0)}, \psi_{mn'k'}^{(0)}> = \int_D \psi_{mnk}^{(0)}(x,y)^* \psi_{mn'k'}^{(0)}(x,y)dxdy = \delta_{mm'}\delta_{kk'}$$

where D denotes the cross section of the guide. The first order transverse equations give

$$E_{\perp,m}^{(1)} = (-\pi m/dh(\omega,m)^2)\nabla_\perp E_{zm}^{(1)}$$

$$-(j\mu_0\omega/h(\omega,m)^2)\nabla_\perp H_{zm}^{(1)} \times \hat{z}$$

$$-(j\mu_0\pi m\delta\omega/dh(\omega,m)^2)\hat{z} \times H_{\perp,m}^{(0)}$$

$$+(\pi m/dh(\omega,m)^2 \sum_p \delta F_1(\omega,x,y,m,p)\hat{z} \times H_{\perp,p}^{(0)}$$

$$-(\omega\mu_0\epsilon_0\delta\omega/h(\omega,m)^2)E_{\perp,m}^{(0)}$$

$$+(j\mu_0\omega/h(\omega,m)^2) \sum_p \delta F_2(\omega,x,y,m,p)E_{\perp,p}^{(0)}$$

and

$$H_{\perp,m}^{(1)} = (\pi m/dh(\omega,m)^2)\nabla_\perp H_{zm}^{(1)}$$

$$+(j\mu_0\omega/h(\omega,m)^2)\nabla_\perp E_{zm}^{(1)} \times \hat{z}$$

$$-(j\epsilon_0\pi m\delta\omega/dh(\omega,m)^2)\hat{z}\times E_{\perp,m}^{(0)} - (\pi m/dh(\omega,m)^2\sum_p \delta F_2(\omega,x,y,m,p)\hat{z}\times E_{\perp,p}^{(0)}$$

$$-(\omega\mu_0\epsilon_0\delta\omega/h(\omega,m)^2)H_{\perp,m}^{(0)}$$

$$-(j\epsilon_0\omega/h(\omega,m)^2)\sum_p \delta F_1(\omega,x,y,m,p)H_{\perp,p}^{(0)}$$

where

$$h(\omega,m)^2 = \omega^2\mu_0\epsilon_0 - (\pi m/d)^2$$

Thus, we get

$$\hat{z}.\nabla_\perp \times E_{\perp,m}^{(1)} = (j\mu_0\omega/h(\omega,m)^2)\nabla_\perp^2 H_{zm}^{(1)}$$

$$-(j\mu_0\pi m\delta\omega/dh(\omega,m)^2)(\nabla_\perp.H_{\perp,m}^{(0)})+(\pi m/dh(\omega,m)^2)\sum_p \nabla_\perp.(\delta F_1(\omega,x,y,m,p)\hat{z}\times H_{\perp,p}^{(0)}(x,y))$$

$$-(\mu_0\epsilon_0\omega\delta\omega/h(\omega,m)^2)\hat{z}.\nabla_\perp \times E_{\perp,m}^{(0)}$$

$$+(j\mu_0\omega/h(\omega,m)^2)\sum_p \hat{z}.\nabla_\perp \times (\delta F_2(\omega,x,y,m,p)E_{\perp,p}^{(0)}(x,y))$$

$$= -j\omega\mu_0 H_{z,m}^{(1)} - j\mu_0\delta\omega H_{z,m}^{(0)} + \sum_p \delta G_1(\omega,x,y,m,p)H_{z,p}^{(0)}(x,y)$$

This equation can be rearranged and simplified using the identities

$$\nabla_\perp.H_{\perp,m}^{(0)}(x,y) + (m\pi/d)H_{z,m}^{(0)} = 0,$$

$$\hat{z}.\nabla_\perp \times E_{\perp,m}^{(0)} = -j\omega\mu_0 H_{z,m}^{(0)}$$

as

$$(\nabla_\perp^2 + h(\omega,m)^2)H_{z,m}^{(1)}(x,y)$$

$$+2\delta\omega.\omega\mu_0\epsilon_0 H_{z,m}^{(0)}+\sum_p \hat{z}.\nabla_\perp\times(\delta F_2(\omega,x,y,m,p)E_{\perp,p}^{(0)})+(jh(\omega,m)^2/\mu_0\omega)\sum_p \delta G_1(\omega,x,y,m,p)H_{z,p}^{(0)} = 0$$

13.8 Optimal control for partial differential equations

.

Examples:

[a] Controlling the electromagnetic field within a box over a given space-time range with a control current density sources so that the controlled em fields are close in distance to a given em field.

[b] Controlling the energy-momentum tensor of matter and radiation in the Einstein field equations of general relativity so that the controlled metric is

close in distance to a given metric over a given space-time range. Both using the approximate linearized Einstein field equations and the fully non-linear Einstein field equations.

[c] Controlling the stirring force in a fluid so that the fluid velocity pattern matches a given pattern over a given space-time interval.

[2] Study of gravitational waves produced by an electromagnetic source.

$$R_{\mu\nu} - (1/2)Rg_{\mu\nu} = -8\pi GS_{\mu\nu}$$

$$S_{\mu\nu} = (-1/4)F_{\alpha\beta}F^{\alpha\beta}g_{\mu\nu} + F_{\mu\alpha}F_\nu^\beta$$

This is the energy-momentum tensor of the radiation field. Assume that the em four potential is

$$A_\mu = A_{(0)} + \delta A_\mu, g_{\mu\nu} = g_{(0)} + \delta g_{\mu\nu} = \eta_{\mu\nu} + h_{\mu\nu}(x)$$

where $A_\mu^{(0)}$ satisfies the flat space-time Maxwell equations, ie, if

$$F_{\mu\nu} = F_{\mu\nu}^{(0)} + \delta F_{\mu\nu},$$

$$F_{\mu\nu}^{(0)} = A_{\nu,\mu}^{(0)} - A_{\mu,\nu}^{(0)},$$

$$\delta F_{\mu\nu} = \delta A_{\nu,\mu} - \delta A_{\mu,\nu}$$

then

$$\delta S_{\mu\nu} = (-1/4)F_{\alpha\beta}^{(0)}F^{(0)\alpha\beta}h_{\mu\nu} + F_{\mu\alpha}^{(0)}F_{\nu\beta}^{(0)}h_{\alpha\beta}$$

$$+ 2\eta_{\alpha\beta}(F_{\mu\alpha}^{(0)}\delta F_{\nu\beta} + F_{\nu\beta}^{(0)}\delta F_{\mu\alpha})$$

This is of the general form

$$\delta S_{\mu\nu}(x) = C_1(\mu\nu\alpha\beta, x)h_{\alpha\beta}(x) + C_2(\mu\nu\alpha\beta, x)\delta F_{\alpha\beta}(x)$$

where the functions C_1, C_2 are completely decided by the unperturbed em wave $F_{\mu\nu}^{(0)}(x)$. We have already seen that the first order perturbation to the Einstein tensor

$$G^{\mu\nu} = R^{\mu\nu} - (1/2)Rg^{\mu\nu}$$

has the form

$$G^{(1)\mu\nu} = \delta G^{\mu\nu} = \delta R^{\mu\nu} - (1/2)\delta R.\eta^{\mu\nu} =$$

$$C_3(\mu\nu\alpha\beta\rho\sigma)h_{\alpha\beta,\rho\sigma}$$

whose ordinary four divergence vanishes, ie

$$C_3(\mu\nu\alpha\beta\rho\sigma)h_{\alpha\beta,\rho\sigma\nu} = 0$$

The first order perturbed Einstein-Maxwell field equations then give

$$C_3(\mu\nu\alpha\beta\rho\sigma)h_{\alpha\beta,\rho\sigma}(x) = C_1(\mu\nu\alpha\beta, x)h_{\alpha\beta}(x)$$

$$+ C_2(\mu\nu\alpha\beta, x)\delta F_{\alpha\beta}(x)$$

The Maxwell equations are

$$(F^{\mu\nu}\sqrt{-g})_{,\nu} = 0$$

which is to be combined with the gauge condition

$$(A^{\mu}\sqrt{-g})_{,\mu} = 0$$

or equivalently,

$$(g^{\mu\nu}\sqrt{-g}A_{\nu})_{,\mu} = 0$$

The unperturbed, ie, flat space-time components of these equations are

$$\eta_{\mu\alpha}\eta_{\nu\beta}F^{(0)}_{\alpha\beta,\nu} = 0, \eta_{\mu\nu}A^{(0)}_{\nu,\mu} = 0$$

Substituting

$$F^{(0)}_{\mu\nu} = A^{(0)}_{\nu,\mu} - A^{(0)}_{\mu,\nu}$$

these lead to the flat-space-time wave equation

$$\Box A^{(0)}_{\mu} = 0, \Box = \partial_{\alpha}\partial^{\alpha} = \eta_{\alpha\beta}\partial_{\alpha}\partial_{\beta}$$

The first order perturbed versions of the Maxwell equations and gauge condition are

$$(\delta(g^{\mu\alpha}g^{\nu\beta}\sqrt{-g})F^{(0)}_{\alpha\beta,\nu})$$
$$+\eta_{\mu\alpha}\eta_{\nu\beta}\delta F_{\alpha\beta,\nu} = 0$$

and

$$\eta_{\mu\nu}\delta A_{\nu,\mu} + (\delta(g^{\mu\nu}\sqrt{-g})A^{(0)}_{\nu})_{,\mu} = 0$$

Chapter 14

Quantization of cavity fields with inhomogeneous media, field dependent media parameters from Boltzmann-Vlasov equations for a plasma, quantum Boltzmann equation for quantum radiation pattern computation, optimal control of classical fields, applications classical nonlinear filtering

14.1 Computing the shift in the characteristic frequencies of oscillation in a cavity resonator due to gravitational effects and the effect of non-uniformity in the medium

Assume first that the unperturbed frequencies are $\omega_0[p, n], n = 1, 2, \ldots$ for a given $p \in \mathbb{Z}_+$. p decides the z-dependence of the fields. For example, the variation of $H_z^{(0)}$ with z is $sin(\pi pz/d)$ while the variation of $E_z^{(0)}$ with z is $cos(\pi pz/d)$. The variation of $E_\perp^{(0)}$ with z is $sin(\pi pz/d)$ and the variation of $H_\perp^{(0)}$ with z is $cos(\pi pz/d)$. This ensures that $H_z^{(0)}$ and $E_\perp^{(0)}$ vanish when $z = 0, d$. Define

$$h(\omega, p)^2 = \omega^2 \mu_0 \epsilon_0 - (\pi p/d)^2$$

Then, the first order perturbed equations for fields and characteristic frequencies are

$$(\nabla_\perp^2 + h(\omega_0[p, n], p)^2)[E_z^{(1)}, H_z^{(1)}]^T +$$

$$2\mu_0 \epsilon_0 \omega_0[p, n] \delta\omega [E_{z,p,n}^{(0)}, H_{z,p,n}^{(0)}]^T$$

$$= [\delta F_1(p, n, x, y), \delta F_2(p, n, x, y)]^T$$

where

$$[\delta F_1(p, n, x, y), \delta F_2(p, n, x, y)]^T$$

is of the form

$$\delta \mathcal{L}(E_{z,p,n}^{(0)}(x, y), H_{z,p,n}^{(0)}(x, y)]^T)$$

with δL being a linear first order partial differential operator depending on the permittivity and permeability perturbations $\epsilon_0 \delta \chi_e(\omega, x, y), \mu_0 \delta \chi_m(\omega, x, y)$ where $\omega = \omega_0[p, n]$. We may assume that the unperturbed modes $\psi_{p,n}(x, y) = [E_{z,p,n}^{(0)}(x, y), H_{z,p,n}^{(0)}(x, y)]^T, p, n = 1, 2, \ldots$ for form a complete onb for $L^2(D)^2$ where D is the cross sectional area of the resonator in the xy plane. This is because these are eigenfunctions of the Laplace operator $\nabla_\perp^2$ which is Hermitian when we apply joint Dirichlet and Neumann boundary conditions, ie E_z and $\frac{\partial H_z}{\partial n}$ vanish on the boundary ∂D. Thus, we infer that the shift in the characteristic frequency $\omega_0[p, n] \, \delta\omega$ is given by

$$\delta\omega = \delta\omega[p, n] =$$

$$(-1/2\mu_0 \epsilon_0 \omega_0[p, n]) \int_D (\bar{E}_{z,p,n}^{(0)}(x, y).\delta F_1(p, n, x, y) + \bar{H}_{z,p,n}^{(0)}(x, y).\delta F_2(p, n, x, y))dxdy$$

Further since the eigenfunctions $\psi_{p,n}$ of $\nabla_\perp^2$ with the stated boundary conditions are orthogonal, we obtain that its first order perturbation is given by

$$\delta\psi_{p,n}(x, y) = \sum_{(p', n') = (p, n)} \psi_{p', n'}(x, y) \int_D \psi_{p', n'}(x, y)^* [\delta F_1(p, n, x, y), \delta F_2(p, n, x, y)]dxdy$$

Here, we are assuming non-degeneracy of the unperturbed modes. If we have a degeneracy of $k(p, n)$ for each p, n, then it means that corresponding to the unperturbed characteristic frequency $\omega_0[p, n]$ or equivalently to the unperturbed modal eigenvalue $h(\omega_0[p, n], p)^2$, we have an orthonormal basis of unperturbed eigenfunctions

$$\psi_{n,p,m}(x, y) = [E_{z,p,n,m}^{(0)}(x, y), H_{z,p,n,m}^{(0)}(x, y)]^T, m = 1, 2, \ldots, k(p, n)$$

and by using the secular determinant theory in time independent perturbation theory developed for solving quantum mechanical problems, the perturbations to the frequency $\omega_0[p,n]$ are $\delta\omega[p,n,m], m = 1,2,...,k(p,n)$, where these are solutions to the determinantal equations

$$det(2\mu_0\epsilon_0\omega_0[p,n]\delta\omega I_k + ((< \psi_{p,n,m}, \delta L\psi_{p,n,m'} >))_{1\leq m,m'\leq k(p,n)}) = 0$$

14.2 Quantization of the field in a cavity resonator having non-uniform permittivity and permeability

We've derived the first order eigen-equations which determine the shift in the cavity resonator frequencies when there is a small perturbation in the uniformity of the medium. These equations were found to have the general form

$$(\nabla_\perp^2 + h(p,n)^2)\delta\psi(x,y) + a(p,n)\delta\omega\psi_{p,n}(x,y) = \delta F(p,n,x,y)$$

These equations could be derived from a variational principle using the action functional

$$S_{p,n}(\delta\psi) = (1/2)\int (|\nabla_\perp\delta\psi(x,y)|^2 - h(p,n)^2|\delta\psi(x,y)|^2)+$$

$$-a(p,n)\int (\delta\psi(x,y)^T\psi_{p,n}(x,y))dxdy + \int \delta\psi(x,y)^T\delta F(p,n,x,y)$$

We assume that the unperturbed system is classical but the perturbed system has quantum fluctuations $\delta\psi$. Although this variational principle is suitable for applying the finite element method, it is not suitable for quantization since time is not explicitly involved. One approach to quantizing this would be to replace $h(p,n)^2$ by $\omega^2\mu_0\epsilon_0$ where $\omega = i\partial/\partial t$ but that would again land us up in difficulty since we cannot give any physical interpretation for $\delta\omega$ in terms of time. So the only way out is start with the basic Maxwell equations with susceptibilities χ_e and χ_m defined as functions of *omega* as $\chi_e(i\omega,x,y), \chi_m(i\omega,x,y)$ and replace $i\omega$ by the operator $\partial/\partial t$ at the end of all calculations. Then we would end up with an equation for $\delta\psi(\omega,x,y)$ as

$$(\nabla_\perp^2 + \omega^2\mu_0\epsilon_0 - \pi^2p^2/d^2)\delta\psi(\omega,x,y) =$$

$$\sum_m \delta F(\omega,p,m,x,y)\psi_{p,m}(\omega,x,y)$$

which can be interpreted in the time domain as

$$(\nabla_\perp^2 - \mu_0\epsilon_0\partial_t^2 - \pi^2p^2/d^2)\delta\psi(t,x,y) =$$

$$\sum_m \delta F(\partial_t, p, m, x, y)\psi_{p,m}(t, x, y)$$

where the operators $F(\partial_t, p, m, x, y)$ are built out of the operators $\chi_e(\partial_t, x, y, z)$ and $\chi_m(\partial_t, x, y, z)$. The above equation can be derived from a variational principle with the action

14.3   Problems in transmission lines and waveguides

[1] Calculate using perturbation theory for partial differential equations, the approximate changes in the line voltage and current when the distributed parameter R of the line gets perturbed by a small non-uniform term $\delta R(z)$, ie, the line equations are

$$v_{,z}(t, z) + (R + \delta R(z))i(t, z) + Li_{,t}(t, z) = 0,$$

$$i_{,z}(t, z) + Gv(t, z) + Cv_{,t}(t, z) = 0$$

Quantize these line equations by first removing the dissipative terms involving $R + \delta R$ and G, and deriving the resulting line equations from an appropriate Hamiltonian after expanding the line equations as a Fourier series in the spatial variable z and then introduce Lindblad noise terms that enable us to model the dissipative effects.

[2] Given a coaxial cable having inner radius a and outer radius b and with the medium in between the two cylinders having parameters (ϵ, μ, σ), calculate the distributed line parameters.

[3] A transmission line at a given frequency has characteristic impedance $Z_0 = R_0 + jX_0$. The load connected to it has impedance $Z_L = R_L + jX_L$ at that frequency and a propagation constant γ. At a distance of d_1 from the load, a stub of characteristic impedance $Z_{01} = R_{01} + jX_{01}$, length l_1 and propagation constant γ_1 is attached. At a distance of $d_2 > d_1$, from the load, another stub of characteristic impedance $Z_{02} = R_{02} + jX_{02}$, length l_2 and propagation constant γ_2 is attached. Find the input impedance of the line at a distance $d_3 > d_2$ from the load. If this input impedance matches a line of characteristic impedance $Z_0' = R_0' + jX_0'$ attached to it at d_3, then calculate the values of d_1 and d_2. Explain how you would solve this problem using the Smith chart.

[4] Explain from first principles how you would calculate the location of the nth voltage maximum and voltage minimum from the load end of a transmission line when the load Z_L and characteristic impedance R_0 of the line are given. Assume that the line is lossless. Also calculate the VSWR of the line. Finally, explain how you would determine the reflection coefficient of the line at the

load end (both magnitude and phase) and the wavelength/propagation constant given the location of the nth voltage maximum, the distance between successive voltage maxima and the VSWR.

[5] Draw the constant r and constant x circles in the $Re(\Gamma) - Im(\Gamma)$ plane from the defining relation for the reflection coefficient

$$\Gamma = \frac{r + jx - 1}{r + jx + 1}$$

14.4   Problems in optimization theory

[1] State and prove Appolonius' theorem in an infinite dimensional Hilbert space $\mathcal{H}$ and use it to establish the orthogonal projection theorem for closed convex subsets of $\mathcal{H}$: If W is a closed and convex subset of $\mathcal{H}$, then for each $x \in \mathcal{H}$, there exists a unique vector $Px \in W$ such that

$$\| x - Px \| = inf_{w \in W} \| x - w \|$$

[2] Let $X = C^2[0, 1]$ denote the normed linear space of all twice continuously differentiable functions on $[0, 1]$. For $f : X \to \mathbb{R}$ and $x \in X$, we say that $Df(x) : X \to X^*$ exists (X^* denotes the Banach space of all bounded linear functions on X), if there exists a $y \in X^*$ such that for all $z \in X$, one has

$$lim_{\epsilon \to 0} |(f(x + \epsilon z) - f(x))/\epsilon - y(z)| = 0$$

In this case, we set $Df(x) = y$. Suppose f is given by

$$f(x) = \int_0^1 L(x(t), x'(t))dt, x \in X$$

where $L : \mathbb{R} \times \mathbb{R} \to \mathbb{R}$ is a twice continuously differentiable function of its arguments. Then prove that $Df(x)$ exists and is given by

$$Df(x)(z) = \int_0^t z(t)(\frac{\partial L(x(t), x'(t))}{\partial x} - \frac{d}{dt}\frac{\partial L(x(t), x'(t))}{\partial x'})dt$$

Justify all the conditions imposed on the spaces X and the function L.

[3] Write down the optimal Bellman-Hamilton-Jacobi equation for minimizing the expected cost

$$C(u) = \mathbb{E} \int_0^T L(x(t), x'(t), u(t))dt$$

where $x(t)$ satisfies the equations

$$dx(t) = x'(t)dt, dx'(t) = -\gamma x'(t)dt - Kx(t) + u(t) + \sigma dB(t)$$

where $B(.)$ is standard Brownian motion and

$$L(x(t), x'(t), u(t)) = x'^2(t)/2 + x^2(t)/2 + u^2(t)/2$$

The control function is $u(t)$ and it is constrained to be of the instantaneous feedback type, ie $u(t) = \chi_t(x(t), x'(t))$. Before doing this problem, you must first prove that the bivariate process $(x(t), x'(t))^T$ is a Markov process and calculate its infinitesimal generator.

[4] Consider a 2-D image field $g(x, y)$ obtained by rotating and translating a given image $f(x, y)$ and further adding an additive white Gaussian noise $w(x, y)$ to it with autocorrelation

$$\mathbb{E}(w(x, y)w(x', y')) = \sigma_w^2 \delta(x - x')\delta(y - y')$$

The transformed image is therefore given by

$$g(x, y) = f((x-a)cos(\theta) + (y-b)sin(\theta), -(x-a)sin(\theta) + (y-b)cos(\theta)) + w(x, y)$$

Using two dimensional Fourier transforms and one dimensional Fourier series, derive an algorithm for estimating the rotation angle θ and the translation vector (a, b) from measurements of $g(x, y)$ and $f(x, y)$ over the entire plane using the maximum likelihood method. Finally, calculate an approximate formula for the covariance matrix of the estimation error:

$$Cov((\hat{\theta}_{ML} - \theta, \hat{a}_{ML} - a, \hat{b}_{ML} - b))$$

in terms of $f, a, b, \theta, \sigma_w$. The approximation involves retaining only linear terms in the noise field.

[5] State the Peter-Weyl theorem. Let χ_1, χ_2 be the characters of two inequivalent irreducible representations of a compact group G. Prove using the Peter-Weyl theorem that

$$\int_G |\chi_1(g)|^2 dg = 1, \int_G \bar{\chi}_1(g)\chi_2(g)dg = 0$$

14.5 Another approach to quantization of wave-modes in a cavity resonator having non-uniform medium based on the scalar wave equation

The wave field $\psi(r)$ satisfies

$$(\nabla^2 + h(\omega, r)^2)\psi(r) = 0, r \in B, \psi(r) = 0, r \in \partial B$$

We write

$$h(\omega, r) = \omega^2/c^2 + \delta\lambda(\omega, r)$$

and

$$\psi(r) = \psi_0(r) + \delta\psi(r), \omega = \omega_0 + \delta\omega$$

Then, application of standard first order perturbation theory gives

$$(\nabla^2 + \omega_0^2/c^2)\psi_0(r) = 0, r \in B, \psi(r) = 0, r \in \partial B$$

$$(\nabla^2 + \omega_0^2/c^2)\delta\psi(r) + (2\omega_0\delta\omega/c^2)\psi_0(r) + \delta\lambda(\omega_0, r)\psi_0(r) = 0, r \in B,$$

$$\delta\psi(r) = 0, r \in \partial B$$

The solutions to the unperturbed equation are

$$\omega_0 = \omega_0[n], n = 1, 2, ..., \psi_0(r) = \psi_{0,n,k}(r), k = 1, 2, ..., d(n), n = 1, 2, ...$$

ie

$$(\nabla^2 + \omega_0[n]^2/c^2)\psi_{0,n,k}(r) = 0, k = 1, 2, ..., d(n)$$

where we may assume that $\{\psi_{0,n,k}, 1 \leq k \leq d(n), n \geq 1\}$ forms an onb for $L^2(B)$ since ∇^2 is a self-adjoint operator. In other words, the unperturbed mode $\omega_0[n]$ has a degeneracy of $d(n)$. Then, it follows by writing

$$\psi_0(r) = \sum_{k=1}^{d(n)} c(n, k)\psi_{0,n,k}(r), \omega_0 = \omega_0[n]$$

that

$$((\omega_0[n]^2 - \omega_0[m]^2)/c^2) < \psi_{0,m,s}, \delta\psi >$$

$$+ (2\omega_0[n]/c^2)\delta\omega \sum_k c(n, k) < \psi_{0,m,s}, \psi_{0,n,k} >$$

$$+ \sum_k c(n, k) < \psi_{0,m,s}, \delta\lambda(\omega_0[n], r)\psi_{0,n,k} >= 0$$

or equivalently,

$$((\omega_0[n]^2 - \omega_0[m]^2)/c^2) < \psi_{0,m,s}, \delta\psi > + (2\omega_0[n]/c^2)\delta\omega c(n, s)\delta_{m,n}$$

$$+ \sum_k c(n, k) < \psi_{0,m,s}, \delta\lambda(\omega_0[n], r)\psi_{0,n,k} >= 0$$

In particular, considering the cases $m = n$ and $m \neq n$ separately gives us

$$det(2\omega_0[n]/c^2)\delta\omega I_{d(n)} + ((< \psi_{0,n,s}, \delta\lambda(\omega)[n], r))\psi_{0,n,k} >))_{1 \leq s, k \leq d(n)}) = 0$$

which has solutions

$$\delta\omega = \delta\omega[n, k], k = 1, 2, ..., d(n)$$

and with the corresponding secular eigenvectors $((c_k(n,s)))_{s=1}^{d(n)}, k = 1, 2, ..., d(n)$, and for $m \neq n$,

$$< \psi_{0,m,s}, \delta\psi >= (\omega_0[m]^2 - \omega_0[n]^2)^{-1} \sum_k c(n,k) < \psi_{0,m,s}, \delta\lambda(\omega)[n], r)\psi_{0,n,k} >$$

where we substitute $c(n,k) = c_l(n,k), l = 1, 2, ..., d(n)$ to get $d(n)$ solutions for $\delta\psi(r)$. Thus, corresponding to the oscillation frequency $\omega_0[n] + \delta\omega[n,k]$, the wave field is

$$\sum_{s=1}^{d(n)} c_k(n,s)\psi_{0,n,s} + \delta\psi(r)$$

where

$$\delta\psi(r) = \sum_{m=\eta, 1 \leq s \leq d(m)} \psi_{0,m,s}(r)(\omega_0[m]^2 - \omega_0[n]^2)^{-1} \sum_l c_k(n,l) < \psi_{0,m,s}, \delta\lambda(\omega_0[n], r)\psi_{0,n,l} >$$

To carry out the quantization, we derive the above generalized wave equation from the action principle

$$\delta_\psi S[\psi] = 0$$

where

$$S[\psi] = \int \psi(t,r)(\nabla^2 + h(-i\partial_t, r)^2)\psi(t,r)dt d^3r = 0$$

Quantization can be performed by considering the Feynman path integral corresponding to this action between two states $|i>$ and $|f>$:

$$< f|S|i >= \int exp(iS[\psi])D\psi$$

To quantize this using Heisinberg operators, we expand $h(\omega,r)^2 = \omega^2/c^2 + \delta\lambda(\omega,r)$ as a power series in $j\omega$:

$$h(i\partial_t, r)^2 = (-1/c^2)\partial_t^2 + \sum_{m \geq 0} \delta\lambda_m(r)\partial_t^m$$

If we truncate this series at $m = N$, then we have the approximation $(c = 1)$

$$h(\partial_t, r)^2 = -\partial_t^2 + \sum_{m=0}^{N} \delta\lambda_m(r)\partial_t^m$$

and our modified wave equation in the time domain becomes

$$(\nabla^2 - \partial_t^2 + \sum_{m=0}^{N} \delta\lambda_m(r)\partial_t^m)\psi(t,r) = 0$$

The Lagrangian density for this equation is conveniently written by introducing a dual field $\phi(t,r)$:

$$\mathcal{L}(\phi, \partial_t^m \psi, 0 \le m \le N, \nabla \psi, \nabla^2 \psi) =$$

$$(1/2)\phi(\nabla^2 - \partial_t^2 + \sum_{m=0}^{N} \delta\lambda_m(r)\partial_t^m)\psi$$

which is equivalent to (ie differs by a partial derivative)

$$(1/2)[\partial_t\phi.\partial_t\psi - (\nabla\phi, \nabla\psi) + \phi\sum_{m=0}^{N}\lambda_m(r)\partial_t^m\psi]$$

The variational equations

$$\delta_\phi \int \mathcal{L}d^4x = 0, \delta_\psi \int \mathcal{L}d^4x = 0$$

give us the generalized wave equation along with its dual:

$$[\partial_t^2 - \nabla^2 - \sum_{m=0}^{N}\delta\lambda(r)\partial_t^m]\psi(tr,r) = 0,$$

$$[\partial_t^2 - \nabla^2 - \sum_{m=0}^{N}\lambda_m(r)(-1)^m\partial_t^m]\phi(t,r) = 0$$

To cast this Lagrangian in a form from which the Hamiltonian can be derived, we must allow only first order partial derivatives w.r.t. time to appear in the picture. To this end, we define the auxiliary fields

$$\psi_k = \partial_t^k\psi, k = 1, 2, ..., N-1$$

and then the field equation can be expressed as a sequence of first order in time equations:

$$\partial_t\psi_k = \psi_{k+1}, k = 0, 1, ..., N-2, \psi_0 = \psi,$$

$$\lambda_N(r)\partial_t\psi_{N-1} + \sum_{m=0}^{N-1}\lambda_m(r)\psi_m(t,r) + \nabla^2\psi(t,r) - \psi_1 = 0$$

These equations can be derived from a Lagrangian density after introducing dual fields $\phi_k, k = 0, 1, ..., N$:

$$\mathcal{L}(\phi_k, \psi_k, \partial_t\psi_k, k = 0, 1, ..., N-1) =$$

$$\sum_{k=0}^{N-2}\phi_k(\partial_t\psi_k - \psi_{k+1})$$

$$+\phi_{N-1}(\lambda_N(r)\partial_t\psi_{N-1} + \sum_{m=0}^{N-1}\lambda_m(r)\psi_m(t,r) + \nabla^2\psi(t,r) - \psi_1)$$

Problem: By applying the Legendre transformation to this Lagrangian density, write down the corresponding Hamiltonian density and discuss the constraints involved.

14.6   Derivation of the general structure of the field dependent permittivity and permeability of a plasma

We start with the Boltzmann equation for the particle distribution function $f(t, r, v)$ with the relaxation time approximation used in place of the collision term:

$$f_{,t}(t,r,v)+(v,\nabla_r)f(t,r,v)+(q/m)(E(t,r)+v\times B(t,r),\nabla_v)f(t,r,v)$$
$$= (f_0(t,r,v)-f(t,r,v))/\tau(v)$$

We write the frequency domain solution as

$$\hat{f}(\omega,r,v) = \tau(v)^{-1}[j\omega+1/\tau(v)+(v,\nabla_r)+(q/m)(E(j\partial/\partial\omega,r)+v\times B(j\partial/\partial\omega,r),\nabla_v)]^{-1}\hat{f}_0(\omega,r,v)$$

and the current density in the frequency domain is given by

$$\hat{J}(\omega,r) = q\int v\hat{f}(\omega,r,v)d^3v$$

while the charge density is

$$\hat{\rho}(\omega,r) = q\int \hat{f}(\omega,r,v)d^3v$$

The Maxwell equations are

$$\nabla.\hat{E}(\omega,r) = \hat{\rho}(\omega,r)/\epsilon_0, \nabla.\hat{B}(\omega,r) = 0,$$

$$\nabla\times\hat{E}(\omega,r) = -j\omega\hat{B}(\omega,r), \nabla\times\hat{B}(\omega,r) = \mu_0\hat{J}(\omega,r) + j\omega\mu_0\epsilon_0\hat{E}(\omega,r)$$

and solving these equations gives us $\hat{J}(\omega,r)$ as a functional of $\hat{E}$ and $\hat{B}$. Equating $\hat{J}(\omega,r)$ to $(\sigma + j\omega\epsilon)\hat{E}(\omega,r)$ gives us σ and ϵ as nonlinear functionals of $\hat{E}$ and $\hat{B}$.

14.7   Other approaches to calculating the permittivity and permeability of a plasma via the use of Boltzmann's kinetic transport equation

Let the unperturbed electrostatic potential be $\Phi(r)$. The corresponding equilibrium particle density in phase space is given by

$$f_0(r,v) = N.exp(-\beta(mv^2/2 + q\Phi(r)))/Z(\beta)$$

where N is the total number of particles and $Z(\beta)$ is the classical partition function

$$Z(\beta) = \int exp(-\beta(mv^2/2 + q\Phi(r)))d^3rd^3v$$

It satisfies the equilibrium kinetic equation

$$[(v, \nabla_r) + (q/m)(-\nabla\Phi(r), \nabla_v)]f_0(r, v) = 0$$

On application of an external electromagnetic field $E(t, r), B(t, r)$ assumed to be small, the particle distribution function gets perturbed to

$$f(t, r, v) = f_0(r, v) + \delta f(t, r, v)$$

which satisfies upto first order terms, the perturbed kinetic equation

$$i\delta f_{,t}(t, r, v) + (v, \nabla_r)\delta f(t, r, v) - (q/m)(\nabla\Phi(r), \nabla_v)\delta f$$
$$+ (q/m)(E(t, r) + v \times B(t, r), \nabla_v)f_0(r, v) = -\delta f(t, r, v)/\tau(v)$$

The term involving the magnetic field cancels out and and in the special case when $\Phi = 0$, this simplifies to

$$i\delta f_{,t} + (v, \nabla_r)\delta f + (q/m)(E(t, r), \nabla_v)f_0 + \delta f/\tau(v) = 0$$

which gives on Fourier transforming,

$$[(i\omega + 1/\tau(v)) + (v, \nabla_r)]\hat{\delta}f(\omega, r, v) = (q\beta)(\hat{E}(\omega, r), v)f_0(r, v)$$

The solution to this equation is given by

$$\delta f(\omega, r, v) = \int K(\omega, v, r - r')(\hat{E}(\omega, r'), v)f_0(r', v)d^3r'$$

where
$$K(\omega, v, r) =$$

$$q\beta(2\pi)^{-3} \int exp(ik.r)[i\omega + 1/\tau(v) + i(v, k)]^{-1}d^3k$$

The current density in the plasma medium is therefore given by

$$q \int v\hat{\delta}f(\omega, r, v)d^3v = \hat{J}(\omega, r) = \mu_0^{-1}\nabla \times \hat{B}(\omega, r) - i\omega\epsilon_0\hat{E}(\omega, r)$$

We have in fact,

$$\hat{J}(\omega, r) = q \int vK(\omega, v, r - r')(v, \hat{E}(\omega, r'))f_0(r', v)d^3r'd^3v$$

from which the complex anisotropic permittivity matrix $\epsilon(\omega, r)$ defined by

$$\hat{J}(\omega, r) = j\omega \int \epsilon(\omega, r - r')E(\omega, r')d^3r'$$

can be read of immediately. This is the linear theory of permittivity. It should be noted that the complex permittivity contains the plasma conductivity as a component. When $\Phi \neq 0$, things are a little more complicated Formally, we can write the solution as

$$\hat{\delta f}(\omega, r, v) = q\beta[i\omega + 1/\tau(v) + (v, \nabla_r) - (q/m)(\nabla\Phi(r), \nabla_v)]^{-1}((\hat{E}(\omega, r), v)f_0(r, v))$$

Formally, denoting the kernel of the operator $(i\omega + 1/\tau(v) + (v, \nabla_r))^{-1}$ by $K = K(\omega, v, r - r')$, we can write the kernel of the operator $(q/m)[i\omega + 1/\tau(v) + (v, \nabla_r) - (q/m)(\nabla\Phi(r), \nabla_v)]^{-1}$ as

$$(q\beta)[K + \sum_{n \geq 1}(KT)^n K]$$

where

$$T = (q/m)(\nabla\Phi(r), \nabla_v)$$

In this way the current density in the frequency domain is

$$\hat{J}(\omega, r) = (q\beta) \int v[K + \sum_{n \geq 1}(KT)^n K](\hat{E}(\omega, r), \nabla_v)f_0(r, v)d^3v$$

from which we easily read out the inhomogeneous frequency dependent permittivity.

Taking non-linearities into account. Let $J(\omega, r)$ be the current density. We write the Boltzmann equation in perturbation theoretic form:

$$\delta f_{,t} + (v, \nabla_r)(f_0 + \delta f) + (q/m)(-\nabla\Phi(r) + \delta E(t, r), \nabla_v)(f_0 + \delta f)$$
$$+(q/m)(\delta E(t, r) + v \times \delta B(t, r), \nabla_v)\delta f + \delta f/\tau(v) = 0$$

Expanding

$$\delta f = \sum_{n \geq 1} \delta_n f$$

we have on equating order zero terms,

$$(v, \nabla_r)f_0 - (q/m)(\nabla\Phi(r), \nabla_v)f_0 = 0$$

which is satisfied by the above Gibbsian f_0. Equating n^{th} order terms gives us

$$\delta_n f_{,t} + (v, \nabla_r)\delta_n f - (q/m)(\nabla\Phi(r), \nabla_v)\delta_n f$$
$$+(q/m)(\delta E + v \times \delta B, \nabla_v)\delta_{n-1}f + \delta_n f/\tau(v) = 0, n \geq 1$$

where

$$\delta_0 f = f_0$$

whose solution can be expressed as

$$\delta_n f(t, r, v) =$$
$$(q/m)[i\partial/\partial t + 1/\tau(v) + (v, \nabla_r) - (q/m)(\nabla\Phi(r), \nabla_v)]^{-1}(\delta E(t, r) + v \times \delta B(t, r),$$
$$\nabla_v)\delta_{n-1}f(t, r, v)], n \geq 1$$

By iterating this equation, we can express $\delta f(t, r, v) = \sum_{n \geq 1} \delta_n f(t, r, v)$ as a Volterra series in the electric and magnetic fields $\delta E(t, r), \delta B(t, r)$ and hence determine the current density as a similar Volterra series in the electric and magnetic fields:

14.8 Derivation of the permittivity and permeability functions using quantum statistics

In the presence of an external electric and magnetic field and taking into account quantum noise in the sense of Hudson and Parthasarathy, the Schrodinger evolution equation is

$$dU(t) = (-(iH(t) + P)dt + L_1 dA(t) + L_2 dA(t)^* + Sd\Lambda(t))U(t)$$

where

$$H(t) = (\alpha, P + eA(t,r)) + \beta m - eV(t,r)$$

Solving this equation gives us the state of the system at time t as

$$\rho_s(t) = Tr_2(U(t)(\rho_s(0) \otimes \rho_{env}(0))U(t)^*)$$

and hence, the average electric dipole moment and magnetic dipole moment can be computed using respectively the equations

$$p(t) = -eTr(\rho_s(t)r), m(t) = (e/2m)Tr(\rho_s(t)(L + g\sigma/2))$$

where $L = r \times P = -ir \times \nabla$.

14.9 Approximate discrete time nonlinear filtering for non-Gaussian process and measurement noise

(Summary of Rohit Singh's Ph.D work).

The process to be estimated on a real time basis is a Markov process $X(n), n \geq 0$ with one step transition probability distribution $P_n(x, dy) = p_n(x, y)dy$, ie,

$$Pr(X(n+1) \in dy|X(n) = x) = p_n(x, y)dy$$

The measurement process is

$$y(n) = h_n(X(n) + v(n))$$

where $v(n)$ is iid noise with pdf p_v. The measurement data upto time n is given by

$$Y_n = \sigma(y(m), m \leq n)$$

and the MAP estimate of the process X at time n is

$$\hat{X}(n) = argmax_{X(n)} p(X(n)|Y_n)$$

We wish to calculate $\hat{X}(n)$ recursively, ie, on a real time basis. Using Bayes' rule, and the Markov property, we get

$$p(X(n+1)|Y_{n+1}) = p(y(n+1), Y_n, X(n+1))/P(y(n+1), Y_n) = A/B$$

$$A = \int p(y(n+1)|X(n+1))p(X(n+1)|X(n))p(X(n)|Y_n)dX(n)$$

$$B = \int AdX(n+1)$$

so the MAP estimate of $X(n+1)$ given Y_{n+1} is given by

$$\hat{X}(n+1) = -argmax_{X(n+1)}A$$

Write

$$X(n+1) = f_n(X(n)) + W(n+1)$$

where W is an iid process. Then,

$$A = \int p_v(y(n+1) - h_{n+1}(X(n+1)))p_w(X(n+1) - f_n(X(n))p(X(n)|Y_n)dX(n)$$

Write

$$X(n+1) = f_n(\hat{X}(n)) + \delta X$$

Then, we have approximately,

$$h_{n+1}(X(n+1)) = h_{n+1}(f_n(\hat{X}(n)) + \delta X) =$$

$$h_{n+1}(f_n(\hat{X}(n))) + h'_{n+1}(f_n(\hat{X}(n))\delta X + (1/2)h''_{n+1}(f_n(\hat{X}(n))(\delta X \otimes \delta X)$$

and writing

$$e(n+1) = y(n+1) - h_{n+1}(f_n(\hat{X}(n)))$$

we get approximately.

$$p_v(y(n+1) - h_{n+1}(X(n+1)) =$$

$$p_v(e(n+1)) + p'_v(e(n+1))h'_{n+1}(f_n(\hat{X}(n))))\delta X + (1/2)\delta X^T [p'_v(e(n+1))h''_{n+1}(f_n(\hat{X}(n)))$$

$$+ h'_{n+1}(f_n(\hat{X}(n))^T p''_v(e(n+1))h'_{n+1}(f_n(\hat{X}(n))))]\delta X$$

Likewise,

$$p_w(X(n+1) - f_n(X(n)) = p_w(f_n(\hat{X}(n)) + \delta X - f_n(X(n)))$$

$$= p_w(f_n(\hat{X}(n)) - f_n(X(n)) + p'_w(f_n(\hat{X}(n)) - f_n(X(n)))\delta X + (1/2)p''_w(f_n(\hat{X}(n))$$

$$- f_n(X(n)))\delta X \otimes \delta X$$

and hence upto $O(|\delta X|^2)$, we have

$$A = \int A(X(n))p(X(n)|Y_n)dX(n)$$

where

$$A(X(n)) = p_v(e(n+1))p_w(f_n(\hat{X}(n)) - f_n(X(n))) +$$

$$[p'_v(e(n+1))h'_{n+1}(f_n(\hat{X}(n)))) + p'_w(f_n(\hat{X}(n)) - f_n(X(n)))]\delta X$$

$$+ (1/2)\delta X^T [[p'_v(e(n+1))h''_{n+1}(f_n(\hat{X}(n))) + p''_w(f_n(\hat{X}(n)) - f_n(X(n)))$$

$$+2p'_w(f_n(\hat{X}(n)) - f_n(X(n)))^T p'_v(e(n+1))h'_{n+1}(f_n(\hat{X}(n))))]\delta X$$

$$= P_1(n, e(n+1), \hat{X}(n), X(n))^T \delta X + (1/2)\delta X^T P_2(n, e(n+1), \hat{X}(n), X(n))\delta X$$

say, plus terms that are independent of δX. Maximizing this over δX gives us

$$\hat{X}(n+1) = f_n(\hat{X}(n)) - P_2(n, e(n+1), \hat{X}(n))^{-1} P_1(n, e(n+1), \hat{X}(n))$$

where

$$P_1(n, e(n+1), \hat{X}(n)) = \int P_1(n, e(n+1), \hat{X}(n), X(n)) p(X(n)|Y_n) dX(n),$$

$$P_2(n, e(n+1), \hat{X}(n)) = \int P_2(n, e(n+1), \hat{X}(n), X(n)) p(X(n)|Y_n) dX(n)$$

We can now also compute the error covariance matrix update equation:

$$X(n+1) - \hat{X}(n+1) = f_n(X(n)) + w(n+1) - f_n(\hat{X}(n)) + \psi(n, e(n+1), \hat{X}(n))$$

where

$$\psi(n, e(n+1), \hat{X}(n)) = -P_2(n, e(n+1), \hat{X}(n))^{-1} P_1(n, e(n+1), \hat{X}(n))$$

and then we get approximately with

$$X(n) - \hat{X}(n) = E(n),$$

$$E(n+1) = f'_n(\hat{X}(n))E(n) + w(n+1) + \psi(n, y(n+1) - h_{n+1}(f_n(\hat{X}(n))), \hat{X}(n))$$

$$= f'_n(\hat{X}(n))E(n) + w(n+1) + \psi(n, h_{n+1}(f_n(X(n)) + w(n+1)) - h_{n+1}(f_n(\hat{X}(n))) + v(n+1), \hat{X}(n))$$

$$= f'_n(\hat{X}(n))E(n) + w(n+1) + \psi(n, (h_{n+1} \circ f_n)'(\hat{X}(n)))E(n) + h'_{n+1}(f_n(X(n))w(n+1) + v(n+1), \hat{X}(n))$$

$$= f'_n(\hat{X}(n))E(n) + w(n+1) + \psi(n, 0, \hat{X}(n)) + \psi_{,2}(n, 0, \hat{X}(n))((h_{n+1} \circ f_n)'(\hat{X}(n))E(n) +$$

$$h'_{n+1}(f_n(X(n))w(n+1) + v(n+1))$$

If we assume that $E(n)$ is orthogonal to the σ algebra generated by Y_n, then we get approximately for the covariance $P(n+1)$ of $E(n+1)$,

$$P(n+1) = Q_1[n]P[n]Q_1[n]^T + Q_2[n]P_v[n]Q_2[n]^T + Q_3[n]P_w[n]Q_3[n]^T$$

where

$$Q_1[n] = f'_n(\hat{X}(n)) + \psi_{,2}(n, 0, \hat{X}(n))(h_{n+1} \circ f_n)'(\hat{X}(n))$$

14.10   Quantum theory of many body systems with application to current computation in a Fermi liquid

The Fermi operator field is $\psi_a(r)$. They satisfy the anticommutation relations

$$[\psi_a(r), \psi_b(r')^*]_+ = \delta_{ab}\delta^3(r - r'),$$

$$[\psi_a(r), \psi_b(r')]_+ = 0, [\psi_a(r)^*, \psi_b(r')^*]_+ = 0$$

The Hartree-Fock Hamiltonian of the fluid taking into account interactions with external fields as well as internal interactions is

$$H = \sum_a \int \psi_a(r)^*(-\nabla^2/2m)\psi_a(r)d^3r + \sum_{ab} \int V_{ab}(r, r')\psi_a(r)^*\psi_b(r')d^3rd^3r'$$

$$+ \sum_{abcd} \int V_{abcd}(r_1, r_2, r_3, r_4)\psi_a(r_1)^*\psi_b(r_2)^*\psi_c(r_3)^*\psi_d(r_4)d^3r_1d^3r_2d^3r_3d^3r_4$$

A special case of this Hamiltonian is

$$H = \sum_a \int \psi_a(r)^*(-\nabla^2/2m)\psi_a(r)d^3r + \sum_a \int V_a(r)\psi_a(r)^*\psi_a(r)d^3r$$

$$+ \sum_{a,b} \int V_{ab}(r, r')\psi_a(r)^*\psi_a(r)\psi_b(r')^*\psi_b(r')d^3rd^3r'$$

The potential terms in this latter Hamiltonian correspond to the number $\psi(r)^*\psi_a(r)d^3r$ of Fermions of type a interacting with an external potential $V_a(r)$ and numbers $\psi_a(r)^*\psi_a(r)d^3r$ of particles of type a at r interacting with numbers $\psi_b(r')^*\psi_b(r')d^3r'$ of particles of type b at r' via an interaction potential $V_{ab}(r, r')$. We shall work with this latter Hamiltonian as it has a nice physical interpretation. The above Fermion anticommutation rules give us on writing

$$\psi_a(t, r) = exp(itH)\psi_a(r).exp(-itH),$$

the same anticommutation rules for the time dependent field operators $\psi_a(t, r)$ for a fixed time t and the fact that H is a constant of the motion, ie, $exp(itH).H.exp(-itH) =$
H so that H is unchanged when $\psi_a(r)$ is replaced by $\psi_a(t, r)$ etc. We may also allow the potentials V_a, V_{ab} to depend explicitly on time in situations describing the application of a time varying voltage to a Fermi liquid or to a superconductor. In such a case, however, the Hamiltonian would not be a constant, it would depend explicitly on time and we write $H(t)$ for it. However the anticommutation relations between the field operators would still remain the same since $\psi_a(t, r) = U(t)^*\psi_a(r)U(t)\forall a, r$ where $U(t)$ is a unitary operator defined by

$$U(t) = T\{exp(-i \int_0^t H(s)ds)\}$$

The aim is to compute the average current in the liquid. For this, we assume that at the start, the liquid is in a Gibbs state $\rho(0) = exp(-\beta H_0)/Z(\beta)$, $Z(\beta) = Tr(exp(-\beta H_0))$ and H_0 is obtained by replacing $V_a(t,r)$ by $V_{a0}(r)$ and $V_{ab}(t,r,r')$ by $V_{ab0}(r,r')$ with $V_a(t,r) = V_{a0}(r) + \delta V_a(t,r)$, $V_{ab}(t,r,r') = V_{ab0}(r,r') + \delta V_{ab}(t,r,r')$. We regard $\delta V_a(t,r)$ and $\delta V_{ab}(t,r,r')$ as being the applied external potentials to the system. In the linear response theory, we compute the average current as a linear functional of these external potentials. The Hamiltonian can thus be expressed as $H(t) = H_0 + \delta H(t)$ and the state of the system as $\rho_0(t) + \delta\rho(t)$ where upto linear orders,

$$\rho_0'(t) = -i[H_0, \rho_0(t)]$$

which is satisfied by $\rho_0(t) = \rho(0)$ and

$$\delta\rho'(t) = -i[H_0, \delta\rho(t)] - i[\delta H(t), \rho(0)]$$

whose solution is

$$\delta\rho(t) = -i\int_0^t exp(-i(t-s)H_0)[\delta H(t), \rho(0)].exp(i(t-s)H_0)ds$$

The Heisenberg field operators $\psi_a(t,r)$ satisfy

$$\psi_{a,t}(t,r) = i[H_0 + \delta H(t), \psi_a(t,r)]$$

and we can also apply first order perturbation theory to this to get

$$\psi_a(t,r) = \psi_{a0}(t,r) + \delta\psi_a(t,r),$$

$$\psi_{a0,t}(t,r) = i[H_0, \psi_{a0}(t,r)]$$

so that

$$\psi_{a0}(t,r) = exp(itH_0)\psi_a(r).exp(-itH_0)$$

$$\delta\psi_a(t,r) = i[H_0, \delta\psi_a(t,r)] + i[\delta H(t), \psi_{a0}(t,r)]$$

with solution

$$\delta\psi_a(t,r) = i\int_0^t exp(i(t-s)H_0).[\delta H(s), \psi_{a0}(s,r)].exp(-i(t-s)H_0)ds$$

The current density operator at time t is

$$J(t,r) = (-ie/2m)(\psi_a(t,r)^*\nabla\psi_a(t,r) - \psi_a(t,r)\nabla\psi_a(t,r)^*)$$

and in the presence of an external magnetic field described by the magnetic vector potential $A(r)$, this current gets modified to

$$J(t,r) = (e/2m)(\psi_a(t,r)^*(-i\nabla + eA(r))\psi_a(t,r) - \psi_a(t,r)(-i\nabla - eA(r))\psi_a(t,r)^*)$$

summation over the repeated index a being implied. The Hamiltonian used in this case should be modified to

$$H(t) = T + V_1(t) + V_2(t)$$

where

$$T = \int \psi_a(r)^* ((-i\nabla + eA(r))^2/2m)\psi_a(r)d^3r +$$

$$\int V_a(t,r)\psi_a(r)^*\psi_a(r)d^3r + \int V_{ab}(r,r')\psi_a(r)^*\psi_a(r)\psi_b(r')^*\psi_b(r')d^3rd^3r'$$

and the average current is

$$<J>(t,r) = Tr(\rho(0)J(t,r))$$

which can be calculated using the linear response theory described above. The Heisenberg equations for

$$\psi_a(t,r) = U(t)^*\psi_a(r)U(t)$$

with

$$U(t) = T\{exp(-i\int_0^t H(s)ds)\}$$

are

$$\psi_{,t}(t,r) = U(t)^* i[H(t),\psi_a(r)]U(t) = i[U(t)^* H(t)U(t), U(t)^*\psi_a(r)U(t)]$$

$$= i[\tilde{H}(t),\psi_a(t,r)]$$

where $\tilde{H}(t)$ is obtained from $H(t)$ by replacing $\psi_a(r)$ with $\psi_a(t,r)$, and the commutation and anticommutation relations between the operators $\psi_a(t,r), \psi_a(t,r)^*$ are the same as those between the $\psi_a(r), \psi_a(r)$ after premultiplying the result with $U(t)^*$ and postmultiplying it with $U(t)$. We therfore compute

$$[H(t),\psi_a(r)] = [T,\psi_a(r)] + [V_1(t),\psi_a(r)] + [V_2(t),\psi_a(r)]$$

$$T\psi_a(r)] = (-1/2)\int \psi_b(r')^*\nabla'^2\psi_b(r')d^3r'\psi_a(r)$$

$$= (-1/2)\int \psi_b(r')^*\nabla'^2\psi_b(r')\psi_a(r)d^3r'$$

$$= (1/2)\int \psi_b(r')^*\psi_a(r)\nabla'^2\psi_b(r')d^3r'$$

$$= (1/2)\int (\delta_{ab}\delta^3(r-r') - \psi_a(r)\psi_b(r')^*)\nabla'^2\psi_b(r')d^3r'$$

$$= (1/2)\nabla^2\psi_a(r) + \psi_a(r)T$$

Equivalently,

$$[T,\psi_a(r)] = (1/2)\nabla^2\psi_a(r)$$

Next,

$$V_1(t)\psi_a(r) = (\int V_b(t,r')\psi_b(r')^*\psi_b(r')d^3r')\psi_a(r) =$$

$$= -\int V_b(t,r')\psi_b(r')^*\psi_a(r)\psi_b(r')d^3r' =$$

$$-\int V_b(t,r')(\delta_{ab}\delta^3(r-r') - \psi_a(r)\psi_b(r')^*)\psi_b(r')d^3r'$$

$$= -V_a(t,r)\psi_a(r) + \psi_a(r)V_1(t)$$

or equivalently,

$$[V_1(t), \psi_a(r)] = -V_a(t,r)\psi_a(r)$$

Thus,

$$[T + V_1(t), \psi_a(r)] = ((1/2)\nabla^2 - V_a(t,r))\psi_a(r)$$

Finally,

$$V_2(t)\psi_a(r) = \int V_{bc}(t,r',r'')\psi_b(r')^*\psi_b(r')\psi_c(r'')^*\psi_c(r'')\psi_a(r)d^3r'd^3r''$$

14.11 Optimal control of gravitational, matter and em fields

The perturbed Einstein-Maxwell equations are

$$\delta R^{\mu\nu} - (1/2)\delta R.\eta^{\mu\nu} = -8\pi G(\delta T^{\mu\nu} + \delta S^{\mu\nu})$$

where

$$g_{\mu\nu} = \eta_{\mu\nu} + \delta g_{\mu\nu}$$

is a small perturbation of the flat space-time metric,

$$\delta T^{\mu\nu} = (1+p'(\rho_0))\delta\rho.V^\mu V^\nu + (\rho_0+p(\rho_0))(V^\mu\delta v^\nu + V^\nu\delta v^\mu) - p'(\rho_0)\delta\rho\eta^{\mu\nu}$$
$$-p(\rho_0)\eta^{\mu\alpha}\eta^{\nu\beta}\delta g_{\alpha\beta}$$

is a small perturbation of the energy-momentum tensor of the matter field and finally

$$\delta S^{\mu\nu} = (1/4)F^{(0)\alpha\beta}F^{(0)}_{\alpha\beta}\eta^{\mu\rho}\eta^{\nu\sigma}\delta g_{\rho\sigma} - (1/2)\eta^{\mu\nu}F^{(0)\alpha\beta}\delta F_{\alpha\beta}$$

is a small perturbation of the energy-momentum tensor of the em field. We can write

$$G^{\mu\nu} = R^{\mu\nu} - (1/2)Rg^{\mu\nu},$$

$$\delta G^{\mu\nu} = C_1(\mu\nu\alpha\beta\rho\sigma)\delta g_{\alpha\beta,\rho\sigma}$$

$$\delta S^{\mu\nu} = C_2(\mu\nu\alpha\beta,x)\delta g_{\alpha\beta}(x) + C_3(\mu\nu\alpha\beta,x)\delta F_{\alpha\beta}(x)$$

where C_2 and C_3 are functions of the unperturbed em field tensor $F^{(0)}_{\alpha\beta}(x)$. Finally, we can write

$$\delta T^{\mu\nu}(x) = C_4(\mu\nu\alpha, x)\delta v^\alpha(x) + C_5(\mu\nu, x)\delta\rho(x) + C_6(\mu\nu\alpha\beta, x)\delta g_{\alpha\beta}(x)$$

As a result of the Einstein field equations, we have the perturbed fluid dynamical equations (MHD)

$$\delta((T^{\mu\nu} + S^{\mu\nu})_{:\nu}) = 0$$

Also the perturbed Maxwell equations are

$$\delta(F^{\mu\nu}\sqrt{-g}))_{,\nu} = \delta(J^\mu\sqrt{-g})$$

which can be expressed in the form

$$(\delta(g^{\mu\alpha}g^{\nu\beta}\sqrt{-g})_{,\nu}F^{(0)}_{\alpha\beta} +$$

$$+\delta(g^{\mu\alpha}g^{\nu\beta}\sqrt{-g})F^{(0)}_{\alpha\beta,\nu} +$$

$$\eta^{\mu\alpha}\eta^{\nu\beta}\delta F_{\alpha\beta,\nu} = J^{(0)\mu}\delta\sqrt{-g} + \delta J^\mu$$

Since

$$S^{\mu\nu}_{:\nu} = F^{\mu\nu}J_\nu$$

we can write the perturbed MHD equations as

$$\delta(T^{\mu\nu}_{:\nu}) = \delta(F^{\mu\nu}J_\nu)$$

Now,

$$T^{\mu\nu}_{:\alpha} = T^{\mu\nu}_{,\alpha} + \Gamma^\mu_{\alpha\rho}T^{\nu\rho} + \Gamma^\nu_{\alpha\rho}T^{\mu\rho}$$

so that since the unperturbed Christoffel symbols are zero, we get

$$\delta(T^{\mu\nu}_{:\nu}) = (\delta T^{\mu\nu})_{,\nu}$$

$$+T^{(0)\nu\rho}\delta\Gamma^\mu_{\alpha\rho} + T^{(0)\mu\rho}\delta\Gamma^\nu_{\alpha\rho}$$

We note that these equations have to be modified if the background space-time, ie, unperturbed space-time is curved. We also note that

$$\delta\Gamma^\mu_{\alpha\beta} = \delta(g^{\mu\nu}\Gamma_{\nu\alpha\beta})$$

$$= \eta^{\mu\nu}\delta\Gamma_{\nu\alpha\beta} - \eta^{\mu\rho}\eta^{\nu\sigma}\Gamma^{(0)}_{\nu\alpha\beta}\delta g_{\rho\sigma}$$

$$= (1/2)\delta g^\mu_{\alpha,\beta} + \delta g^\mu_{\beta,\alpha} - \delta g^{,\mu}_{\alpha\beta})$$

since

$$\Gamma^{(0)}_{\nu\alpha\beta} = 0$$

because the unperturbed space-time is flat. We note that

$$\delta F_{\alpha\beta} = \delta A_{\beta,\alpha} - \delta A_{\alpha,\beta}$$

Now, denoting the set of fields $\delta g_{\mu\nu}(x), \delta A_\mu(x), \delta v^\mu(x), \delta\rho(x)$ by the symbols $\phi_k(x)$ and the control current sources $\delta J^\mu(x)$ by $s_m(x)$, the above linearized field equations have the general structure

$$C_1(r, k, \mu\nu, x)\phi_{k,\mu\nu}(x) + C_2(r, k, \mu, x)\phi_{k,\mu}(x) + C_3(r, k, x)\phi_k(x)$$
$$= s_r(x), r = 1, 2, ..., p$$

with the summation being over the repeated indices. In arriving at this general form, we have chosen a specific coordinate system so that only six out of the ten metric coefficients $\delta g_{\mu\nu}(x)$ are independent and further that only three out of the four $\delta v^\mu(x)$ are independent, these being related by

$$0 = \delta(g_{\mu\nu}v^\mu v^\nu) = 2\eta_{\mu\nu}V^\mu \delta v^\nu + V^\mu V^\nu \delta g_{\mu\nu}$$

Subject to these equations of motion, we wish to select the control input fields $s_r(x)$ so that the response field functions ϕ_k are as close as possible in distance to given fields. Using Lagrange multiplier fields $\lambda_r(x)$, we may therefore consider the problem of minimizing

$$L(\phi_k, s_k, \lambda_k) = \sum_k \int w_k(x)(\phi_k(x) - \phi_k^d(x))^2 d^4x$$

$$-\sum_k \int (C_1(r, k, \mu\nu, x)\phi_{k,\mu\nu}(x) + C_2(r, k, \mu, x)\phi_{k,\mu}(x) + C_3(r, k, x)\phi_k(x)$$
$$-s_r(x))\lambda_r(x)d^4x$$

with summation over the repeated indices being implied.

14.12 Calculating the modes in a cylindrical cavity resonator with a partition in the middle

. The cylinder is over the length $0 \leq z \leq d = d_1 + d_2$ and the partition is at $z = d_1$. The radius of the cylinder is R and the parameters of the medium for $0 \leq z \leq d_1$ are (ϵ_1, μ_1) and those of the medium for $d_1 < z < d_1 + d_2$ are (ϵ_2, μ_2). The side walls at $\rho = R$ and the top and bottom walls at $z = 0, d$ are perfect conductors. The standard formulae relating the transverse components of the electric and magnetic fields to the longitudinal components are

$$E_{k\perp} = (1/h_k^2)\partial/\partial z(\nabla_\perp E_{kz}) - (j\omega\mu_k/h^2)\nabla_\perp H_{kz} \times \hat{z},$$

$$H_{k\perp} = (1/h_k^2)\partial/\partial z(\nabla_\perp H_{kz}) + (j\omega\epsilon_k/h_k^2)\nabla_\perp E_{kz} \times \hat{z}$$

$k = 1, 2$. $k = 1$ stands for the bottom medium and $k = 2$ for the top medium. The boundary conditions are that H_z and $E_\perp$ vanish at $z = 0, d$, $\mu H_z, \epsilon E_z, H_\perp, E_\perp$ are continuous at $z = d_1$ and E_z, H_ρ vanish at $\rho = R$. These give us

$$(\nabla_\perp^2 + h_k^2)(E_{kz}, H_{kz}) = 0, k = 1, 2$$

with different $h'_k s$ for the electric and magnetic field. The solutions that match these boundary conditions are

$$E_{1z} = A_1 J_m(h_1\rho)(P_1 cos(m\phi) + P_2 sin(m\phi))cos(\alpha_1 z)$$

$$E_{2z} = A_2 J_m(h_2\rho)(P_1 cos(m\phi) + P_2 sin(m\phi))cos(\alpha_2(d - z))$$

$$H_{1z} = B_1 J_m(h'_1\rho)(Q_1 cos(m\phi) + Q_2 sin(m\phi))sin(\beta_1 z),$$

$$H_{2z} = B_2 J_m(h'_2\rho)(Q_1 cos(m\phi) + Q_2 sin(m\phi))sin(\beta_2(d - z))$$

with

$$\epsilon_1 A_1 cos(\alpha_1 d_1) = \epsilon_2 A_2 cos(\alpha_2 d_2),$$

$$h_k = \alpha_m[n], h'_k = \beta_m[n], k = 1, 2$$

where $\alpha_m[n]$ are the roots of $J_m(x)$ while $\beta_m[n]$ are the roots of $J'_m(x)$. Matching of $E_\perp$ at $z = d_1$ gives us

$$\alpha_1 A_1 sin(\alpha_1 d_1) = -\alpha_2 A_2 sin(\alpha_2 d_2)$$

so we get

$$\alpha_2\epsilon_1 cos(\alpha_1 d_1)sin(\alpha_2 d_2) + \alpha_1\epsilon_2 sin(\alpha_1 d_1)cos(\alpha_2 d_2) = 0$$

Further,

$$\mu_1 B_1 sin(\beta_1 d_1) = \mu_2 B_2 sin(\beta_2 d_2),$$

$$\beta_1\mu_1 B_1 cos(\beta_1 d_1) = -\beta_2\mu_2 B_2 cos(\beta_2 d_2)$$

which gives

$$\beta_2\mu_1 sin(\beta_1 d_1)cos(\beta_2 d_2) + \beta_1\mu_2 cos(\beta_1 d_1)sin(\beta_2 d_2) = 0$$

Thus, if we know α_1, β_1, we can determine α_2, β_2. But what are the possible values of α_1, β_1 that guarantee oscillations ? For the TM mode, we have using the wave equation for E_z,

$$\omega^2\mu_1\epsilon_1 - \alpha_1^2 = \alpha_m[n]^2,$$

$$\omega^2\mu_2\epsilon_2 - \alpha_2^2 = \alpha_m[n]^2$$

Note that h_1 and h_2 are the same for a given mode because of the radial part of fields ϵE_z at the interface $z = d_1$, ie, $J_m(h_1\rho) = J_m(h_2\rho), 0 \le \rho \le R$ implies $h_1 = h_2 = \alpha_m[n]$. Likewise, considering the TE mode we have

$$\omega^2\mu_1\epsilon_1 - \beta_1^2 = \beta_m[n]^2,$$

$$\omega^2\mu_2\epsilon_2 - \beta_2^2 = \beta_m[n]^2$$

These equations imply that we have two additional equations relating $\alpha_k, \beta_k, k = 1, 2$ which can be solved to determine the characteristic frequencies of oscillations of the TE and TM modes:

$$(\alpha_1^2 + \alpha_m[n]^2)/\mu_1\epsilon_1 = (\alpha_2^2 + \alpha_m[n]^2)/\mu_2\epsilon_2,$$

$$(\beta_1^2 + \beta_m[n]^2)/\mu_1\epsilon_1 = (\beta_2^2 + \beta_m[n]^2)/\mu_2\epsilon_2,$$

Exercise: Using the above formulas, write down the complete expansions for the electric and magnetic fields in the time domain for the TE and TM modes of oscillation in a cylindrical DRA.

Quantization of the fields inside the cylindrical DRA with the two partitions. We can write

$$E_{1z}(t,r) = Re \sum_{nmp} J_m(\alpha_m[n]\rho/R)(C_1(n,m,p)\cos(m\phi)+D_1(n,m,p)\sin(m\phi))\sin(\alpha_1[n,m,p]z)exp(j\omega[n,m,p]t),$$

$$= \sum_{n,m,p} \psi_{1nmp}(t,r), 0 < z < d_1$$

say and

$$E_{2z}(t,r) = Re \sum_{nmp} J_m(\alpha_m[n]\rho/R)(C_2(n,m,p)\cos(m\phi)+D_2(n,m,p)\sin(m\phi))\sin(\alpha_2[n,m,p](d-z))exp(j\omega[n,m,p]t)$$

$$= \sum_{nmp} \psi_{2nmp}(t,r), d_1 < z < d$$

say. It should be noted that

$$C_1(n,m,p) : D_1(n : m,p) = C_2(n,m,p) : D_2(n,m,p)$$

Assume that the TM modes only are sustained by the resonator. Then, the transverse components of the electric and magnetic fields are given by

$$E_{k\perp}(t,r) = \sum_{nmp} \alpha_m[n]^{-2}\nabla_\perp\partial\psi_{k,nmp}(t,r)/\partial z, k = 1, 2$$

$$H_{k,\perp}(t,r) = \sum_{nmp} \alpha_m[n]^{-2}\epsilon_k\nabla_\perp\partial\psi_{k,nmp}(t,r)/\partial t \times \hat{z}, k = 1, 2$$

The average energy stored within the resonator is

$$H(C_1, D_1, C_2) = (\epsilon_1/2T) \int_0^T dt \int_{0<z<d_1,(\rho,\phi)\in S} (E_{1z}(t,r)^2 + |E_{1\perp}(t,r)|^2)d^3r$$

$$+(\epsilon_2/2T) \int_0^T dt \int_{d_1<z<d,(\rho,\phi)\in S} (E_{2z}(t,r)^2 + |E_{2\perp}(t,r)|^2)d^3r$$

$$+(1/2\mu T) \int_0^T dt \int_{d_1<z<d,(\rho,\phi)\in S} (H_{2\perp}(t,r)|^2d^3r$$

after taking the limit $T \to \infty$. We can evaluate the time averages by the following device. If the frequencies $\omega[n], n = 1, 2, ...$ are all distinct, then

$$T^{-1} \int_0^T dt \int_B |\sum_n C(n)Phi_n(r)exp(j\omega[n]t)|^2d^3r$$

converges as $T \to \infty$ to

$$\sum_n |C(n)|^2 \int_B |\Phi_n(r)|^2 d^3 r$$

In the second quantized picture, we would regard $C(n), \bar{C}(n)$ as annihilation and creation operators.

14.13 Summary of the algorithm for nonlinear filtering in discrete time applied to fan rotation angle estimation:

State model:
$$x[n+1] = f_n(x[n]) + w[n+1]$$

Measurement model:
$$z[n] = h_n(x[n]) + v[n]$$

w, v are iid non-Gaussian processes with pdf's p_w and p_v respectively. The measurement data collected upto time n is given by

$$Z_n = \{z[k] : k \le n\}$$

We have using the Markov property,

$$p(x[n+1]|Z_{n+1}) = \int p(z[n+1]|x[n+1])p(x[n+1]|x[n])p(x[n]|Z_n)dx[n]/p(z[n+1]|Z_n)$$

So the MAP estimate of $x[n+1]$ given Z_{n+1} is given by

$$\hat{x}[n+1] = argmax_{x[n+1]} p(x[n+1]|Z_{n+1})$$

$$= argmax_{x[n+1]} \int p(z[n+1]|x[n+1]p(x[n+1]|x[n])p(x[n]|Z_n)dx[n]$$

Now,

$$\int p(z[n+1]|x[n+1]p(x[n+1]|x[n])p(x[n]|Z_n)dx[n]$$

$$= \int p_v(z[n+1] - h_{n+1}(x[n+1]))p_w(x[n+1] - f_n(x[n]))p([n]|Z_n)dx[n]$$

Write
$$e[n] = x[n] - \hat{x}[n]$$

If we assume that the MAP estimate approximately coincides with the MMSE estimate, then we have

$$\int e[n]p(x[n]|Z_n)dx[n] = 0$$

Now, we have approximately,

$$f_n(x[n]) = f_n(\hat{x}[n]) + f_n'(\hat{x}[n])e[n] + (1/2)f_n''(\hat{x}[n])(e[n] \otimes e[n])$$

and so approximately,

$$p_w(x[n+1]-f_n(x[n])) = p_w(x[n+1]-f_n(\hat{x}[n]))-p_w'(x[n+1]-f_n(\hat{x}[n]))f_n'(\hat{x}[n])e[n]$$

$$-(1/2)p_w''(x[n+1]-f_n(\hat{x}[n]))(f_n'(\hat{x}[n]) \otimes f_n'(\hat{x}[n]))(e[n] \otimes e[n])$$

$$-(1/2)p_w'(x[n+1]-f_n(\hat{x}[n]))f_n''(\hat{x}[n])(e[n] \otimes e[n])$$

and hence writing

$$p[n] = \int e[n] \otimes e[n]p(x[n]|Z_n)dx[n]$$

we get

$$\int p_w(x[n+1]-f_n(x[n]))p(x[n]|Z_n)dx[n] =$$

$$p_w(x[n+1]-f_n(\hat{x}[n]))-(1/2)[p_w''(x[n+1]$$

$$-f_n(\hat{x}[n]))(f_n'(\hat{x}[n]) \otimes f_n'(\hat{x}[n]))+p_w'(x[n+1]-f_n(\hat{x}[n]))f_n''(\hat{x}[n])]p[n]$$

With this approximation, we get

$$\hat{x}[n+1] =$$

$$argmax_{x[n+1]}[p_v(z[n+1] - h_{n+1}(x[n+1]))(p_w(x[n+1] - f_n(\hat{x}[n]))$$

$$-(1/2)[p_w''(x[n+1]-f_n(\hat{x}[n]))(f_n'(\hat{x}[n]) \otimes f_n'(\hat{x}[n]))+p_w'(x[n+1]-f_n(\hat{x}[n]))f_n''(\hat{x}[n])]p[n])]$$

This computation can be further approximated by writing

$$\hat{x}[n+1] = f_n(\hat{x}[n]) + \delta x$$

and expanding the above expression to be maximized upto quadratic orders in δx and then carrying out the maximization.

14.14 Classical filtering theory applied to Levy process and Gaussian measurement noise. Developing the EKF for such problems

(Summary of Rohit Singh's Ph.D work)

$X(t)$ the state is a Markov process with generator K_t. For example, for the sde

$$dX(t) = f_t(X(t))dt + \int g(t, X(t), x)_{x \in E}dN(t, x)$$

with N a Poisson field having rate measure $\lambda(t, dx)$,

$$K_t\phi(x) = f_t(x)^T \nabla\phi(x) + \int_E (\phi(x + g(t, x, y)) - \phi(x))\lambda(t, dy)$$

The measurement process is

$$dz(t) = h_t(X(t))dt + \sigma_v V(t)$$

where V is vector valued standard Brownian motion. We define for an observable $\psi(x)$ on the system space,

$$\pi_t(\psi) = \mathbb{E}(\psi(X(t))|Z_t), Z_t = \sigma(z(s) : s \leq t)$$

Then, we have the Kushner-Kallianpur filter

$$d\pi_t(\phi) = \pi_t(K_t\phi)dt + \sigma_v^{-2}(\pi_t(h_t\phi) - \pi_t(h_t)\pi_t(\phi))^T(dz(t) - \pi_t(h_t)dt)$$

We wish to make an EKF like approximation to this infinite dimensional sde. Let $\pi_t(x) = \hat{X}(t)$ and $P(t) = cov(X(t) - \hat{X}(t)|Z_t)$. Then, we have approximately,

$$d\hat{X}(t) = \hat{(K_t x)}dt + \sigma_v^{-2}(\pi_t(h_t x)) - \pi_t(h_t)\hat{X}(t))^T(dz(t) - h_t(\hat{X}(t))dt)$$

Writing

$$K_t\phi(x) = \int K_t(x,y)\phi(y)dy$$

we have

$$K_t x = \int K_t(x,y)ydy$$

and hence

$$\hat{K_t x} \approx \int K_t(\hat{X}(t),y)ydy + (1/2)\int Tr(\nabla_x \nabla_x^T K_t(\hat{X}(t),y)P(t))ydy$$

$$h_t(X(t))X(t) \approx h_t(\hat{X}(t))\hat{X}(t) + h_t(\hat{X}(t))(X(t) - \hat{X}(t))$$
$$+h_t'(\hat{X}(t))(X(t) - \hat{X}(t))\hat{X}(t) + h_t'(\hat{X}(t))(X(t) - \hat{X}(t))(X(t) - \hat{X}(t))$$
$$+(1/2)h_t''(\hat{X}(t))((X(t) - \hat{X}(t)) \otimes (X(t) - \hat{X}(t)))\hat{X}(t)$$

So

$$\pi_t(h_t x) \approx h_t(\hat{X}(t))\hat{X}(t) + P(t)h_t'(\hat{X}(t))^T$$
$$+(1/2)h_t''(\hat{X}(t)) < (X(t) - \hat{X}(t)) \otimes (X(t) \otimes \hat{X}(t))\hat{X}(t)$$

On the other hand,

$$\pi_t(h_t)\hat{X}(t) \approx$$
$$h_t(\hat{X}(t))\hat{X}(t) + (1/2)h_t''(\hat{X}(t)) < (X(t) - \hat{X}(t)) \otimes (X(t) - \hat{X}(t)) > \hat{X}(t)$$

Taking the difference gives

$$\pi_t(h_t x) - \pi_t(h_t)\hat{X}(t) \approx P(t)H_t^T$$

where

$$H_t = h_t'(\hat{X}(t))$$

This gives us the first part of the EKF

$$d\hat{X}(t) \approx \pi_t(K_t x)dt + \sigma_v^{-2}P(t)H_t^T(dz(t) - h_t(\hat{X}(t))dt)$$

$$\approx dt \int K_t(\hat{X}(t), y)y\,dy + \sigma_v^{-2}P(t)H_t^T(dz(t) - h_t(\hat{X}(t))dt)$$

For the second part, we need a differential equation for $P(t)$. Let

$$e(t) = X(t) - \hat{X}(t), \int K_t(x, y)y\,dy = F_t(x).$$

We have

$$P(t) = cov(e(t)|Z_t)$$

Now,

$$de(t) = dX(t) - d\hat{X}(t)$$

$$d(e(t)e(t)^T) = de(t).e(t)^T + e(t)de(t)^T + de(t)de(t)^T$$

Now,

$$\mathbb{E}(de(t).e(t)^T|Z_t) = \mathbb{E}[(dX(t)-F_t(\hat{X}(t))dt-\sigma_v^{-2}P(t)H_t^T(h_t(X(t))dt+\sigma_v dV(t)$$
$$-h_t(\hat{X}(t))dt))e(t)^T]$$

$$= \mathbb{E}[dX(t)(X(t) - \hat{X}(t))^T|Z_t] - \sigma_v^{-2}P(t)H_t^T H_t \mathbb{E}[e(t)e(t)^T|Z_t]dt$$
$$(\pi_t((K_t x)x^T) - \pi_t(K_t x)\hat{X}(t)^T)dt - \sigma_v^{-2}P(t)H_t^T H_t P(t)dt$$

The second term $\mathbb{E}[e(t)de(t)^T|Z_t]$ is the transpose of this. Finally, by Ito's formula,

$$\mathbb{E}[de(t).de(t)^T|Z_t] =$$

$$\mathbb{E}[(dX(t) - \pi_t(K_t x)dt + \sigma_v^{-2}P(t)H_t^T(dz(t) - h_t(\hat{X}(t))dt)).$$

$$.(dX(t) - \pi_t(K_t x)dt + \sigma_v^{-2}P(t)H_t^T(dz(t) - h_t(\hat{X}(t))dt))^T|Z_t]$$

$$= \mathbb{E}[dX(t)dX(t)^T|Z_t] + \sigma_v^{-2}P(t)H_t^T H_t P(t)dt$$

Combining all these, we get finally the generalized Riccati equation

$$dP(t)/dt = \pi_t((K_t x)(x - \hat{X}(t))^T + (x - \hat{X}(t))(K_t x)^T) - \sigma_v^{-2}P(t)H_t^T H_t P(t)$$

$$+dt^{-1}\mathbb{E}[dX(t).dX(t)^T|Z_t]$$

Now writing $F_t(x) = K_t x$ gives

$$\pi_t((K_t x)(x - \hat{X}(t))^T) = \pi_t(F_t(x)(x - \hat{X}(t))^T)$$

$$\approx F_t'(\hat{X}(t))P(t)$$

Thus, our generalized Riccati equation further approximates to

$$dP(t)/dt = F_t'(\hat{X}(t))P(t) + P(t)F_t'(\hat{X}(t))^T + dt^{-1}\mathbb{E}[dX(t).dX(t)^T|Z_t]$$

$$-\sigma_v^{-2}P(t)H_t^T H_t P(t)$$

Now, we observe that

$$\mathbb{E}(dX(t)dX(t)^T|X(t)=x) = \mathbb{E}(X(t+dt)X(t+dt)^T+xx^T-xX(t+dt)^T$$
$$-X(t+dt)x^T|X(t)=x)$$

$$= \mathbb{E}(X(t+dt)X(t+dt)^T-xx^T-x(X(t+dt)-x)^T-(X(t+dt)-x)x^T|X(t)=x)$$

$$= dt(K_t(xx^T)-x(K_tx)^T-(K_tx)x^T)$$

So

$$\mathbb{E}(dX(t).dX(t)^T|Z_t) = dt\pi_t(K_t(xx^T) - x(K_tx)^T - (K_tx)x^T)$$

Now we've defined

$$F_t(x) = K_tx = \int K_t(x,y)y dy$$

We also define

$$C_t(x) = K_t(xx^T) = \int K_t(x,y)yy^T dy$$

Then, we can write

$$\mathbb{E}(dX(t)dX(t)^T|Z_t) = dt\pi_t(C_t(x) - xF_t(x)^T - F_t(x)x^T)$$

This approximately equals

$$dt[C_t(\hat{X}(t)) - \hat{X}(t).F_t(\hat{X}(t))^T - F_t(\hat{X}(t))\hat{X}(t)^T]$$

so that our approximate EKF becomes

$$dP(t)/dt = F_t'(\hat{X}(t))P(t) + P(t)F_t'(\hat{X}(t))^T$$

$$+[C_t(\hat{X}(t)) - \hat{X}(t).F_t(\hat{X}(t))^T - F_t(\hat{X}(t))\hat{X}(t)^T] - \sigma_v^{-2}P(t)H_t^T H_t P(t)$$

14.15 Quantum Boltzmann equation for calculating the radiation fields produced by a plasma

Consider an ensemble of N identical particles with Hamiltonian

$$H = \sum_{a=1}^{N} H_a + \sum_{1\leq a<b\leq N} V_{ab}$$

where the H_a' are identical one particle Hamiltonians acting on Hilbert spaces $\mathcal{H}_a$ and $V_{ab}'s$ are identical two particle Hamiltonians acting the tensor product space $\mathcal{H}_a \otimes \mathcal{H}_b$. The Schrodinger-Liouville-Von-Neumann equation for the N particle density operator ρ is

$$i\rho' = [H,\rho] = \sum_a [H_a,\rho] + \sum_{a<b} [V_{ab},\rho]$$

Thus, assuming that all the marginal densities of any given order are identical, we get on partial tracing,

$$i\rho_1' = iTr_{23...N}\rho' = [H_1, \rho_1] + (N-1)Tr_2[V_{12}, \rho_{12}]$$

and

$$i\rho_{12}' = iTr_{34...N}\rho' = [H_1 + H_2 + V_{12}, \rho_{12}] + (N-2)Tr_3[V_{13} + V_{23}, \rho_{123}]$$

We write

$$\rho_{12} = \rho_1 \otimes \rho_1 + g_{12}$$

where g_{12} is small and likewise,

$$\rho_{123} = \rho_1 \otimes \rho_1 \otimes \rho_1 + g_{123}$$

where g_{123} is small. We have then approximately assuming V_{ab} to be small, ie, of the same order as g_{12}, the following equations obtained using perturbation theory:

$$i\rho_{12}' = [H_1 + H_2, \rho_{12}] + (N-1)Tr_2[V_{12}, \rho_1 \otimes \rho_1]$$

so that

$$\rho_{12}(t) = exp(-itad(H_1+H_2))(\rho_1(0)\otimes\rho_1(0)) - i(N-1)$$
$$\int_0^t exp(-i(t-s)ad(H_1+H_2))(Tr_2[V_{12}, \rho_1(s), \rho_1(s)])ds$$

This expression may be substituted into the equation for ρ_1. Alternately, since $[V_{12}, g_{12}]$ is of the second order of smallness, we get on retaining only first order of smallness terms in the equation of evolution of ρ_1, the following quantum Boltzmann equation:

$$i\rho_1'(t) = [H_1, \rho_1(t)] + (N-1)Tr_2[V_{12}, \rho_1(t) \otimes \rho_1(t)]$$

Solving this perturbatively gives

$$\rho_1(t) = exp(-itad(H_1))(\rho_1(0)) - i(N-1)\int_0^t exp(-i(t-s)ad(H_1))$$

$$.(Tr_2[V_{12}, exp(-isad(H_1)) \otimes exp(-isad(H_1))(\rho_1(0) \otimes \rho_1(0))])ds$$

Likewise, using first order perturbation theory, we get

$$ig_{12}' = [H_1 + H_2, g_{12}] + (N-1)Tr_2[V_{12}, \rho_1(t) \otimes \rho_1(t)]$$

from which, we get

$$g_{12}(t) = -i(N-1)\int_0^t exp(-i(t-s)(H_1+H_2))(Tr_2[V_{12}, \rho_1(s) \otimes \rho_1(s)])ds$$

and we get

$$\rho_{12}(t) = \rho_1(t) \otimes \rho_1(t) + g_{12}(t)$$

Likewise, using higher order perturbation theory, we can solve for all the marginals $\rho_{12...r}(t), r = 1, 2, ..., N$. We write

$$\rho(t) = \rho_{12..N}(t)$$

We now wish to describe the current produced by such a system of interacting charged quantum particles. Let m denote the mass of any one particle and $p_1, ..., p_N$ their three momenta. Let $-e$ denote the charge on any one particle. Then the total charge density operator is

$$-e \sum_{k=1}^{N} \delta^3(r - r_k)$$

and the total current density operator is

$$(-e/2m) \sum_{k=1}^{N} (p_k \delta^3(r - r_k) + \delta^3(r - r_k)p_k$$

$$= (-e/m) \sum_{k} ((i\nabla\delta^3(r - r_k))/2 + \delta^3(r - r_k)p_k)$$

where $(r_k, p_k = -i\nabla_{r_k})$ are respectively the position and momentum operators of the kth particle. Let

$$\rho_t(r_1, ..., r_N | r_1', ..., r_N')$$

denote the kernel in position space of the density operator $\rho(t)$. Then, the quantum averaged charge density at time t is

$$\sigma(t, r) = -e \sum_{k=1}^{N} Tr(\rho.\delta^3(r - r_k))$$

$$= -e \sum_{k=1}^{N} \int \rho_t(r_1, ..., r_N | r_1, .r_k.., r_N)\delta^3(r - r_k)d^3r_1...d^3r_N$$

$$= -e \sum_{k=1}^{N} \rho_t(r_1, .., r.., r_N | r_1, ..., r, ..., r_N)d^3r_1...d^3\hat{r}_k..d^3r_N$$

where the hat above r_k means omission of that variable in the integration. Likewise, the quantum averaged current density at time t is

$$J(t, r) = -(e/m) \sum_{k=1}^{N} Tr(\rho((-i\nabla\delta^3(r - r_k))/2 - i\delta^3(r - r_k)\nabla_k))$$

$$= (ie/m) \sum_{k} \int [\nabla^{(2)}\rho(r_1, ..., r, ..., r_N | r_1, ..., r, ..., r_N)/2$$

$$+ \nabla_k^{(1)}\rho(r_1, ..., r, ..., r_N | r_1, ..., r, ..., r_N)]d^3r_1...d^3\hat{r}_k...d^3r_N$$

From these expressions for the quantum averaged charge and current densities, the far field averaged radiation patterns can be computed. More generally, suppose we wish to calculate the higher moments of the far field radiation patterns.

Chapter 15

Classical and quantum drone design

15.1 Project proposal on drone design for the removal of pests in a farm

The project would involve the following: First, we have to design a flying drone like a miniature aeroplane that carries pesticide. The flight and movement of the drone will be controlled by propellers whose angular velocities $\omega_k(t), k = 1, 2, ..., d$ can be controlled by a remote base station through electromagnetic waves and transmitter-received antennas. The state of the drone at any given time t is specified by six variables; three position variables $r = (x, y, z)$ for the centre of mass position and three Euler angle rotation variables $\xi = (\phi, \theta, \psi)$. The equations of motion of the drone are of the form

$$r''(t) = F(r(t), r'(t), \xi(t), \xi'(t), \omega(t)) + w(t)$$

$$\xi''(t) = G(r(t), r'(t), \xi(t), \xi'(t), \omega(t)) + v(t)$$

where the first equation is Newton's second law of motion in the form force equals mass times acceleration and the second equation is again derived from Newton's second law in the form, rate of change of angular momentum is torque. The external forces and torques on the drone come from the gravitational field and the propeller angular velocities. The lift of the drone comes from Bernoulli's principle, namely, that the top surface of a wing has a longer line length that the bottom surface. Therefore, if the propellers give a forward thrust to the drone, then air above the wing surface will cover a longer distance than air below the wing for a fixed time duration. This would mean that the speed of air is more on the top surface than on the bottom surface. Consequently, by Bernoulli's principle, the air pressure would be more on the bottom surface than on the top surface of the wing causing a lift of the drone. Suppose that we desire the drone to follow a desired trajectory $r_d(t)$ over a time duration $[0, T]$. Then, we would have to give an instantaneous feedback to the angular velocities $\omega(t)$ of propellers based on the difference between estimated trajectory and the desired trajectory so that some cost function like

$$\mathbb{E}\int_0^T E(r(t), r_d(t), \omega(t))dt$$

is minimized. When the drone is thus controlled to move along a given trajectory, it will sprinkle pesticide on the crop that occur along its trajectory to remove the insects. Another problem faced is that during the motion of the drone, there may be certain obstacles like trees and pillars which it would have to read by means of a camera and then we must provide a feedback force that depends on the obstacle's position and the current position of the drone so that collision does not take place. The problem of drone design along with the controller at the base station is thus a complex one involving concepts from stochastic optimal control theory, extended Kalman observers/filters and feedback control laws.

We can also design quantum drones that are miniature quantum aeroplanes having dimensions of the order of Angstroms. The dynamics of such projectiles will once again be described by the observables $r = (x, y, z)$ and $\xi = (\phi, \theta, \psi)$. We would have to write down the Hamiltonian of such a quantum drone in the form

$$H(t) = (1/2m)\mathbf{P}^T\mathbf{P} + mgz + (1/2)\mathbf{P}_\xi^T J(\xi)\mathbf{P}_\xi + H_I(\mathbf{r}, \xi, \omega(t))$$

where H_I is the interaction Hamiltonian between the propeller angular velocity pseudo-vector and the position and angular variables of the drone. The wave function of the drone $\psi(t, r, \xi)$ satisfies the Schrodinger equation

$$i\psi_{,t}(t, r, \xi) = H(t)\psi(t, r, \xi)$$

where $H(t)$ is obtained by replacing $\mathbf{P}$ by $-i\nabla_r$ and P_ξ by $-i\nabla_\xi$. At time t, the probability density of the drone's position and angular variables (r, ξ) is given by $|\psi(t, r, \xi)|^2$ and the objective is that this pdf should track a given fuzzy trajectory specified by a desired pdf $f_d(t, r, \xi)$, for example,

$$\int_0^T \int_{\mathbb{R}^3 \times [0, 2\pi)^3} (|\psi(t, r, \xi)|^2 - f_d(t, r, \xi))^2 d^3r d^3\xi dt$$

may be minimized w.r.t $\omega(t), 0 \le t \le T$.

15.2   Quantum drones based on Dirac's relativistic wave equation

The motion of the quantum drone having a charge of $-e$ in an external em field described by the electromagnetic four potential $A_\mu(x)$ is described by the Dirac equation for the wave function $\psi(x) \in \mathbb{C}^4, x \in \mathbb{R}^4$:

$$[\gamma^\mu(i\partial_\mu + eA_\mu(x)) - m]\psi(x) = 0$$

This can be rearranged as

$$i\partial_0\psi(x) = [-eA_0(x) + (\alpha, -i\nabla + eA(x)) + \beta m]\psi(x)$$

or equivalently, taking quantum noise into account,

$$idU(t) = (-eA_0(t,r)dt + c_b^a(r) \otimes d\Lambda_a^b(t) + (\alpha, -i\nabla + eA(t,r))dt - \beta m dt)U(t)$$

where

$$\alpha^r = \gamma^0\gamma^r, \beta = \gamma^0$$

The unperturbed Hamiltonian of the free projectile is

$$H_0 = (\alpha, -i\nabla) + \beta m$$

and the perturbed Hamiltonian in the sense of white noise calculus is

$$H(t) = H_0 + V(t,r)$$

where the interaction potential is

$$V(t,r) = -eA_0(t,r) + c_b^a(r) \otimes d\Lambda_a^b(t)/dt + e(\alpha, A(t,r)) - \beta m$$

The unperturbed evolution operator is

$$U_0(t) = exp(-itH_0)$$

and the perturbed evolution operator is

$$U(t) = U_0(t)W(t)$$

where

$$W(t) = I + \sum_{n \geq 1}(-i)^n \int_{0 < t_n < ... < t_1 < t} \tilde{V}(t_1)...V(t_n)dt_1..dt_n$$

with

$$\tilde{V}(t) = U_0(-t)V(t)U_0(t) =$$

$$-eU_0(-t)A_0(t,r)U_0(t) + U_0(-t)c_b^a(r)U_0(t) \otimes d\Lambda_a^b(t)/dt + eU_0(-t)(\alpha, A(t,r))U_0(t) - \beta m$$

The objective is to calculate the average scattering matrix when the bath is in a given coherent state $|\phi(u) >$. The scattering matrix will tell us the probability of the projectile getting scattered within a given solid angle when its initial state is a given wave. An example of such a computation based on first order perturbation theory is as follows. Write

$$\psi(x) = \psi_i(x) + \psi_s(x)$$

where the incident wave $\psi_0(x)$ satisfies the free Dirac equation:

$$i\partial_0\psi_0(x) = H_0\psi_0(x)$$

This gives us assuming

$$\psi_0(x) = u(P)exp(-ip.x), p = (p^0, P), p^0 = E(P) = \sqrt{m^2 + P^2}$$

the algebraic free Dirac equation

$$E(P)u(P) = [(\alpha, P) + \beta m]u(P)$$

For a fixed P, this equation has two linearly independent solutions which may be assumed to be orthogonal: $u(P, \sigma), \sigma = 1, 2$. The scattered wave $\psi_s(x)$ by first order relativistic noisy Born scattering theory is given by

$$(i\partial_0 - H_0)\psi_s(x) = -V(x)\psi_0(x)$$

Writing

$$S(x) = (2\pi)^{-4}\int (q^0 - (\alpha, Q) - \beta m)^{-1}exp(iq.x)d^4q$$

where

$$q = (q^0, Q)$$

we get

$$\psi_s(x) = -\int S(x - x')V(x')\psi_0(x')d^4x'$$

The second order correction $\psi_{2s}(x)$ satisfies

$$(i\partial_0 - H_0)\psi_{2s}(x) = -V(x)\psi_s(x)$$

so that

$$\psi_{2s}(x) = -\int S(x - x')V(x')\psi_s(x')d^4x' =$$

$$\int S(x - x')V(x')S(x' - x'')V(x'')\psi_0(x'')d^4x'd^4x''$$

The noise averaged scattering amplitude at from the initial state $\psi_0(x)$ to the final state $\psi_f(x)$ is then

$$< \psi_f \otimes \phi(u)|(\psi_s + \psi_{2s})\otimes]phi(u) >$$

$$=< \phi(u)| - \int \psi_f(x)^*(S(x - x')V(x')\psi_0(x')d^4xd^4x'+$$

$$\int \psi_f(x)^*S(x - x')V(x')S(x' - x'')V(x'')\psi_0(x'')d^4xd^4x'd^4x''|\phi(u) >$$

Chapter 16

Current in a quantum antenna

16.1 Hartree-Fock equations for obtaining the approximate current density produced by a system of interacting electrons

$$H = \sum_{a=1}^{N} H_a + \sum_{a<b} V_{ab}$$

$H_a, a = 1, 2, ..., N$ are identical copies of a one particle Hamiltonian acting in different Hilbert spaces and $V_{ab}, a < b$ are identical copies of the interaction Hamiltonian between two particles We try a wave function

$$|\psi_t> = \otimes_{k=1}^{N} |\psi_{kt}>$$

Substituting this into the Schrodinger equation gives

$$\sum_k \psi_{1t} \otimes ... \otimes (id\psi_{kt}/dt) \otimes ... \otimes \psi_{nt} =$$

$$= \sum_k \psi_{1t} \otimes ... \otimes H_k \psi_{kt} \otimes ... \otimes \psi_{nt} +$$

$$+ \sum_{a<b} \psi_{1t} \otimes ... \otimes V_{ab} \psi_{at} ... \otimes \psi_{bt} \otimes ... \otimes \psi_{nt}$$

Taking the inner product on both sides with $\psi_{1t} \otimes .. \hat{\psi}_{kt} \otimes ... \otimes \psi_{nt}$ where a hat above a symbol means omission of that symbol gives the approximate equations

$$id|\psi_{kt}>/dt = H_k|\psi_{kt}> + \sum_{m \neq k} <I \otimes \psi_{mt}|V_{km}|\psi_{kt} \otimes \psi_{mt}>$$

In terms of kernels, we can express this as

$$id\psi_{kt}(r)/dt = H_k\psi_{kt}(r) + (\sum_{m\neq k}\int V(r,r')|\psi_{mt}(r')|^2 d^3r')\psi_{kt}(r)$$

This equation however does not take into account the Pauli exclusion principle. For taking that into account, we must try an antisymmetrized wave function

$$|\psi_t> = \sum_{\sigma\in S_n}|\psi_{\sigma 1,t}> \otimes... \otimes |\psi_{\sigma n,t}>$$

Substituting and taking the inner products after assuming orthonormality of the component wave functions gives us

$$id|\psi_{kt}>/dt = H_k|\psi_{kt}> + \sum_{m\neq k}<I\otimes\psi_{mt}|V|\psi_{kt}\otimes\psi_{mt}>$$

$$-\sum_{m\neq k}<\psi_{mt}\otimes I|V|\psi_{kt}\otimes\psi_{mt}>$$

which in coordinate form can be expressed as

$$id\psi_{kt}(r)/dt = H_k\psi_{kt}(r) + \sum_{m\neq k}(\int V(r,r')|\psi_{mt}(r')|^2 d^3r')\psi_{kt}(r)$$

$$-\sum_{m\neq k}(\int V(r,r')\bar{\psi}_{mt}(r')\psi_{kt}(r')d^3r')\psi_{mt}(r) = 0$$

These equations are special cases of nonlinear Schrodinger equations of the form

$$id\psi_{kt}(r)/dt = -\nabla^2\psi_{kt}(r)/2m + V_0(r)\psi_{kt}(r) + \sum_m V_{km}(r,\psi_{1t},...,\psi_{Nt})\psi_{mt}(r), k = 1,2,...,N$$

The charge density is

$$\rho(t,r) = -e\sum_k |\psi_{kt}(r)|^2$$

and its rate of increase is given by

$$\rho_{,t}(r) = -e\sum_k(\bar{\psi}_{kt,t}(r).\psi_{kt}(r) + \bar{\psi}_{kt}(r).\psi_{kt,t}(r))$$

$$= (ie/2m)\sum_k[\psi_{kt}(r)\nabla^2\bar{\psi}_{kt}(r) - \bar{\psi}_{kt}(r)\nabla^2\psi_{kt}(r)]$$

$$= div((ie/2m)\sum_k[\psi_{kt}(r)\nabla\bar{\psi}_{kt}(r) - \bar{\psi}_{kt}(r)\nabla\psi_{kt}(r)])$$

$$= -divJ(t,r)$$

where the current density $J(t,r)$ is given by

$$J(t,r) = (-ie/2m) \sum_k [\psi_{kt}(r)\nabla\bar{\psi}_{kt}(r) - \bar{\psi}_{kt}(r)\nabla\psi_{kt}(r)]$$

It should be noted that this expression for the current density is the same as that derived using the linear Schrodinger equations but the wave functions ψ_{kt} are calculated by solving the nonlinear Schrodinger equation. It should be noted that in deriving the above formula for the current density, we have made used of the relation

$$\bar{V}_{km}(r,\psi_1,...,\psi_N) = V_{mk}(r,\psi_1,...,\psi_N)$$

which is readily verified.

16.2 Controlling the current produced by a single quantum charged particle quantum antenna

The wave function is $\psi(t,r)$. The external magnetic vector potential is $A(t,r)$ and the external electric potential is $V(t,r)$. The Hamiltonian of the electron after it interacts with the external em field is given by

$$H(t) = (1/2m)(\nabla + ieA(t,r))^2\psi(t,r) + (V_0(r) - eV(t,r))$$

Writing down the Schrodinger equation

$$i\psi_{,t}(t,r) = H(t)\psi(t,r)$$

and calculating the rate of change of the smeared charge density

$$\rho(t,r) = -e|\psi(t,r)|^2$$

$$\rho_{,t} = -e\bar{\psi}'\psi - e\bar{\psi}\psi'$$

$$= ie(\bar{\psi}\bar{\psi}H\psi - \psi(\bar{H}\bar{\psi})) = -divJ$$

gives us

$$J(t,r) = (-ie/2m)(\bar{\psi}(\nabla + ieA)\psi - \psi(\nabla - ieA)\bar{\psi})$$

$$= (-ie/2m)(\bar{\psi}\nabla\psi - \psi\nabla\bar{\psi}) + (e^2/2m)A|\psi|^2$$

The external em field potentials $A(t,r), V(t,r)$ are controlled by an external current source $J_c(t,r)$. Apart from this external current source, there is the current J produced by the charged particle after quantum smearing. Thus, Maxwell's equations for the em four potential are

$$\nabla^2 A - (1/c^2)A_{,tt} = -\mu_0(J + J_c),$$

$$\nabla^2 V - (1/c^2)V_{,tt} = -(\rho + \rho_c)/\epsilon_0$$

where

$$\rho_c = -\int_0^t div\,J_c\,dt$$

The solutions to these equations are obtained by the retarded potential formula

$$A(t,r) = (\mu_0/4\pi)\int (J(t - |r - r'|/c, r') + J_c(t - |r - r'|/c, r'))d^3r'/|r - r'|$$

$$V(t,r) = (1/4\pi\epsilon_0)\int (\rho(t - |r - r'|/c, r') + \rho_c(t - |r - r'|/c, r'))d^3r'/|r - r'|$$

The aim is to design the control current density $J_c(t,r)$ over a given time range $t \in [0,T]$ within a box $[0,L]^3$, so that the generated quantum current density $J(t,r)$ tracks a desired current density $J_d(t,r)$ in the sense that

$$\int_{[0,T]\times[0,L]^3} |J_d(t,r) - J(t,r)|^2 d^3r\,dt$$

is a minimum.

The second quantized picture:

Assume that there are two species of Fermions described by the Fermionic operator wave fields $\psi_a(t,r), a = 1,2$ satisfying the canonical anticommutation relations

$$[\psi_a(t,r), \psi_b(t,r')^*]_+ = \delta_{ab}\delta^3(r - r')$$

The unperturbed Hamiltonian operator of this Fermionic field is

$$H_0 = (-1/2m)\int \psi_a(r)^*\nabla^2\psi_a(r)d^3r + \int V_{ab}(r,r')\psi_a(r)^*\psi_b(r')d^3r\,d^3r'$$

The quantum statistical Gibbsian density operator is then

$$\rho_G = exp(-\beta H_0)/Z(\beta), Z(\beta) = Tr(exp(-\beta H_0))$$

In the presence of an external electromagnetic field described by a magnetic vector potential $A(t,r)$ and scalar electric potential $V(t,r)$, the second quantized Hamiltonian of the Fermi liquid assumes the form

$$H(t) = (-1/2m)\int \psi_a(r)^*(\nabla + ieA(t,r))^2\psi_a(r)d^3r - e\int V(t,r)\psi_a(r)^*\psi_a(r)d^3r$$

$$+ \int V_{ab}(r,r')\psi_a(r)^*\psi_b(r')d^3r\,d^3r'$$

and we may also add to this a "Cooper pair" term

$$\int U_{ab}(r,r')\psi_a(r)^*\psi_a(r)\psi_b(r')^*\psi_b(r')d^3r\,d^3r'$$

coming from the interaction between $\psi_a(r)^*\psi_a(r)d^3r$ Fermions (number operator) at d^3r with $\psi_b(r')^*\psi_b(r')d^3r'$ Fermions at d^3r'. In this, case, the unperturbed Hamiltonian for the Fermionic field would be taken as

$$H_0 =$$

$$(-1/2m)\int \psi_a(r)^*\nabla^2\psi_a(r)d^3r + \int V_{ab}(r,r')\psi_a(r)^*\psi_b(r')d^3rd^3r'$$

$$\int U_{ab}(r,r')\psi_a(r)^*\psi_a(r)\psi_b(r')^*\psi_b(r')d^3rd^3r'$$

while the interaction Hamiltonian between the Fermionic field and the external electromagnetic field is

$$H_I(t) = (-ie/2m)\int \psi_a(r)^*(2(A(t,r),\nabla) + div A(t,r))\psi_a(r)d^3r$$

$$-e\int V(t,r)\psi_a(r)^*\psi_a(r)d^3r$$

where the canonical anticommutation relations are satisfied by the Fermionic field operators:

$$[\psi_a(r),\psi_b(r')^*]_+ = \delta_{ab}\delta^3(r-r'),$$

$$[\psi_a(r),\psi_b(r')]_+ = 0, [\psi_a(r)^*,\psi_b(r')^*]_+ = 0$$

The Fermionic fields satisfy Heisenberg's equation of motion:

$$\psi_{a,t}(t,r) = i[H(t),\psi_a(t,r)]$$

where $H(t)$ is obtained from the previous $H(t) = H_0 + H_I(t)$ by replacing $\psi_a(r),\psi_a(r)^*$ with $\psi_a(t,r),\psi_a(t,r)^*$ respectively and using the same anticommutation relations for the time dependent field operators. The Fermionic current density operator is

$$J(t,r) = (-ie/2m)\int [\psi_a(t,r)^*(\nabla+ieA(t,r))\psi_a(t,r)$$

$$-\psi_a(t,r)^*(\nabla-ieA(t,r))\psi_a(t,r)d^3r]$$

The average current density is then

$$J_{av}(t,r) = Tr(\rho_G J(t,r))$$

and the objective of our Fermi-liquid quantum antenna is to make this current density track a desired current density $J_d(t,r)$ by controlling the external electromagnetic fields A, V.

Chapter 17

photons in a gravitational field with gate design applications and image processing in electromagnetics

17.1 Some remarks on quantum blackhole physics

[1] Time always flows in the forward direction, the entropy of a blackhole always increases and the mass of a blackhole always increases since any particle is always attracted towards itself and absorbed by the massive gravitational field of the blackhole. This is why computing the entropy of a blackhole becomes important. At time $t = 0$, assume that a system of particles in the vicinity of a blackhole is in the pure state $|\psi_m>$. We assume that $|\psi_m>$ in position space is concentrated within the critical radius of the blackhole. Let H_m denote the Hamiltonian of the system of particles, H_G, the Hamiltonian of the gravitational field of the blackhole and H_I the interaction Hamiltonian between the particles and the gravitational field of the blackhole. H_m acts in the system Hilbert space $\mathcal{H}_m$, H_G acts in the gravitational field Hilbert space $\mathcal{H}_G$ while H_I acts in the tensor product space $\mathcal{H}_m \otimes \mathcal{H}_G$. The initial state of the blackhole gravitational field and the system particles is a pure state $|\psi(0)>=|\psi_m(0) \otimes \psi_G(0)>$. After time t, it evolves to the state

$$|\psi(t)>= exp(-it(H_m + H_G + H_I))|\psi(0)>$$

The states after time t of the system of particles and the gravitational field of the blackhole are both mixed and are respectively given by

$$\rho_m(t) = Tr_2(|\psi(t)><\psi(t)|), \rho_G(t) = Tr_1(|\psi(t)><\psi(t)|)$$

The initial entropy of the system is zero and so also of the initial gravitational field of the blackhole since both of these states are pure. The final entropies of these are non-zero in general and are respectively given by

$$S_m(t) = -Tr(\rho_m(t)log(\rho_m(t)), S_G(t) = -Tr(\rho_G(t)log(\rho_G(t)))$$

In particular, this shows that by interacting with material particles, the entropy of the blackhole increases. We can also understand Hawking radiation via the tunneling phenomena in quantum mechanics. Consider the case of a Schwarzchild blackhole. Classically, a particle within the critical radius $m = 2GM/c^2$ of this blackhole cannot escape outside in finite coordinate time. Quantum mechanically there is a small probability of such an escape taking place. This can be computed as follows. The KG equation for a particle in the metric $g_{\mu\nu}$ is given by

$$(g^{\mu\nu}\psi_{,\nu}))_{:\mu} + (2\pi mc^2/h)^2\psi = 0$$

or writing

$$\beta = 2\pi mc^2/h$$

this equation becomes

$$(g^{\mu\nu}\sqrt{-g}\psi_{,\nu})_{,\mu} + \beta^2\sqrt{-g}\psi = 0$$

In the case of a Schwarzchild metric and when the wave function depends only on the radial coordinate r, this equation reduces to

$$g^{00}\sqrt{-g}\psi_{,00} + (g^{11}\sqrt{-g}\psi_{,1})_{,1} + \beta^2\sqrt{-g}\psi = 0$$

or equivalently,

$$\alpha(r)^{-1}r^2\psi_{,00}(t,r) - (\alpha(r)r^2\psi_{,1}(t,r))_{,1} + \beta^2 r^2\psi(t,r) = 0$$

This equation is solved using separation of variables. If we take the initial KG wave function as $\psi(0,r) = K\chi_{r<r_c}$ where $r_c = 2m$ and $4\pi r_c^3 K^2/3 = 1$, which corresponds to the situation that initially the KG particle is uniformly distributed within the Schwarzchild radius, then after time t, we will find that $\psi(t,r) \neq 0$ for $r > r_c$. This result means that although classically the particle cannot tunnel through the critical radius, yet quantum mechanically, there is a small probability of this happening. Instead of KG particles, we could also work with photons which are governed by the Maxwell equations in curved space-time. Let $A_\mu(x)$ denote the four potential of the photon. Assume that at time $t = 0$, all the photons are contained within the Schwarzchild radius. For example we may take

$$A_\mu(0,r) = \psi_\mu(r), A_{\mu,0}(0,r) = \phi_\mu(r)$$

Then the $A_\mu(t,r)$ satisfy the Maxwell equations in the Schwarzchild metric

$$(g^{\mu\alpha}g^{\nu\beta}\sqrt{-g}F_{\alpha\beta})_{,\nu} = 0$$

with the gauge condition

$$(A^\mu\sqrt{-g})_{,\mu} = 0$$

We find therefore that A_μ satisfies a second order pde in the space-time variables and the two specified initial conditions suffice to solve for A_μ for all values of space-time. After rearrangement, the Maxwell equations give us

$$(g^{\mu\alpha}g^{\nu\beta}\sqrt{-g})_{,\nu}F_{\alpha\beta} + g^{\mu\alpha}g^{\nu\beta}\sqrt{-g}F_{\alpha\beta,\nu} = 0$$

The gauge condition gives us

$$(g^{\mu\alpha}\sqrt{-g})_{,\mu}A_\alpha + g^{\mu\alpha}\sqrt{-g}A_{\alpha,\mu} = 0$$

A better approach to this problem is to work directly with covariant derivatives, ie,

$$F^{\mu\nu}_{:\nu} = 0, \ A^\mu_{:\mu} = 0$$

and then get

$$A^{\nu:\mu}_{:\nu} - A^{\mu:\nu}_{:\nu} = 0$$

Now,

$$A^{\nu:\mu}_{:\nu} = g^{\mu\alpha}A^\nu_{:\alpha:\nu} =$$
$$g^{\mu\alpha}(A^\nu_{:\alpha:\nu} - A^\nu_{:\nu:\alpha})$$

(in view of the gauge condition $A^\nu_{:\nu} = 0$)

$$= g^{\mu\alpha}g^{\nu\beta}R^\rho_{\beta\alpha\nu}A_\rho$$

Thus, the Maxwell equations reduce to

$$\Box A^\mu = g^{\mu\alpha}g^{\nu\beta}R^\rho_{\beta\alpha\nu}A_\rho$$

where $\Box$ denotes the Laplace-Beltrami wave operator of curved space-time acting on four vector fields. We can write

17.2 EM field pattern produced by a rotated and translated antenna with noise deblurring

A transmitter antenna is completely specified by its current density $J(\omega, r)$ at the frequency ω. If the antenna is rotated and translated by $(R, a) \in SO(3) \times \mathbb{R}^3$, the resulting current density is $J_1(\omega, r) = J(\omega, R^{-1}(r - a))$. The electric field patterns in space due to the original antenna and the transformed antenna are respectively given by

$$F(\omega, r) = \int K(\omega, r - r') J(\omega, r') d^3 r',$$

and

$$F_1(\omega, r) = \int K(\omega, r - r') J_1(\omega, r') d^3 r' + w(r)$$

where K is the vector valued Green's function defined by

$$F(\omega, r) = \nabla \times \nabla \times A(\omega, r) / j\omega\mu\epsilon$$

where

$$A(\omega, r) = (mu/4\pi) \int J(\omega, r') exp(-j\omega|r - r'|/c) d^3 r' / |r - r'|, c = (\mu\epsilon)^{-1/2}$$

Here, we are assuming the medium to be linear, homogeneous and isotropic. The aim is to estimate the rotation-translation pair (R, a) from measurements on F and F_1. More generally, if the medium is nonlinear, inhomogeneous and anisotropic, then we can write with J, J_1, F, F_1 being 3×1 vector fields,

$$F(\omega, r) = \sum_{n \geq 1} \int K_n(\omega, \omega_1, ..., \omega_n, r, r_1, ..., r_n)(\otimes_{k=1}^n J(\omega_k, r_k) d^3 r_1 ... d^3 r_n,$$

$$F_1(\omega, r) = \sum_{n \geq 1} \int K_n(\omega, \omega_1, ..., \omega_N, r, r_1, ..., r_n)(\otimes_{k=1}^n J_1(\omega_k, r_k)) d^3 r_1 ... d^3 r_n,$$

$$= \sum_{n \geq 1} \int K_n(\omega, \omega_1, ..., \omega_n, r, r_1, ..., r_n)(\otimes_{k=1}^n J(\omega_k, R^{-1}(r_k - a)) d^3 r_1 ... d^3 r_n,$$

$$= \sum_{n \geq 1} \int K_n(\omega, \omega_1, ..., \omega_n, r, Rr_1 + a, ..., Rr_n + a) \otimes_{k=1}^n J(\omega_k, r_k)) d^3 r_1 ... d^3 r_n$$

In these expressions, K_n is a matrix valued complex function of size $3 \times 3^n, n \geq 1$. By measuring the untransformed em field F at different frequencies and at different spatial locations, we can get to estimate the kernels $K_n(\omega, \omega_1, ..., \omega_n, r, r_1, ..., r_n)$ and by measuring the transformed em field F at different frequencies and at different spatial locations, we can get to estimate the kernels

$$K'_n(\omega, \omega_1, ..., \omega_n, r, r_1, ..., r_n) =$$

$$= K_n(\omega, \omega_1, ..., \omega_n, r, Rr_1 + a, ..., Rr_n + a)$$

From the knowledge of these two kernels, the rotation translation pair (R, a) may be determined by applying a combination of 3-D spatial Fourier transforms and the Peter-Weyl theory for $SO(3)$ based on spherical harmonic expansions.

17.3 Estimation the 3-D rotation and translation vector of an antenna from electromagnetic field measurements

Original antenna current density: $J(\omega, r)$

Rotated and translated antenna current density: $J_1(\omega, r) = J(\omega, R^{-1}(r-a))$.

Original electromagnetic field:

$$F(\omega, r) = \int G_1(\omega, r-r_1) J(\omega, r_1) d^3 r_1$$

$$+ \int G_2(\omega_1, \omega-\omega_1, r-r_1, r-r_2) J(\omega_1, r_1) J(\omega-\omega_1, r_2) d^3 r_1 d^3 r_2$$

Electromagnetic field after rotating and translating the antenna:

$$F_1(\omega, r) = \int G_1(\omega, r-r_1) J_1(\omega, r_1) d^3 r_1$$

$$+ \int G_2(\omega_1, \omega-\omega_1, r-r_1, r-r_2) J_1(\omega_1, r_1) J_1(\omega-\omega_1, r_2) d^3 r_1 d^3 r_2$$

$$= \int G_1(\omega, r - a - Rr_1) J_1(\omega, r_1) d^3 r_1 +$$

$$\int G_2(\omega_1, \omega-\omega_1, r-a-Rr_1) G_2(\omega_1, \omega-\omega_1, r-a-Rr_2) J_1(\omega_1, r_1) J_1(\omega-\omega_1, r_2) d^3 r_1 d^3 r_2$$

Assuming that G_1, G_2 depend on r only via $|r|$, it follows that after taking noise into account,

$$F_1(\omega, r) = F(\omega, R^{-1}(r - a)) + w(\omega, r)$$

where w is the noise field. (R, a) can thus be identified from measurements of the original and final em field patterns using Fourier transforms in $\mathbb{R}^3$ and the Peter-Weyl theory for $SO(3)$.

More generally, consider a set of N antennae having current densities $J_k(\omega, r), k = 1, 2, ..., N$. These antennae are permuted, rotated and translated so that the resulting sequence of current densities becomes $J_{\sigma k}(\omega, R^{-1}(r-a)), k = 1, 2, ..., N$. The resulting em field pattern before applying this set of transformations (R, a, σ) is

$$F(\omega, r) = F(\omega, r|r_1, ..., r_N)$$

and the em field pattern after applying the rotation, translation and permutation transformation is given by

$$F_1(\omega, r|r_1, ..., r_N) = F(\omega, R^{-1}(r - a)|r_{\sigma 1}, ..., r_{\sigma N}) + w(\omega, r)$$

We wish to estimate the group element (R, a, σ) from measurements of F_1 and F_2 at different $r's$. Let π_1 denote a unitary representation of S_n-the permutation group of n objects and π_2 a unitary representation of $SO(3)$. We have assuming zero noise,

$$\int F_1(\omega, r|r_1, ..., r_N).exp(-jk.r) d^3 r =$$

$$\int F(\omega, r|r_{\sigma 1}, ..., r_{\sigma N}) exp(-jk.(Rr + a)) d^3 r$$

$$= exp(-jk.a) \int F(\omega, r|r_{\sigma 1}, ..., r_{\sigma N}) exp(-j(R^T k).r) d^3 r$$

We express this identity as

$$\hat{F}_1(\omega, k|r_1, ..., r_N) = exp(-jk.a)\hat{F}(\omega, R^T k|r_{\sigma 1}, ..., r_{\sigma N})$$

from which, we deduce that

$$|\hat{F}_1(\omega, k|r_1, ..., r_N)| = |\hat{F}(\omega, R^T k|r_{\sigma 1}, ..., r_{\sigma N})|$$

Calculating the spherical harmonic coefficients on both sides on the sphere $k \in k_0 S^2$ gives us

$$\int |\hat{F}_1(\omega, k_0 \hat{n}|r_1, ..., r_N)\bar{Y}_{lm}(\hat{n})d\Omega(\hat{n})$$

$$= \int |\hat{F}(\omega, k_0 \hat{n}|r_{\sigma 1}, ..., r_{\sigma N})|\bar{Y}_{lm}(R\hat{n})d\Omega(\hat{n})$$

$$= \int |\hat{F}(\omega, k_0 \hat{n}|r_{\sigma 1}, ..., r_{\sigma N})| \sum_{m'} (\bar{\pi}_l)_{m',m}(R^{-1})\bar{Y}_{lm'}(\hat{n})d\Omega(\hat{n})$$

This equation can be expressed in matrix form after noting that $\pi_l(R^{-1}) = (\pi_l(R))^*$,

$$|\hat{F}_1(\omega, k_0|r_1, ..., r_N)|_l = \pi_l(R)|\hat{F}(\omega, k_0|r_{\sigma 1}, ..., r_{\sigma N})|_l$$

17.4   Mackey's theory of induced representations applied to estimating the Poincare group element from image pairs

The Poincare group is the semidirect product of $\mathbb{R}^4$ (space-time translations) with the proper orthochronous Lorentz group (spatial rotations and boosts). Essentially, this group can be expressed as

$$\mathcal{P} = \mathbb{R}^4 \otimes_s SL(2, \mathbb{C})$$

where we identify $SL(2, \mathbb{C})$ as the double cover of the proper orthochronous Lorentz group via the map $X \rightarrow gXg^*$ with X being a Hermitian matrix that represents the space-time coordinates and $g \in SL(2, \mathbb{C})$. For a given character χ_0 of $\mathbb{R}^4$ and the corresponding stability subgroup H_0 of $SL(2, \mathbb{C})$ with a given irreducible representation L of H_0, we wish to determine the irreducible representation of $\mathcal{P}$ obtained by inducing $\chi_0 \times L$ from $\mathbb{R}^4 \otimes_s H_0$ to $\mathcal{P} = \mathbb{R}^4 \times_s SL(2, \mathbb{C})$. It is easy to see that if $n \in \mathbb{R}^4$ and $h \in SL(2, \mathbb{C})$, then the representation U of $\mathcal{P}$ induced by χ_0 and L is defined by

$$U(nh)f(\chi) = \chi(n)L(\gamma(\chi)^{-1}h\gamma(h^{-1}\chi))f(h^{-1}\chi), \chi \in O_{\chi_0}$$

where the representation space of U consists of all functions f defined on the orbit O_{χ_0} of χ_0 under $SL(2,\mathbb{C})$ with values in the representation space V of L. Here, γ is a cross-section map in the sense that for each $\chi \in O_{\chi_0}$, $\gamma(\chi) \in SL(2,\mathbb{C})$ is such that $\gamma(\chi)\chi_0 = \chi$ and that if χ, χ' are two distinct elements of O_{χ_0}, then $\gamma(\chi) \neq \gamma(\chi')$. The representation property is easily checked: For $h_1, h_2 \in SL(2,\mathbb{C})$,

$$(U(h_1)(U(h_2)f))(\chi) =$$
$$L(\gamma(\chi)^{-1}h_1\gamma(h_1^{-1}\chi))(U(h_2)f)(h_1^{-1}\chi)$$
$$= L(\gamma(\chi)^{-1}h_1\gamma(h_1^{-1}\chi))L(\gamma(h_1^{-1}\chi)^{-1}h_2\gamma(h_2^{-1}h_1^{-1}\chi))f(h_2^{-1}h_1^{-1}\chi)$$
$$= L(\gamma(\chi)^{-1}h_1h_2\gamma((h_1h_2)^{-1}\chi))f((h_1h_2)^{-1}\chi)$$
$$= U(h_1h_2)f(\chi)$$

We now take an image field $f_1(x)$ defined on a manifold $x \in \mathcal{M}$ on which the group $\mathcal{P}$ acts. Here, $\mathcal{M} = \mathbb{R}^4$. The transformed image field is $f_2(x) = f_1(g^{-1}x) + w(x)$ where $g = nh \in \mathcal{P}$. The left invariant Haar measure on $\mathcal{P}$ is $dndh$ where dn is the standard Lebesgue measure on $\mathbb{R}^4$ while dh is the left invariant Haar measure on $SL(2,\mathbb{C})$. Our aim is to estimate $g = nh$ from measurements of both f_1 and f_2. We have in the absence of noise,

$$\int f_2(g'x_0)U(g')dg' = \int f_1(g^{-1}g'x_0)U(g')dg' = \int f_1(g'x_0)U(gg')dg'$$

$$= U(g)\int f_1(g'x_0)U(g')dg'$$

which can be written in the language of group theoretic Fourier transforms as

$$\hat{f}_2(U) = U(g)\hat{f}_1(U)$$

From this equation, $U(g)$ can be estimated accurately using a linear least squares algorithm provided that we have a sufficient large number of image field pairs (f_1, f_2), all related through U. We now observe that for ψ defined on the orbit O_{χ_0} of χ_0 under $SL(2,\mathbb{C})$, we have

$$\hat{f}_1(U)\psi(\chi) = \int f_1(g x_0)U(g)\psi(\chi)dg =$$

$$\int_{\mathbb{R}^4 \times SL(2,\mathbb{C})} f_1(nhx_0)\chi(n)L(\gamma(\chi)^{-1}h\gamma(h^{-1}\chi))\psi(h^{-1}\chi)dndh$$

where $\chi \in O_{\chi_0}$ is arbitrary. This expression by left invariance of the measure dh on $SL(2,\mathbb{C})$, can equivalently be expressed by the change of variables $h \to \gamma(\chi)h$ as

$$\int f_1(n\gamma(\chi)hx_0)\chi(n)L(h\gamma(h^{-1}\chi_0))\psi(h^{-1}\chi_0)dh$$

We therefore have the following relations for determining the operator $U(g)$. We write

$$\psi_{1k}(\chi) = \hat{f}_1(U)\psi_k(\chi), \psi_{2k}(\chi) = \hat{f}_2(U)\psi_k(\chi), k = 1, 2, ..., n$$

Then,

$$U(g)\psi_{1k}(\chi) = \psi_{2k}(\chi), k = 1, 2, ..., n$$

or equivalently, with $g = nh$,

$$\psi_{2k}(\chi) = \chi(n)L(\gamma(\chi)^{-1}h\gamma(h^{-1}\chi))\psi_{1k}(h^{-1}\chi), k = 1, 2, ..., n --- (1)$$

This gives us on taking the norm in the representation space V in which the unitary irreducible representation L of the stability group H_0 of χ_0 acts,

$$\| \psi_{2k}(\chi) \| = \| \psi_{1k}(h^{-1}\chi) \|, k = 1, 2, ..., n$$

from which h may be determined by taking the ordinary group theoretic Fourier transform on $SL(2, \mathbb{C})$. Assuming that h has thus be determined, n is easily determined using the equations (1) for different $\chi \in O_{\chi_0}$.

A remark on the different orbits in $\mathbb{R}^4$ of $SL(2, \mathbb{C})$ action.

(a) Corresponding to $(1, 0, 0, 0)^T$ (mass $m = 1$), the corresponding representing matrix in $\mathcal{H}$, the space of 2×2 Hermitian matrices on which $SL(2, \mathbb{C})$ acts by adjoint action, is

$$X_0 = I_2$$

The stability subgroup H_0 for X_0 is the set of all $g \in SL(2, \mathbb{C})$ for which $gX_0g^* = X_0$, ie, $gg^* = I_2$. This means that $g \in SU(2)$, ie, $H_0 = SU(2)$.

(b) Corresponding to $(0, 0, 0, 1)^T$ (imaginary mass $m = i$), the element of $\mathcal{H}$ is

$$X_1 = \begin{pmatrix} 1 & 0 \\ 0 & -1 \end{pmatrix}$$

The stability group of X_1 consists of all $g \in SL(2, \mathbb{C})$ for which

$$gX_1g^* = X_1$$

or equivalently,

$$gX_1 = X_1g^{*-1}, g \in SL(2, \mathbb{C})$$

Writing

$$g = \begin{pmatrix} a & b \\ c & d \end{pmatrix}, ad - bc = 1$$

we get

$$g^{-1} = \begin{pmatrix} d & -b \\ -c & a \end{pmatrix}$$

$$g^{*-1} = \begin{pmatrix} \bar{d} & -\bar{c} \\ -\bar{b} & \bar{a} \end{pmatrix}$$

Thus,

$$gX_1 = \begin{pmatrix} a & -b \\ c & -d \end{pmatrix},$$

$$X_1 g^{*-1} = \begin{pmatrix} \bar{d} & -\bar{c} \\ \bar{b} & -\bar{a} \end{pmatrix}$$

So the stability group H_0 of X_1 is the set of all $g \in SL(2,\mathbb{C})$ for which

$$d = \bar{a}, c = \bar{b}$$

Thus, H_0, consists of all matrices of the form

$$\begin{pmatrix} a & b \\ \bar{b} & \bar{a} \end{pmatrix}, |a|^2 - |b|^2 = 1$$

We now consider the stability group H_0 of $(1,0,0,1)$ (zero mass). The element X_2 of $\mathcal{H}$ corresponding this is

$$X_2 = 2 \begin{pmatrix} 1 & 0 \\ 0 & 0 \end{pmatrix}$$

Thus, the condition for

$$g = \begin{pmatrix} a & b \\ c & d \end{pmatrix} \in SL(2,\mathbb{C})$$

to leave this element fixed, ie, that $gX_2 g^* = X_2$ is that g have the form

$$g = \begin{pmatrix} exp(i\theta) & z \\ 0 & exp(-i\theta) \end{pmatrix}, \theta \in [0, 2\pi), z \in \mathbb{C}$$

Denoting this element by $g(z, \theta)$, we see that

$$g(z,\theta).g(z',\theta') = g(z'.exp(i\theta) + z.exp(-i\theta'), \theta + \theta')$$

The group H_0 is isomorphic to the semidirect product $\mathbb{C} \otimes_s \mathbb{T}$ where the isomorphism takes $g(z,\theta)$ to the element $(z.exp(i\theta), \theta)$ with the action of $\mathbb{T}$ on $\mathbb{C}$ being defined by

$$\theta[z] = exp(2i\theta).z$$

Indeed, we then have for this semidirect product

$$(z.exp(i\theta), \theta).(z'.exp(i\theta'), \theta') =$$

$$(z'.exp(i(\theta' + 2\theta)) + z.exp(i\theta), \theta + \theta') =$$

$$(z'.exp(i\theta) + z.exp(-i\theta')).exp(i(\theta + \theta')), \theta + \theta')$$

which corresponds to the matrix $g(z'.exp(i\theta) + z.exp(-i\theta'), \theta + \theta')$, thereby demonstrating the required group isomorphism.

Remark: We wish to demonstrate that the two definitions of the induced representation of the semidirect product of an Abelian group N with another group H such that H normalizes N are equivalent. The first definition is based on the following construction: Choose a $\chi_0 \in \hat{N}$ and let H_0 be its stability subgroup in H. Let O_{χ_0} be the orbit of χ_0 under H. Let L be an irreducible representation of H_0 in a vector space V. It is easy to see that $\tilde{L} : nh \to \chi_0(n)L(h)$ is an irreducible representation of $G_0 = N \otimes_s H_0$ in V where $n \in N, h \in H_0$. The conventional method of constructing $U = Ind_{G_0}^G \tilde{L}$ is to define the representation space Y of U to be the set of all maps $f : G \to V$ for which $f(gg_0) = \tilde{L}(g_0)^{-1}f(g)$ for all $g \in G, g_0 \in G_0$ and then define

$$U(g)f(x) = f(g^{-1}x), g, x \in G, g \in Y$$

We now give another construction and prove that the two methods of constructing the induced representation are isomorphic. For $f \in Y$, define $\psi_f : O_{\chi_0} \to V$ by $\psi_f(\chi) = f(\gamma(\chi))$ where $\gamma(\chi) \in H$ is such that $\gamma(\chi)\chi_0 = \chi$ for each $\chi \in \hat{N}$. Now, define

$$W(nh)\psi_f(\chi) = \chi(n)L(\gamma(\chi)^{-1}h\gamma(h^{-1}\chi))\psi_f(h^{-1}\chi), n \in N, h \in H$$

We observe that $\psi_{f_1} = \psi_{f_2}$ for $f_1, f_2 \in Y$ implies $f(\gamma(\chi)) = 0$ for all $\chi \in O_{\chi_0}$ where $f = f_1 - f_2 \in Y$. This implies in turn that

$$0 = \tilde{L}(nh)^{-1}f(\gamma(\chi)) = f(\gamma(\chi)nh), n \in N, h \in H_0, \chi \in O_{\chi_0}$$

This in turn implies that $f(g) = 0$ for all $g \in G$ since $\gamma(\chi)$ runs over one element from each coset of $G/G_0 = H/H_0$ as χ runs over O_{χ_0} Thus, the map $f \to \psi_f$ is a bijection from Y onto the set of all functions on O_{χ_0}. Note that if $\psi : O_{\chi_0} \to V$ is a map, we define $f(nh) = \psi(h.\chi_0), n \in N, h \in H$. We then have

$$\psi_f(\chi) = f(\gamma(\chi)) = \psi(\gamma(\chi)\chi_0) = \psi(\chi)$$

which demonstrates the bijection once we observe that for $n \in N, h \in H, h_0 \in H_0$, we have

$$f(nh(n_0h_0)^{-1}) = \psi(hh_0^{-1}\chi_0) = \psi(h\chi_0) = f(nh)$$

implying that $f \in Y$.

If $\psi : O_{\chi_0} \to \mathbb{C}$, define $T\psi : G \to \mathbb{C}$ by

$$(T\psi)(nh) = A(n,h).\psi(h.\chi_0), n \in N, h \in H$$

It is easily seen that $T\psi = 0$ iff $\psi = 0$. Further, for $n_0, n \in N, h_0 \in H_0, h \in H$, we have

$$(T\psi)(nhn_0h_0) = (T\psi)(nhn_0h_0) =$$

$$(T\psi)(nhn_0h^{-1}hh_0) = A(nhn_0h^{-1}, hh_0)\psi(h.\chi_0)$$

In order that $T\psi \in Y$, we require that

$$A(nhn_0h^{-1}, hh_0) = \bar{\chi}_0(n_0).L(h_0^{-1})A(n,h)$$

This happens provided that

$$A(n,h) = \bar{h}[\chi_0](n)L(h^{-1}\gamma(h\chi_0))$$

for then,

$$A(nhn_0h^{-1}, hh_0) = \bar{h}h_0[\chi_0](nh[n_0])L(h_0^{-1}h^{-1}\gamma(h[\chi_0]))$$
$$= \bar{h}[\chi_0](nh[n_0])L(h_0^{-1})L(h^{-1}(\gamma(h[\chi_0]))$$
$$= \bar{\chi}_0(n_0)L(h_0^{-1})A(n,h)$$

We further note that any $f \in Y$ is of the form $T\psi$ for some function ψ on O_{χ_0}, the reason being that if $f \in Y$, then we define $\psi(\chi) = f(\gamma(\chi))$. Then,

$$T\psi(nh) = A(n,h)f(\gamma(nh\chi_0)) = A(n,h)f(\gamma(hh^{-1}nh\chi_0)) = A(n,h)f(\gamma(h\chi_0)) =$$
$$A(n,h)f(h.\gamma(\chi_0)\gamma(\chi_0)^{-1}.h^{-1}\gamma(h\chi_0))$$
$$= A(n,h)L(\gamma(h\chi_0)^{-1}h)f(h.\gamma(\chi_0))$$
$$= \bar{h}[\chi_0](n)L(h^{-1}\gamma(h\chi_0))L(\gamma(h\chi_0)^{-1}h)f(h)$$
$$= \bar{h}[\chi_0](n)f(h) = \chi_0(h^{-1}n^{-1}h)f(h) = f(nh)$$

Finally, we show that the two definitions of the induced representation are equivalent. To se this, we note that

$$U(n_1h_1)T\psi(nh) = T\psi(h_1^{-1}n_1^{-1}nh) = T\psi(h_1^{-1}n_1^{-1}nh_1h_1^{-1}h)$$
$$= A(h_1^{-1}n_1^{-1}nh_1, h_1^{-1}h)\psi(h_1^{-1}h\chi_0)$$
$$= h_1^{-1}h[\chi_0](h_1^{-1}n^{-1}n_1h_1)L(h^{-1}h_1\gamma(h_1^{-1}h\chi_0))\psi(h_1^{-1}h\chi_0)$$
$$= h_1^{-1}h[\chi_0](h_1^{-1}[n^{-1}n_1])L(h^{-1}h_1\gamma(h_1^{-1}h\chi_0))\psi(h_1^{-1}h\chi_0)$$
$$= h[\chi_0](n^{-1}n_1)L(h^{-1}h_1\gamma(h_1^{-1}h\chi_0))\psi(h_1^{-1}h\chi_0)$$

while on the other hand,

$$W(n_1h_1)\psi(\chi) = \chi(n_1)L(\gamma(\chi)^{-1}h_1\gamma(h_1^{-1}\chi))\psi(h_1^{-1}\chi)$$

and therefore,

$$TW(n_1h_1)\psi(nh) = A(n,h)(W(n_1h_1)\psi)(h.\chi_0)$$
$$= A(n,h)h[\chi_0](n_1)L(\gamma(h[\chi_0])^{-1}h_1\gamma(h_1^{-1}h[\chi_0]))\psi(h_1^{-1}h[\chi_0])$$
$$= \bar{h}[\chi_0](n)L(h^{-1}\gamma(h\chi_0))h[\chi_0](n_1)L(\gamma(h[\chi_0])^{-1}h_1\gamma(h_1^{-1}h[\chi_0]))\psi(h_1^{-1}h[\chi_0])$$
$$= h[\chi_0](n^{-1}n_1)L(h^{-1}h_1\gamma(h_1^{-1}h[\chi_0]))\psi(h_1^{-1}h.\chi_0)$$

thus proving that

$$U(n_1h_1)T = TW(n_1h_1), n_1 \in N, h_1 \in H$$

ie,

$$U(g) = TW(g)T^{-1}$$

thereby establishing the equivalence of the two definitions of the induced representation for the semidirect product.

17.5   Effect of electromagnetic radiation on the expanding universe

First the unperturbed em field is calculated by solving Maxwell's equations in the Robertson-Walker metric:

$$g_{00} = 1, g_{11} = -S^2(t)f(r), f(r) = 1/(1-kr^2), g_{22} = -S^2(t)r^2, g_{33} = -S^2(t)r^2 sin^2(\theta)$$

The solution to Maxwell's equations in this metric will result in a generalized wave equation for the unperturbed em four potential $A_\mu(x)$. The coefficients of this wave equation will depend upon t, r, θ, with the dependence on t coming via the term $S(t)$. This wave equation will be a second order partial differential equation in the space and time variables. Hence, if we know the em four potential A_μ as well as its time derivative $A_{\mu,0}$ at time $t = 0$, then we can in principle get a unique solution for all $t \geq 0$. We now assume that at time $t = 0$, A_μ and $A_{\mu,0}$ are random functions of the spatial variable and we calculate the ensemble averaged energy-momentum tensor

$$S_{\mu\nu}(t, r, \theta, \phi) = (-1/4) < F_{\alpha\beta}F^{\alpha\beta} > g_{\mu\nu} + g^{\alpha\beta} < F_{\mu\alpha}F_{\nu\beta} >$$

The probability distribution of

$$A_\mu(0, r, \theta, \phi), A_{\mu,0}(0, r, \theta, \phi), \mu = 0, 1, 2, 3$$

must be chosen in such a way that $S_{\mu\nu}$ is a homogeneous and isotropic $(0, 2)$ tensor. Once this has been done, we solve the perturbed Einstein field equations by considering the background to be Robertson-Walker and a small perturbation to this metric begin given by $\delta g_{\mu\nu}(x)$ which satisfies the electromagnetically perturbed Einstein field equations:

$$\delta R_{\mu\nu} - (1/2)R\delta g_{\mu\nu} - (1/2)g_{\mu\nu}\delta R = -8\pi G S_{\mu\nu}$$

The question therefore arises is that what should be the general form, ie, spacetime dependence of $S_{\mu\nu}$ for it to be regarded as a homogeneous and isotropic tensor? It is easy to see that for this to be the case, $S_{\mu\nu}$ must have the form $F(t,r)g_{\mu\nu}$ where $F(t,r)$ is a scalar dependent only on time and the comoving radial coordinate. The condition that $g^{\nu\alpha}S_{\mu\nu:\alpha} = 0$ then implies

$$F_{,0}(t,r)g_{\mu0} + F(t,r)g_{\mu0:0} + F(t,r)g^{km}g_{\mu k:m} + g^{11}F_{,1}(t,r)g_{\mu1}$$

which gives since $g_{\mu\nu:\alpha} = 0$ the equation

$$F(t,r) = F_0 = constt.$$

Such a solution is not of much value since it is equivalent to adding a cosmological constant term to the Einstein field equations. So we drop the condition that the unperturbed em field have an averaged energy-momentum tensor that is homogeneous and isotropic. More generally, solving the Maxwell equations in the Robertson-Walker space-time gives with the above stated initial conditions,

$$A_\mu(t,r) = \int K_\mu^\nu(t,r,r')A_\nu(0,r')d^3r' + \int L_\mu^\nu(t,r,r')A_{\nu,0}(0,r')d^3r'$$

where K and L are uniquely determined kernels expressible in terms of $S(t)$ and k. Now assuming

$$< A_\mu(0,r)A_\nu(0,r') > = P_{\mu\nu}(r,r'),$$

$$< A_\mu(0,r)A_{\nu,0}(0,r') > = Q_{\mu\nu}(r,r'),$$

$$< A_{\mu,0}(0,r)A_{\nu,0}(0,r') > = M_{\mu\nu}(r,r')$$

We have

$$< A_\mu(t,r)A_\nu(t',r') > = int K_\mu^\alpha(t,r,r_1)K_\nu^\beta(t',r',r_1')P_{\alpha\beta}(r_1,r_1')d^3r_1 2d^3r_1'$$

$$= N_{\mu\nu}^{(1)}(t,r,t',r')$$

say. Next,

$$< A_{\mu,\nu}(t,r)A_{\alpha,\beta}(t',r') > =$$

$$\frac{\partial^2}{\partial x^\nu \partial x^{\beta'}}(< A_\mu(x).A_\alpha(x') >$$

$$= \frac{\partial^2}{\partial x^\nu \partial x^{\beta'}}N_{\mu\alpha}^{(1)}(x,x')$$

$$= N_{\mu\alpha\nu\beta}^{(2)}(x,x')$$

say, where

$$x = (t,r), x' = (t',r')$$

Now, the ensemble averaged energy-momentum tensor of the unperturbed em field is given by

$$S_{\mu\nu}(x) = (-1/4)g^{\alpha\rho}(x)g^{\beta\sigma}(x)g_{\mu\nu}(x) < F_{\alpha\beta}(x)F_{\rho\sigma}(x) >$$

$$+ < F_{\mu\alpha(x)}F_{\nu\beta}(x) > g^{\alpha\beta}(x)$$

17.6   Photons inside a cavity

. Assume that there are N photons, so that, for example, a state of the em field within the box may be $|k_1, s_1, ..., k_N, s_N >$ corresponding to the fact that the lth photon has a four momentum k_l and spin/helicity s_l. This photon field interacts with the electron positron field coming from a probe inserted within the cavity. The state of the electron-positron field within the box is then described by $|p_1, \sigma_1, ..., p_M, \sigma_M, p'_1, \sigma'_1, ..., p'_K, \sigma'_K >$ corresponding to the fact that there are M electrons with four momenta p_k and spin σ_k, $k = 1, 2, ..., M$ and K positrons with four momenta p'_k and spin $\sigma'_k s$, $k = 1, 2, ..., K$. The interaction Hamiltonian between the photon field and the electron-positron field is then

$$H_I(t) = \int J^\mu A_\mu d^3 r = -e \int \psi(x)^* \alpha^\mu \psi(x) A_\mu(x) d^3 r$$

where $x = (t, r)$. After interaction, we wish to describe the final state of the electrons, positrons and photons after time T. This state is given in the interaction picture by

$$|\phi(T) >= T\{exp(-i \int_0^T H_I(t)dt)\}|\phi(0) >$$

where

$$|\phi(0) >= |k_j, s_j, j = 1, 2, ..., N, p_j, \sigma_j, j = 1, 2, ..., M, p'_j, \sigma'_j, j = 1, 2, ..., K >$$

is the initial state of the photons, electrons and positrons. We note the expansions

$$\psi(x) = \int (u(P, \sigma)a(P, \sigma)exp(-ip.x) + \bar{v}(P, \sigma)b(P, \sigma)^* exp(ip.x))d^3 P$$

$$A_\mu(x) = \int (2|K|)^{-1/2}[e_\mu(K, s)c(K, s).exp(-ik.x) + \bar{e}_\mu(K, s)c(K, s)^* exp(ik.x)]d^3 K$$

After time T, the current operator in the interaction picture within the box is

$$J^\mu(T, r) = -e\psi(T, r)^* \alpha^\mu \psi(T, r)$$

where

$$\psi(T, r) = exp(-iH_D T)\psi(0, r)exp(iH_D T)$$

with

$$H_D = \int \psi(0, r)^*((\alpha, -i\nabla) + \beta m)\psi(0, r)d^3 r,$$

$$\psi(0, r) = \int (u(P, \sigma)a(P, \sigma)exp(iP.r) + \bar{v}(P, \sigma)b(P, \sigma)^* exp(-iP.r))d^3 P$$

The radiation field produced by this current density outside the box in the far-field zone is given by

$$A_1^\mu(t, r) = (\mu_0/4\pi r) \int J^\mu(t - r/c + \hat{r}.r'/c, r')d^3 r'$$

Apart from this, there is a surface current density operator on the walls of the cavity resonator. This surface current density operator is $J_S = -\hat{n} \times H$ where H is the magnetic field on the boundary. This is determined from the em four potential within the box in the interaction picture:

$$A_\mu(T, r) = exp(iTH_{em})A_\mu(0, r).exp(-iTH_{em})$$

where

$$A_\mu(0, r) = \int (2|K|)^{-1/2}[e_\mu(K, s)c(K, s).exp(iK.r) + \bar{e}_\mu(K, s)c(K, s)^* exp(-iK.r)]d^3K$$

and

$$H_{em} = \int |K|c(K, s)^* c(K, s)d^3K$$

The far field em four potential operator produced by the surface electric current density on the resonator boundaries is given by

$$A_{2\mu}(T, r) = (\mu/4\pi r) \int J_S(t - r/c + \hat{r}.r'/c, r')d^3r'$$

It follows that the far-field em four potential operator $(A_1^\mu + A_2^\mu)T, r)$ is expressible in terms of the creation and annihilation operators of the electrons, positrons and photons. Denoting this operator by $A_3^\mu(t, r)$, we can compute the far field Poynting vector operator $\mathbf{S}(t, r)$ in terms of the electron-positron-photon creation and annihilation operators in the interaction picture. Thus, the mean, mean square and more generally, any higher order moment of the far field Poynting vector or more generally of the far field electromagnetic field which we denote by $F(A_3^\mu, \mathbf{r})$ can be evaluated as

$$< \phi(T)|F(A_3^\mu, \mathbf{r})|\phi(T) >$$

We can also consider other initial states of the photon field like the coherent states and calculate the higher moments of the far field em potentials in such a state.

Remark: When we solve the Maxwell equations with the boundary conditions required by the walls of the cavity resonator, the solution for the four potential will not be expressed as linear combinations of $exp(\pm ik.x)$ but rather in terms of certain eigenfunctions $\eta_k(t, r) = \eta_k(x), k = 1, 2, ...$ that satisfy the boundary conditions. Likewise the solution of the free Dirac equation with wave function vanishing on the boundary will cause the solution to this confined Dirac equation to be expandable in terms of certain vector valued eigenfunctions $\chi_{1k}(x), \chi_{2k}(x), k = 1, 2, ...$ So, ideally speaking, we should express the em four potential operator field $A_\mu(x)$ and the Dirac electron-positron vector operator field as

$$A_\mu(x) = \sum_k c(k)\eta_k(x) + c(k)^* \bar{\eta}_k(x)$$

$$\psi(x) = \sum_k a(k)\chi_{1k}(x) + b(k)^* \chi_{2k}(x)$$

where

$$[c(k), c(m)^*] = \delta[k - m], [c(k), c(m)] = 0,$$
$$[a(k), a(m)^*]_+ = \delta[k - m], [a(k), a(m)]_+ = 0,$$
$$[b(k), b(m)^*]_+ = \delta[k - m], [a(k), b(m)]_+ = 0,$$
$$[a(k), b(m)^*]_+ = 0, [c(k), a(m)] = 0, [c(k), a(m)^*] = 0,$$
$$[c(k), b(m)] = 0, [c(k), b(m)^*] = 0$$

It follows that the Dirac current density field operator within the cavity at time t given by

$$J^\mu(x) = -e\psi(x)^*\alpha^\mu\psi(x)$$

is expressible in terms of $a(k), a(k)^*, b(k), b(k)^*, k = 1, 2, \ldots$. We can calculate the far field em four potential operator in terms of these operators using the standard retarded potential formula. Further, the magnetic field operator on the walls and hence the surface current density field operator J_s on the walls can be computed as $\hat{n} \times H$ where the magnetic field is calculated as components of $A_{r,s} - A_{s,r}$. Thus, J_s on the walls is expressible in terms of $c(k), c(k)^*$ and hence the far field four potential operator generated by the surface current density can be computed using the standard retarded potential method. The state of the electron-positron-photon field within the cavity after time T with the evolution in the interaction picture taking place in accordance with the interaction Hamiltonian $H_I(t) = -e \int \psi(x)^*\alpha^\mu\psi(x)A_\mu(x)d^3r$ is given by

$$\rho(t) = T\{exp(-i\int_0^t H_I(s)ds)\}.\rho(0).T\{exp(-i\int_0^t H_I(s)ds)\})^*$$

and hence we can calculate the moment

$$Tr(\rho(T).F(A_{3\mu}))$$

where $A_{3\mu}$ is the far field em four potential generated by the current density within the cavity and the surface current density on the walls of the cavity and F is some functional of this far field em four potential.

17.7   Justification of the Hartree-Fock Hamiltonian using second order quantum mechanical perturbation theory

First consider a first quantized system with unperturbed Hamiltonian H_0 and perturbation V. Let $E_n^{(0)}, n = 1, 2, \ldots$ denote the unperturbed energy levels and $|\psi_n^{(0)} >$ the corresponding unperturbed eigenfunctions. The perturbation of these levels and eigenfunctions upto second order in V is given by

$$E_n = E_n^{(0)} + E_n^{(1)} + E_n^{(2)}, |\psi_n >= |\psi_n^{(0)} > +|\psi_n^{(1)} > +|\psi_n^{(2)} >$$

Substituting these into the stationary state Schrodinger equation

$$(H + V)(|\psi_n> = E_n|\psi_n>$$

and equating terms of the same order of magnitude gives us

$$(H_0 - E_n^{(0)})|\psi_n^{(0)}> = 0,$$

$$(H_0 - E_n^{(0)})|\psi_n^{(1)}> +V|\psi_n^{(0)}> -E_n^{(1)}|\psi_n^{(0)}> = 0,$$

$$(H_0 - E_n^{(0)})|\psi_n^{(2)}> +V|\psi_n^{(1)}> -E_n^{(1)}|\psi_n^{(1)}> -E_n^{(2)}|\psi_n^{(0)}> = 0$$

Using the orthonormality of the $\psi_n^{(0)}$, $n = 1, 2, ...$, we get

$$E_n^{(1)} = <\psi_n^{(0)}|V|\psi_n^{(0)}>,$$

$$|\psi_n^{(1)}> = \sum_{m \neq n} |\psi_m^{(0)}> \frac{<\psi_m^{(0)}|V|\psi_n^{(0)}>}{E_n^{(0)} - E_m^{(0)}}$$

and

$$E_n^{(2)} = E_n^{(1)} <\psi_n^{(0)}|\psi_n^{(1)}> - <\psi_n^{(0)}|V|\psi_n^{(1)}>$$

$$= \sum_{m \neq n} \frac{|<\psi_n^{(0)}|V|\psi_m^{(0)}>|^2}{E_m^{(0)} - E_n^{(0)}}$$

Thus, $E_n^{(1)}$ is a quadratic function of the wave functions $\psi_n^{(0)}$ and their complex conjugates while $E_n^{(2)}$ is a fourth degree function of the same wave functions and their complex conjugates. This means that in the second quantization process, $E_n^{(0)}$ will be replaced by

$$<\psi_n^{(0)}|H_0|\psi_n^{(0)}> = \int \psi_n^{(0)*}(r)H_0\psi_n^{(0)}(r)d^3r$$

where $\psi_n^{(0)}(r)'s$ are now quantum fields and $E_n^{(1)}$ will be replaced by the operator

$$\int \psi_n^{(0)*}(r)V(r)\psi_n^{(0)}(r)d^3r$$

while finally, $E_n^{(2)}$ will be replaced by a fourth degree polynomial function in the operator wave fields and their conjugates

17.8 Tetrad formulation of the Einstein-Maxwell field equations

Tetrad: $e_a^\mu(x)$. Metric is

$$g = \eta_{ab}\omega^a \otimes \omega^b, \omega^a = e_\mu^a dx^\mu$$

where
$$(\eta_{ab})) = diag[1, -1, -1, -1]$$
Thus,
$$dx^\mu = e^\mu_a \omega^a$$

Let ∇ denote the metrical connection. Then its torsion is zero. We write
$$\nabla_X e_a = \omega^b_a(X) e_b$$
for any vector field X. ω^a_b is a one form. We have
$$\nabla_{e_a} g = 0$$

Since the torsion is zero, we have
$$\nabla_{e_a} e_b - \nabla_{e_b} e_a - [e_a, e_b] = 0$$

From this, we can easily deduce Cartan's first equation of structure:
$$d\omega^a + \omega^a_b \wedge \omega^b = 0$$

Likewise, the curvature tensor R has tetrad R^a_{bcd} given by
$$R^a_{bcd} e_a = [\nabla_{e_b}, \nabla_{e_c}] e_d - \nabla_{[e_b, e_c]} e_d$$

and we easily deduce using this Cartan's second equation of structure:
$$R^a_{bcd} e^c \wedge e_d = d\omega^a_b + \omega^a_c \wedge \omega^c_b$$

So in order to determine R^a_{bcd}, we must first express the one forms $\{\omega^a_b\}$ in terms of $\{\omega^a\}$. The equation $\nabla_X g = 0$ gives
$$X(g(e_b, e_c)) + g(\nabla_X e_b, e_c) - g(e_b, \nabla_X e_c) = 0$$

for any vector field X and since $\eta_{bc} = g(e_b, e_c)$ are constants (by the definition of tetrad), $X(g(e_b, e_c)) = 0$. Thus, we get
$$g(\omega^d_b(X) e_d, e_c) + g(e_b, \omega^d_c(X) e_d) = 0$$

or equivalently,
$$\omega^d_b(X)\eta_{dc} + \omega^d_c(X)\eta_{bd} = 0$$

which is experssed as
$$\omega_{cb} + \omega_{bc} = 0$$

where ω_{ab} is the one form defined by
$$\omega_{ab} = \eta_{ac}\omega^c_b$$

Now writing
$$\omega^a = e^a{}_\mu dx^\mu$$

we get

$$dw^a = e^a_{\mu,\nu}dx^\nu \wedge dx^\mu = e^a_{\mu,\nu}e^\nu_b e^\mu_c \omega^b \wedge \omega^c$$

and hence, the first equation of structure gives

$$e^a_{\mu,\nu}e^\nu_b e^\mu_c \omega^b \wedge \omega^c + \omega^a_c \wedge \omega^c = 0$$

Comparing the coefficient of ω^c on both sides gives us

$$\omega^a_c = -e^a_{\mu,\nu}e^\nu_b e^\mu_c \omega^b + \lambda^a_c \omega^c$$

where in the last term, there is no summation over c. Lowering the index a using the η-metric then gives us

$$\omega_{ac} = -e_{a\mu,\nu}e^\nu_b e^\mu_c \omega^b + \lambda_{ac}\omega^c$$

It follows that the rhs must be skew-symmetric in the indices (a,c). In other words

$$\lambda_{ac}\omega^c + \lambda_{ca}\omega^a = e_{c\mu,\nu}e^\nu_b e^\mu_a \omega^b$$
$$+ e_{a\mu,\nu}e^\nu_b e^\mu_c \omega^b$$
$$= (e_{c\mu,\nu}e^\mu_a + e_{a\mu,\nu}e^\mu_c)e^\nu_b \omega^b$$

Now,

$$e_{a\mu,\nu}e^\mu_c = -e_{a\mu}e^\mu_{c,\nu}$$
$$= -e^\rho_a g_{\mu\rho}e^\mu_{c,\nu}$$
$$= -e^\rho_a(e_{c\rho,\nu} - e^\mu_c g_{\mu\rho,\nu})$$

Thus, we get

$$\lambda_{ac}\omega^c + \lambda_{ca}\omega^a = g_{\mu\rho,\nu}e^\rho_a e^\mu_c e^\nu_b \omega^b$$

for all a,c. Equivalently,

$$\lambda_{ac}\omega^c + \lambda_{ca}\omega^a = -g_{\mu\rho}(e^\rho_a e^\mu_c)_{,\nu}e^\nu_b \omega^b$$
$$= -g_{\mu\rho}(e^\rho_{a,\nu}e^\mu_c + e^\rho_a e^\mu_{c,\nu})e^\nu_b \omega^b$$
$$= -(e^\rho_{a,\nu}e_{c\rho} + e_{a\mu}e^\mu_{c,\nu})e^\nu_b \omega^b$$
$$-2e_{a\mu}e^\mu_{c,\nu}e^\nu_b \omega^b$$

These equations may be used to calculate the one forms ω^a_b in terms of the one forms ω^a. The same logic may be applied to the determine the tetrad components of the curvature tensor using Cartan's second equation of structure:

$$R^a_{bcd}\omega^c \wedge \omega^d = d\omega^a_b + \omega^a_c \wedge \omega^c_d$$

We've seen that

$$\omega^a_b = f(a,b,c,x)\omega^c,$$
$$d\omega^a = g(a,b,c,x)\omega^b \wedge \omega^c$$

for some appropriate functions f, g determined by the tetrad. Thus,

$$dw_b^a = f_{,mu}(a, b, c, x)dx^\mu \wedge \omega^c$$

$$+f(a, b, c, x)dw^c$$

$$= f_{,\mu}(a, b, c, x)e_d^\mu \omega^d \wedge \omega^c$$

$$+f(a, b, c, x)g(c, k, m, x)\omega^k \wedge \omega^m$$

$$= h(a, b, c, d, x)\omega^c \wedge \omega^d$$

where the function h is easily identified in terms of f, g. This gives

$$R_{bcd}^a \omega^c \wedge \omega^d = h(a, b, c, x)\omega^c \wedge \omega^d + \omega_c^a \wedge \omega_b^c$$

$$= h(a, b, c, x)\omega^c \wedge \omega^d + f(a, c, k, x)f(c, b, m, x)\omega^k \wedge \omega^m$$

$$= P(a, b, c, d, x)\omega^c \wedge \omega^d$$

where the function P is easily identified in terms of h, f. From this identity, it easily follows that

$$R_{bcd}^a = (P(a, b, c, d, x) - P(a, b, d, c, x))/2$$

Exercise: Compute the tetrad components of the curvature tensor for the generalized Kerr metric defined by

$$g = \eta_{ab}\omega^a \wedge \omega^b,$$

where

$$\omega^0 = f_0(x)dx^0, \omega^1 = f_1(x)(dx^1 - a_0(x)dx^0 - a_2(x)dx^2 - a_3(x)dx^3),$$

$$\omega^2 = f_2(x)dx^2, \omega^3 = f_3(x)dx^3$$

17.9 Optimal quantum gate design in the presence of an electromagnetic field propagating in the Kerr metric

The Kerr metric has the form

$$d\tau^2 = a_0(r, \theta)dt^2 - a_1(r, \theta)dr^2 - a_2(r, \theta)dr^2 - a_3(r, \theta)(d\phi - \omega(r, \theta)dt)^2$$

so that

$$g_{00} = a_0 - a_3\omega^2, g_{11} = -a_1, g_{22} = -a_2, g_{33} = -a_3, g_{03} = g_{30} = a_3\omega$$

We first write down the Maxwell equations in this metric in the tetrad formalism:

$$A^\mu e_a^\mu = A_a, A^\mu = A_a e_\mu^a$$

$$e_\mu^a e_b^\nu A_{:\nu}^\mu = e_\mu^a e_b^\nu (e_c^\mu A^c)_{:\nu}$$

$$= e_\mu^a e_b^\nu (e_c^\mu A_{,\nu}^c + e_{c:\nu}^\mu A^c)$$

$$= e_b^\nu A_{,\nu}^a + e_\mu^a e_{c:\nu}^\mu e_b^\nu A^c$$

$$= A_{,b}^a + \gamma_{cb}^a A^c$$

where γ_{cb}^a are the spin coefficients. Note that that A^a are scalars. We now write down the Maxwell equations in the tetrad basis. Here $X_{,b}$ means $e_b^\mu X_{,\mu}$.

$$F_{ab} = e_a^\mu e_b^\nu F_{\mu\nu}$$

are the Maxwell scalars. We have

$$F_{:\nu}^{\mu\nu} = 0$$

for the Maxwell equations which give

$$(F^{ab} e_a^\mu e_b^\nu)_{:\nu} = 0$$

or equivalently,

$$F_{,\nu}^{ab} e_a^\mu e_b^\nu + F^{ab} e_{a:\nu}^\mu e_b^\nu + F^{ab} e_a^\mu e_{b:\nu}^\nu = 0$$

or,

$$F_{,b}^{ab} e_a^\mu + F^{ab} e_{a:\nu}^\mu e_b^\nu + F^{ab} e_a^\mu e_{b:\nu}^\nu = 0$$

Thus,

$$F_{,b}^{ab} + F^{cb} e_\mu^a e_{c:\nu}^\mu e_b^\nu + F^{ab} e_{b:\nu}^\nu = 0$$

or in terms of the spin coefficients,

$$F_{,b}^{ab} + F^{cb} \gamma_{cb}^a + F^{ab} e_\nu^d e_{c:\rho}^\nu e_d^\rho = 0$$

or

$$F_{,b}^{ab} + F^{cb} \gamma_{cb}^a + F^{ab} \gamma_{cd}^d$$

This is the first part of the Maxwell equations in tetrad formalism. The second part is to express the Maxwell equation

$$F_{\mu\nu,\rho} + F_{\nu\rho,\mu} + F_{\rho\mu,\nu} = 0$$

in tetrad formalism. The Maxwell equation

$$F_{\mu\nu,\rho} + F_{\nu\rho,\mu} + F_{\rho\mu,\nu} = 0$$

is equivalent to

$$F_{\mu\nu:\rho} + F_{\nu\rho:\mu} + F_{\rho\mu:\nu} = 0$$

We have

$$F_{\mu\nu:\rho} = (F_{ab}e_\mu^a e_\nu^b)_{:\rho} =$$

$$F_{ab,\rho}e_\mu^a e_\nu^b + F_{ab}(e_{\mu:\rho}^a e_\nu^b + e_\mu^a e_{\nu:\rho}^b)$$

Thus,

$$e_a^\mu e_b^\nu e_c^\rho F_{\mu\nu:\rho} =$$

$$F_{ab,c} + F_{nm}e_a^\mu e_b^\nu e_c^\rho (e_{\mu:\rho}^n e_\nu^m + e_\mu^n e_{\nu:\rho}^m)$$

$$= F_{ab,c} + F_{nb}e_a^\mu e_{\mu:\rho}^n e_c^\rho + F_{am}e_b^\nu e_{\nu:\rho}^m e_c^\rho$$

$$= F_{ab,c} + F_{nb}\gamma_{ac}^n + F_{am}\gamma_{bc}^m$$

or more precisely, this is written as

$$F_{ab,c} + F_{nb}\eta^{nm}\gamma_{amc} + F_{am}\eta^{mn}\gamma_{bnc}$$

$$= F_{ab,c} + \eta^{nm}(\gamma_{amc}F_{nb} + \gamma_{bnc}F_{am})$$

where the spin coefficients are defined by

$$\gamma_{abc} = e_a^\mu e_{b\mu:\nu}e_c^\nu$$

Thus, the homogeneous components of the Maxwell equations can be expressed in tetrad notation as

$$F_{ab,c} + F_{bc,a} + F_{ca,b} + \eta^{nm}(\gamma_{amc}F_{nb} + \gamma_{bnc}F_{am} + \gamma_{bma}F_{nc} + \gamma_{cna}F_{bm}$$

$$+\gamma_{cmb}F_{na} + \gamma_{anb}F_{cm}) = 0$$

Calculating the spinor connection of the gravitational field for the Schwarzchild and Kerr metrics. Let $V_{a\mu}$ be the tetrad and

$$J^{ab} = (1/4)[\gamma^a, \gamma^b]$$

The spinor connection of the gravitational field is

$$\Gamma_\mu = (1/2)J^{ab}V_\nu^a V_{\nu:\mu}^b$$

An easy calculation shows that by taking the local Lorentz matrix $\Lambda(x)$ to be infinitesimal, ie, $I + \omega(x)$ where $\omega_{ab} = -\omega_{ba}$, we have

$$D(\Lambda) = I + dD(\omega) = I + \omega_{ab}J^{ab}$$

Then, we require that if Γ_μ and Γ_μ' are respectively the spinor connections in the original and in locally Lorentz transformed frames, we must have

$$D(\Lambda)\gamma^a V_a^\mu(\partial_\mu + \Gamma_\mu)D(\Lambda)^{-1}$$

$$= V_a^\mu D(\Lambda)\gamma^a D(\Lambda)^{-1}D(\Lambda)(\partial_\mu + \Gamma_\mu)D(\Lambda)^{-1}$$

$$= V_a^\mu \Lambda_b^a \gamma^b (D(\Lambda)(\partial_\mu D(\Lambda)^{-1}) + D(\Lambda)\Gamma_\mu D(\Lambda)^{-1} + \partial_\mu)$$
$$= \Lambda_b^a V_a^\mu \gamma^b (\partial_\mu + \Gamma_\mu')$$

where

$$\Gamma_\mu' = D(\Lambda)(\partial_\mu D(\Lambda)^{-1}) + D(\Lambda).\Gamma_\mu.D(\Lambda)^{-1}$$

or equivalently when $\Lambda = I + \omega$ is inifintesimal, we must have

$$\Gamma_\mu' - \Gamma_\mu = -\omega_{ab,\mu} J^{ab} + \omega_{ab}[J^{ab}, \Gamma_\mu]$$

This is satisfied by choosing Γ_μ as above and using the fact that the tetrad under local Lorentz transformations transforms as

$$V_a^\mu \to V_a^\mu + \omega_{ab} V^{\mu b}$$

We leave this as an exercise for the reader to prove. Now we compute the connection for the Kerr metric in the form

$$d\tau^2 = a_0(x^1, x^2)^2 dx^{02} - a_1(x_1, x_2)^2 dx^{12} - a_2(x^1, x^2)^2 dx^{22} - a_3(x^1, x^2)^2 (dx^3 - \omega(x^1, x^2) dx^0)^2$$

where

$$x^0 = t, x^1 = r, x^2 = \theta, x^3 = \phi$$

The tetrad basis for this problem with the Minkowski metric $((\eta_{ab})) = diag[1, -1, -1, -1]$ is given by

$$\omega^0 = V_\mu^0 dx^\mu = a_0 dx^0,$$
$$\omega^1 = V_\mu^1 dx^\mu = a_1 dx^1,$$
$$\omega^2 = V_\mu^2 dx^\mu = a_2 dx^2,$$
$$\omega^3 = V_\mu^3 dx^\mu = a_3(dx^3 - \omega dx^0)$$

Thus, in terms of components,

$$V_0^0 = a_0, V_r^0 = 0, r = 1, 2, 3,$$
$$V_1^1 = a_1, V_\mu^1 = 0, \mu = 0, 2, 3,$$
$$V_2^2 = a_2, V_\mu^2 = 0, \mu = 0, 1, 3,$$
$$V_0^3 = -\omega a_3, V_3^3 = a_3, V_\mu^3 = 0, \mu = 1, 2$$
$$V_{a\nu:\mu} = V_{a\nu,\mu} - \Gamma_{\nu\mu}^\rho V_{a\rho}$$

Treating the gravitational field and the electromagnetic field as a perturbation to the Dirac equation and noting that the Dirac equation can be expressed as

$$[\gamma^a V_a^\mu (i\partial_\mu + i\Gamma_\mu + eA_\mu) - m]\psi = 0$$

we have the following perturbation equations:

$$\psi = \psi_0 + \psi_1 + \ldots + \psi_n + \ldots$$

where

$$\gamma^\mu(i\partial_\mu - m)\psi_0 = 0,$$

$$[i\gamma^\mu\partial_\mu - m]\psi_1 + i\gamma^a(V_a^\mu - \delta_a^\mu)\partial_\mu + i\gamma^\mu\Gamma_\mu + e\gamma^\mu A_\mu]\psi_0 = 0$$

upto first order. In general, taking higher order terms into consideration, we have

$$[i\gamma^\mu\partial_\mu - m]\psi_{n+1} +$$

$$i\gamma^a(V_a^\mu - \delta_a^\mu)\partial_\mu + i\gamma^\mu\Gamma_\mu + e\gamma^\mu A_\mu]\psi_n = 0, n \geq 0$$

or equivalently, if

$$S(x) = (2\pi)^{-4}\int[i\gamma^\mu p_\mu - m]^{-1}exp(ip.x)d^4p$$

is the electron propagator, we can write

$$\psi_{n+1}(x) =$$

$$-\int S(x - x')[i\gamma^a(V_a^\mu(x') - \delta_a^\mu)\partial'_\mu + i\gamma^\mu\Gamma_\mu(x') + e\gamma^\mu A_\mu(x')]\psi_n(x')d^4x'$$

We assume that the em four potential A_μ satisfies the Maxwell equation in curved space-time with metric $g_{\mu\nu}(x)$ which is weak perturbation of flat space-time. So we can write taking into account the gauge condition,

$$(g^{\mu\nu}A_\nu\sqrt{-g})_{,\nu} = 0, (g^{\mu\alpha}g^{\nu\beta}\sqrt{-g}F_{\alpha\beta})_{,\nu} = 0$$

We write

$$g_{\mu\nu} = \eta_{\mu\nu} + h_{\mu\nu}(x)$$

where $h_{\mu\nu}$ is of the first order of smallness and likewise expand A_μ as

$$A_\mu = A_\mu^{(0)} + A_\mu^{(1)}$$

Then, since upto first order,

$$\sqrt{-g} = 1 + h/2, h = h_\mu^\mu = \eta_{\mu\nu}h_{\mu\nu},$$

$$g^{\mu\nu} = \eta_{\mu\nu} - h^{\mu\nu}, h^{\mu\nu} = \eta_{\mu\alpha}\eta_{\nu\beta}h_{\alpha\beta}$$

we get upto the first order, the following Maxwell equations:

$$(\eta_{\mu\alpha}\eta_{\nu\beta}F_{\alpha\beta}^{(0)})_{,\nu} = 0,$$

$$[(-h^{\mu\alpha}\eta_{\nu\beta} - h^{\nu\beta}\eta_{\mu\alpha} + h\eta^{\mu\alpha}\eta^{\nu\beta}/2)F_{\alpha\beta}^{(0)}]_{,\nu} +$$

$$(\eta_{\mu\alpha}\eta_{\nu\beta}F_{\alpha\beta}^{(1)})_{,\nu} = 0$$

Considerable simplification of this equation is achieved by using the gauge condition upto first order:

$$((\eta_{\mu\nu} - h^{\mu\nu})(1 + h/2)(A_\nu^{(0)} + A_\nu^{(1)}))_{,\mu} = 0$$

On equating zeroth and first order of smallness terms, this gauge condition gives

$$(\eta_{\mu\nu}A_\nu^{(0)})_{,\mu} = 0,$$

$$(\eta_{\mu\nu}A_\nu^{(1)})_{,\mu} + ((h\eta_{\mu\nu}/2 - h^{\mu\nu})A_\nu^{(0)})_{,\mu} = 0$$

Now,

$$F_{\mu\nu} = F_{\mu\nu}^{(0)} + F_{\mu\nu}^{(1)},$$

where

$$F_{\mu\nu}^{(0)} = A_{\nu,\mu}^{(0)} - A_{\mu,\nu}^{(0)},$$

$$F_{\mu\nu}^{(1)} = A_{\nu,\mu}^{(1)} - A_{\mu,\nu}^{(1)},$$

Now using the first order gauge condition,

$$(\eta_{\mu\alpha}\eta_{\nu\beta}F_{\alpha\beta}^{(1)})_{,\nu} =$$

$$[\eta_{\mu\alpha}\eta_{\nu\beta}(A_{\beta,\alpha}^{(1)} - A_{\alpha,\beta}^{(1)})]_{,\nu} =$$

$$-\eta_{\mu\alpha}\eta_{\nu\beta}A_{\alpha,\beta\nu}^{(1)}$$

$$-\eta_{\mu\alpha}((h\eta_{\nu\beta}/2 - h^{\nu\beta})A_\nu^{(0)})_{,\alpha\beta}$$

For the Schwarzchild space-time, the tetrad can be chosen as

$$V_\mu^0 dx^\mu = \sqrt{\alpha(r)}dt, \alpha(r) = 1 - 2m/r$$

$$V_\mu^1 dx^\mu = \alpha(r)^{-1/2}dr,$$

$$V_\mu^2 dx^\mu = rd\theta,$$

$$V_\mu^3 dx^\mu = r.\sin(\theta)d\phi$$

To compute the spinor gravitational connection, we need to evaluate the co-variant derivatives of the tetrad. These are as follows. First we evaluate the connection components $\Gamma_{\alpha\beta}^\mu$:

$$\Gamma_{00}^0 = 0, \Gamma_{10}^0 = \Gamma_{01}^0 = (1/2)g^{00}g_{00,1} = \alpha'(r)/2\alpha(r) = (1/2)(1/(r-2m) - 1/r)$$

$$\Gamma_{11}^0 = 0, \Gamma_{22}^0 = 0, \Gamma_{33}^0 = 0,$$

$$\Gamma_{rs}^0 = 0, r, s = 1, 2, 3$$

$$\Gamma_{02}^0 = \Gamma_{20}^0 = (1/2\alpha(r))g_{00,2} = 0,$$

$$\Gamma_{03}^0 = \Gamma_{30}^0 = 0$$

$$\Gamma_{10}^1 = \Gamma_{01}^1 = (1/2)g^{11}g_{11,0} = 0$$

$$\Gamma_{20}^2 = \Gamma_{02}^2 = 0,$$

$$\Gamma_{00}^1 = (-\alpha(r))g_{00,1} = -\alpha'(r)\alpha(r)$$

17.10   Maxwell's equations in the Kerr metric in the tetrad formalism

The tetrad of the Kerr metric is

$$\omega^0 = a_0(r,\theta)dt = e^0_\mu dx^\mu, \omega^1 = a_1(r,\theta)dr = e^1_\mu dx^\mu,$$

$$\omega^2 = a_2(r,\theta)d\theta = e^2_\mu dx^\mu$$

$$\omega^3 = a_3(r,\theta)(d\phi - \omega(r,\theta)dt) = e^3_\mu dx^\mu$$

so that the metric is

$$(\omega^0)^2 - (\omega^1)^2 - (\omega^2)^2 - (\omega^3)^2 = \eta_{ab}\omega^a \otimes \omega^b$$

We have

$$e^a_\mu e^\mu_b = \delta^a_b$$

or equivalently, if the top index denotes the row index and the bottom index the column index, then

$$((e^a_\mu)) = ((e^\mu_a))^{-1}$$

We shall use the notations $e^\mu_{(a)}$ and $e^{(a)}_\mu$ respectively in place of e^μ_a and e^a_μ in order to avoid confusion. Then, we have

$$a_0 e^0_{(b)} = \delta^0_b, a_1 e^1_{(b)} = \delta^1_b, a_2 e^2_{(b)} = \delta^2_b,$$

$$a_3 e^3_{(b)} - a_3\omega e^0_{(b)} = \delta^3_b$$

from which, we deduce that

$$(e^0_{(0)}, e^0_{(1)}, e^0_{(2)}, e^0_{(3)}) = (1/a_0, 0, 0, 0),$$

$$(e^1_{(0)}, e^1_{(1)}, e^1_{(2)}, e^1_{(3)}) = (0, 1/a_1, 0, 0),$$

$$(e^2_{(0)}, e^2_{(1)}, e^2_{(2)}, e^2_{(3)}) = (0, 0, 1/a_2, 0),$$

$$(e^3_{(0)}, e^3_{(1)}, e^3_{(2)}, e^3_{(3)}) = (\omega/a_0, 0, 0, 1/a_3)$$

The Maxwell scalars $F_{(a)(b)}$ are related to the Maxwell tensor $F_{\mu\nu}$ by the relations

$$F_{(a)(b)} = e^\mu_a e^\mu_b F_{\mu\nu}$$

We thus evaluate:

$$F_{(0)(1)} = e^\mu_{(0)} e^\nu_{(1)} F_{\mu\nu} =$$

$$(1/a_0 a_1)F_{01},$$

$$F_{(0)(2)} = e^\mu_{(0)} e^\nu_{(2)} F_{\mu\nu} = (1/a_0 a_2)F_{02},$$

$$F_{(0)(3)} = e^\mu_{(0)} e^\nu_{(3)} F_{\mu\nu} = (1/a_0 a_3)F_{03}$$

$$F_{(1)(2)} = e^\mu_{(1)} e^\nu_{(2)} F_{\mu\nu} = (1/a_1 a_2) F_{12}$$

$$F_{(1)(3)} = e^\mu_{(1)} e^\nu_{(3)} F_{\mu\nu} = (1/a_1 a_3) F_{13} + (\omega_1/a_1 a_0) F_{10}$$

$$F_{(2)(3)} = e^\mu_{(2)} e^\nu_{(3)} F_{\mu\nu} = (1/a_2 a_3) F_{23} + (\omega_1/a_2 a_0) F_{20}$$

The Maxwell complex scalars in the Newman-Penrose formalism.
The Newman-Penrose tetrad of null geodesics $l, n, m, \bar{m}$ satisfy

$$l.n = 1, m.\bar{m} = -1, l.l = n.n = m.m = \bar{m}.\bar{m} = l.m = l.\bar{m} = 0$$

where by $a.b$ we mean $g_{\mu\nu} a^\mu b^\nu$. We define the complex Maxwell scalars by

$$\phi_1 = F_{\mu\nu} l^\mu m^\mu, \phi_2 = F_{\mu\nu} l^\mu \bar{m}^\mu,$$

$$\phi_3 = F_{\mu\nu}(l^\mu n^\nu + m^\mu \bar{m}^\nu)$$

The six components of $F_{\mu\nu}$ can be recovered from these three complex scalars and their complex conjugates.

The em field when passed through the Kerr-metric:Design of quantum gates based on this idea The em four potential is

$$A_\mu = A^{(0)}_\mu(x) + A^{(1)}_\mu(x)$$

The metric is

$$g_{\mu\nu}(x) = \eta_{\mu\nu} + h_{\mu\nu}(x)$$

The zeroth order Maxwell equations are after taking into account the Lorentz gauge condition for the unperturbed potential,

$$\Box A^{(0)}_\mu = 0, \Box = \partial_\alpha \partial^\alpha$$

The first order Maxwell equations are after taking into account the first order perturbed gauge condition derived from

$$(g^{\mu\nu}\sqrt{-g}A_\mu)_{,\nu} = 0$$

of the form

$$\Box A^{(1)}_\mu = C_1(\mu\nu\rho\sigma\alpha\beta)A^{(0)}_\nu h_{\rho\sigma,\alpha\beta}$$

$$+ C_2(\mu\nu\rho\sigma\alpha\beta)A^{(0)}_{\nu,\alpha\beta} h_{\rho\sigma}$$

$$+ C_3(\mu\nu\rho\sigma\alpha\beta)A^{(0)}_{\nu,\alpha} h_{\rho\sigma,\beta}$$

If the Dirac current field

$$J^\mu = -e\psi(x)^* \alpha^\mu \psi(x)$$

is taken into account as a first order perturbation, then the above equation gets modified to

$$\Box A^{(1)}_\mu = C_1(\mu\nu\rho\sigma\alpha\beta)A^{(0)}_\nu h_{\rho\sigma,\alpha\beta}$$

$$+C_2(\mu\nu\rho\sigma\alpha\beta)A^{(0)}_{\nu,\alpha\beta}h_{\rho\sigma}$$

$$+C_3(\mu\nu\rho\sigma\alpha\beta)A^{(0)}_{\nu,\alpha}h_{\rho\sigma,\beta}$$

$$+\mu_0 e\psi(x)^*\alpha^\mu\psi(x)$$

The Dirac field taking into account gravity and electromagnetic interactions satisfies the first order perturbed equation:

$$V^\mu_a(x) = \delta^\mu_a + f^\mu_a(x)$$

where

$$(\eta_{\mu\nu} + h_{\mu\nu}) = \eta_{ab}(\delta^a_\mu + f^a_\mu)(\delta^b_\nu + f^b_\nu)$$

so that

$$h_{\mu\nu} = \eta_{\mu b}f^b_\nu + \eta_{\nu a}f^a_\mu = f_{\mu\nu} + f_{\nu\mu}$$

So upto first order, we can take

$$f_{\mu\nu} = h_{\mu\nu}/2$$

or equivalently,

$$f^\mu_a = h^\mu_a/2$$

Likewise, the spinor gravitational connection upto first order terms is

$$\Gamma_\mu = (1/2)V^\nu_a V_{b\nu:\mu}J^{ab} =$$

$$(1/2)(\delta^\nu_a + f^\nu_a)(\eta_{b\nu} + f_{b\nu})_{:\mu}J^{ab}$$

$$= (1/2)(\delta^\nu_a + f^\nu_a)(-\Gamma^\alpha_{\nu\mu}\eta_{b\alpha} + f_{b\nu,\mu})J^{ab}$$

$$= (1/2)(-\Gamma_{ba\mu} + f_{ba,\mu})J^{ab}$$

$$= (1/2)((-1/2)(h_{ba,\mu} + h_{b\mu,a} - h_{a\mu,b}) + f_{ba,\mu})J^{ab}$$

$$= (1/2)(-f_{b\mu,a} + f_{a\mu,b})J^{ab}$$

The Dirac equation taking em field interactions and gravitational field interactions into consideration after neglecting nonlinear terms in the metric perturbations $h_{\mu\nu}$ is therefore

$$[\gamma^a(\delta^\mu_a + f^\mu_a)(i\partial_\mu + eA_\mu + (i/2)J^{ab}(f_{a\mu,b} - f_{b\mu,a})) - m]\psi = 0$$

which further simplifies on neglecting quadratic terms in the $f'_{a\mu}$s to

$$[\gamma^\mu(i\partial_\mu + eA_\mu + (i/2)J^{ab}(f_{a\mu,b} - f_{b\mu,a})) + i\gamma^a f^\mu_a\partial_\mu - m]\psi = 0$$

or equivalently, written in the form of first order perturbation theory,

$$(\gamma^\mu(i\partial_\mu - m)\psi =$$

$$-[\gamma^\mu(eA_\mu + (i/2)J^{ab}(f_{a\mu,b} - f_{b\mu,a})) + i\gamma^a f^\mu_a\partial_\mu]\psi$$

Chapter 18

Quantum fluid antennas interacting with media

18.1 Quantum MHD antenna in a quantum gravitational field

The MHD equations in the curved space-time of general relativity are derived using

$$(T^{\mu\nu})_{:\nu} = F^{\mu\nu}J_\nu - \Delta T^{\mu\nu}_{:\nu} = \sigma F^{\mu\nu}F_{\nu\alpha}v^\alpha - \Delta T^{\mu\nu}_{:\nu}$$

where $\Delta T^{\mu\nu}$ is the contribution to the energy-momentum tensor of the fluid matter coming from viscous and thermal effects. Here,

$$T^{\mu\nu} = (\rho + p)v^\mu v^\nu - pg^{\mu\nu}$$

is the energy-momentum tensor of the fluid matter without taking into account viscous and thermal effects. These equations may also be derived from the Einstein-Maxwell field equations:

$$R^{\mu\nu} - (1/2)Rg^{\mu\nu} = -8\pi G(T^{\mu\nu} + S^{\mu\nu}),$$

$$S^{\mu\nu} = (-1/4)F^{\alpha\beta}F_{\alpha\beta}g^{\mu\nu} + g_{\alpha\beta}F^{\mu\alpha}F^{\nu\beta},$$

$$F^{\mu\nu}_{:\nu} = -\mu_0 J^\mu, J^\mu = \sigma F^{\mu\nu}v_\nu$$

Exercise: The coefficient of $\sqrt{-g}\delta g_{\mu\nu}$ in the variation of the action of a field in background curved space-time gives the energy-momentum tensor of the field. For the Dirac field in a background curved space-time with an electromgnetic field, the action is given by

$$S[\psi] = \int Re[\psi(x)^*[V_a^\mu(x)\gamma^0\gamma^a(i\partial_\mu + i\Gamma_\mu + eA_\mu) - m\gamma^0]\psi(x)]\sqrt{-g}d^4x$$

where $V_a^\mu(x)$ is a tetrad for the metric $g_{\mu\nu}$ and Γ_μ is the spinor connection of the gravitational field:

$$\Gamma_\mu = (1/2)J^{ab}V_\nu^a V_{:\mu}^{b\nu}, J^{ab} = (1/4)[\gamma^a, \gamma^b]$$

Using the above discussion, evaluate the energy-momentum tensor of the Dirac field and denote this by $T_{\mu\nu}$. Hence set up the Einstein-Maxwell-Dirac field equations:

$$R_{\mu\nu} - (1/2)Rg_{\mu\nu} = -8\pi G(T_{\mu\nu} + S_{\mu\nu})$$

where

$$S_{\mu\nu} = (-1/4)F_{\alpha\beta}F^{\alpha\beta}g_{\mu\nu} + g^{\alpha\beta}F_{\mu\alpha}F_{\nu\beta}$$

is the energy-momentum tensor of the electromagnetic field. The vanishing four divergence of the Einstein tensor implies the field equations

$$(T^{\mu\nu} + S^{\mu\nu})_{:\nu} = 0$$

which can be derived from the Maxwell equations with Dirac current:

$$F_{:\nu}^{\mu\nu} = -\mu_0 J^\mu$$

where J^μ is the coefficient of δA_μ in the variation of the above Dirac action functional w.r.t A_μ, ie,

$$J^\mu(x) = V_a^\mu(x)\psi(x)^*\gamma^0\gamma^a\psi(x)$$

We now explain how a perturbation theoretic analysis of these equations can be performed to obtain dispersion relations for the perturbed quantities, ie, we linearize the Einstein-Maxwell-Dirac equations around fixed values of these fields and study oscillations in these perturbed quantities.

18.2   Applications of scattering theory to quantum antennas

The basic idea here is the following. We have a projectile coming from time $t = -\infty$ at an infinite distance from the scatterer with an energy of $H_0 = P^2/2m = -\nabla^2/2m$ in the non-relativistic case or $H_0 = (\alpha, P) + \beta m$ in the relativistic case. This projectile interacts with the scatterer with an interaction potential of V so that during the interaction period, the total energy of the projectile becomes $H = H_0 + V$. After interaction, the projectile goes at time $t = \infty$ to infinity. According to first order Born scattering theory, the incident state of the projectile of a steady projectile current is given by

$$\psi_i(r) = C.exp(-ik.r)$$

so that

$$H_0\psi_i = E(k)\psi_i, E(k) = k^2/2m$$

in the non-relativistic case and in the relativistic case, $C = C(k)$ becomes a 4×1 complex vector satisfying

$$((\alpha, k) + \beta m)C(k) = E(k)C(k)$$

where $E(k) = \sqrt{k^2 + m^2}$. Let $\psi_f(r)$ denote the final state of the projectile after it gets scattered. By first order perturbation theory, since the energy of the projectile is conserved, this final state will satisfy the exact equation

$$(H_0 + V)\psi_f = E(k)\psi_f(r)$$

which approximates to

$$H_0(\psi_f - \psi_i) + V\psi_i = E(k)(\psi_f - \psi_i)$$

or equivalently,

$$(H_0 - E(k))(\psi_f - \psi_i) = -V\psi_i$$

In the non-relativistic case, this equation is the same as

$$(\nabla^2 + k^2)(\psi_f - \psi_i) = 2mV(r)\psi_i(r)$$

which has the solution

$$\psi_f(r) = \psi_i(r) - (m.exp(ikr)/4\pi r) \int V(r')\psi_i((r')exp(-ik\hat{r}.r')d^3r'$$

Equivalently, defining

$$F(\hat{r}) = F(\theta, \phi) = (-m/4\pi) \int V(r')exp(-ik.r)exp(-ik\hat{r}.r')d^3r'$$

we can write approximately,

$$\psi_f(r) = C(exp(-ik.r) + F(\hat{r})exp(ikr)/r)$$

in the non-relativistic case. We may assume that the incident flux of particles is parallel to the z axis and directed from $z = -\infty$ to $z = 0$. In this case $k.r = -kz = -kr cos(\theta)$ and we have

$$F(\hat{r}) = (-m/4\pi) \int V(r', \theta', \phi')exp(ik(rcos(\theta) - \hat{r}.r')r^2 sin(\theta)drd\theta d\phi$$

The number of particles getting scattered per unit solid angle at ∞ is therefore proportional to $|F(\hat{r})|^2 = |F(\theta, \phi)|^2$. The smeared out scattered charge density is therefore proportional to $\rho(r, \hat{r}) = (q/r^2)|F(\hat{r})|^2$ and the smeared out current density corresponding to the scattered particles is proportional to

$$J(r, \hat{r}) = q|F(\hat{r})|^2 k\hat{r}/mr^2$$

Note that the smeared out charge and current densities corresponding to a wave function $\psi(r)$ are respectively given by

$$\rho(r) = q|\psi(r)|^2,$$

$$J(r) = (-i/2m)(\psi(r)^*(\nabla + ieA(r))\psi(r) - \psi(r)(\nabla - ieA(r))\psi(r)^*)$$

The magnetic vector potential the electric scalar potentials produced by the scattered charges are respectively given by

$$A(t,r) = (\mu/4\pi)\int cos(\omega(t - |r - r'|/c))J(r')d^3r'/|r - r'|$$

and

$$\Phi(t,r) = (1/4\pi\epsilon)\int cos(\omega(t - |r - r'|/c))\rho(r')d^3r'/|r - r'|$$

The question is how to control the scattering potential V subject to constraints such that the resulting EM field pattern generated by the scattered charged particles is as close as possible to a desired pattern. More, generally, we can pose the following problem. Control the electromagnetic field $A_\mu(x)$ falling on a charged quantum particle with Hamiltonian $H(Q,P)$ so that if $\psi(t,Q)$ denotes the Schrodinger wave function satisfying

$$i\psi_{,t}(t,Q) = [H(Q, -i\nabla - qA(t,Q)) + q\Phi(t,Q)]\psi(t,Q)$$

then we require the smeared out charge and current densities corresponding to this wave function to generate an EM field pattern as close as possible to a desired one.

18.3   Wave function of a quantum field with applications to writing down the Schrodinger equation for the expanding universe

Main idea: Assume that we have a set of fields $\phi_n(x), n = 1, 2, ..., N$ whose classical dynamics is described by a Lagrangian density

$$L(x, \phi_n, \phi_{n,\mu}, n = 1, 2, ..., N, \mu = 0, 1, 2, 3)$$

We write

$$\phi_n(x) = \phi_{n0}(x) + \delta\phi_n(x)$$

where $\phi_{n0}(x)$ are the unperturbed classical fields and $\delta\phi_n(x)$ are the perturbed quantum field fluctuations. We expand L around ϕ_{n0} and obtain thereby an infinite series for the Lagrangian density:

$$L = L_0(x) + \sum_n a_1(n, x)\delta\phi_n(x) + \sum_{n,\mu} b_1(n, \mu, x)\delta\phi_{n,\mu}(x)+$$

$$...+ \sum_{n_1,...,n_k,r_1,...,r_m,\mu_1,...,\mu_m} a_{km}(n_1,n_2,...,n_k,r_1,\mu_1,...,r_m,\mu_m,x)\delta\phi_{n_1}(x)$$

$$...\delta\phi_{n_k}(x)\delta\phi_{n_1,\mu_1}(x)...\delta\phi_{n_k,\mu_k}(x)+...$$

If we expand around the stationary value of the Lagrangian then the first order terms do note appear, ie, $a_1 = 0, b_1 = 0$ and then we have for example upto third order terms,

$$L = L_0(x) + \sum_{n_1,n_2} a_{20}(n_1,n_2,x)\delta\phi_{n_1}(x)\delta\phi_{n_2}(x) + a_{11}(n_1,r_1,\mu_1,x)\delta\phi_{n_1}(x)\delta\phi_{r_1,\mu_1}(x)$$

$$+ a_{02}(r_1,\mu_1,r_2\,mu_2)\delta\phi_{r_1,\mu_1}(x)\delta\phi_{r_2,\mu_2}(x)$$

$$+ \sum_{n_1,n_2,n_3} a_{30}(n_1,n_2,n_3,x)\delta\phi_{n_1}(x)\delta\phi_{n_2}(x)\delta\phi_{n_3}(x)$$

$$+ \sum_{n_1,n_2,r_1,\mu_1} a_{21}(n_1,n_2,r_1,\mu_1,x)\delta\phi_{n_1}(x)\delta\phi_{n_2}(x)\delta\phi_{r_1,\mu_1}(x)$$

$$+ \sum_{n_1,r_1,\mu_1,r_2,\mu_2} a_{12}(n_1,r_1,\mu_1,r_2,\mu_2,x)\delta\phi_{n_1}(x)\delta\phi_{r_1,\mu_1}(x)\delta\phi_{r_2,\mu_2}(x)$$

$$+ \sum_{r_1,\mu_1,r_2,\mu_2,r_3,\mu_3} a_{03}(r_1,mu_1,r_2,\mu_2,r_3,\mu_3,x)\delta\phi_{r_1,\mu_1}(x)\delta\phi_{r_2,\mu_2}(x)\delta\phi_{r_3,\mu_3}(x)$$

The Hamiltonian density corresponding to this Lagrangian density can be written down by performing the Legendre transformation. Retaining only upto cubic terms in the joint position and momentum fields, we have the following approximation for the Hamiltonian density:

$$H = \sum_{n_1,n_2} b_{20}(n_1,n_2)\delta\phi_{n_1}(x)\delta\phi_{n_2}(x) +$$

$$+ \sum_{n_1,r_1,\mu_1} b_{11}(n_1,r_1)\delta\phi_{n_1}(x)\delta\pi_{r_1}(x) +$$

$$\sum_{n_1,k_1,n_2} c_{11}(n_1,k_1,n_2)\delta\phi_{n_1,k_1}(x)\delta\phi_{n_2}(x)$$

$$+ \sum_{n_1,k_1,n_2,k_2} d_{11}(n_1,k_1,n_2,k_2)\delta\phi_{n_1,k_1}(x)\delta\phi_{n_2,k_2}(x)$$

$$+ \sum_{r_1,r_2} f_{11}(r_1,r_2)\delta\pi_{r_1}(x)\delta\pi_{r_2}(x)$$

$$+ \sum_{n_1,k_1,r_1} g_{11}(n_1,k_1,r_1)\delta\phi_{n_1,k_1}(x)\delta\pi_{r_1}(x)$$

+cubic terms involving the products

$$\delta\phi_{n_1}\delta\phi_{n_2}\delta\phi_{n_3}, \delta\phi_{n_1}\delta\phi_{n_2}\delta\phi_{n_3,k_3},$$

$$\delta\phi_{n_1}\delta\phi_{n_2,k_2}\delta\phi_{n_3,k_3}, \delta\phi_{n_1}\delta\phi_{n_2}\delta\pi_{r_3},$$

$$\delta\phi_{n_1}\delta\phi_{n_2,k_2}\delta\pi_{r_3}, \delta\phi_{n_1,k_1}\delta\phi_{n_2,k_2}\delta\pi_{r_3},$$

$$\delta\phi_{n_1}\delta\pi_{r_2}\delta\pi_{r_3}, \delta\phi_{n_1,k_1}\delta\pi_{r_2}\delta\pi_{r_3},$$

$$\delta\pi_{r_1}\delta\pi_{r_2}\delta\pi_{r_3}$$

After this, we replace $\delta\pi_r(x)$ by the variational differential $-i\delta/\delta(\delta\phi_n(x))$ in the Hamiltonian density $\int H d^3x$ and write the Schrodinger equation for the quantum field fluctuations as

$$\left[\int H(t, r\delta\phi_n(r), \delta\phi_{n,k}(r), -i\delta/\delta(\delta\phi_n(r))d^3r\right]\psi(t, \delta\phi_n())$$

$$= i\partial\psi(t, \delta\phi_n(.))/\partial t$$

By applying this idea to the Einstein-Hilbert-Maxwell-Dirac Lagrangian density described in terms of the quantum fields $\delta g_{\mu\nu}(x), \delta A_\mu(x), \delta\psi_l(x)$, we can arrive at the Schrodinger equation for the universe. A slight variant of this idea has been proposed by Hawking (See Hawking and Penrose "The nature of space and time", Oxford,1996.) The idea used by Hawking and Penrose is to expand all the metric tensor coefficients as linear combinations of tensor harmonics, expand the coordinates (which may be chosen arbitrarily) or more specifically, the gauge functions specifying the coordinate system and the velocity field of the matter and the electromagnetic potentials as linear combinations of vector harmonics and finally expand the density field as linear combinations of scalar harmonics. The coefficients in these expansions will be functions of time only while the harmonic functions will be functions of the spatial coordinates only. After substituting these expansions into the overall spatial integral of the Lagrangian density for the gravitational field interacting with the matter field and the EM field, we obtain the Lagrangian of this total field as a nonlinear function of the linear combination coefficients as well as their time derivatives. From this expression, by performing a Legendre transformation, we can obtain the Hamiltonian of the total field as a nonnlinear function of the linear combination coefficients and the corresponding canonical momenta which are simply the partial derivatives of the Lagrangian w.r.t the time derivatives of the linear combination coefficients. Once this discrete form of the Hamiltonian has been set up, the Schrodinger equation for the universe can be immediately set up.

18.4   Simple exclusion process and antenna theory

The generator of a simple exclusion process $\eta_t : \mathbb{Z}_N^3 \to \{0, 1\}$ is given by

$$Lf(\eta) = \sum_{x \neq y} \eta(x)(1 - \eta(y))(f(\eta^{(x,y)}) - f(\eta))$$

The velocity of a particle at site x when it jumps to y in time dt may be defined as $(y - x)/dt \in \mathbb{R}^3$. This is provided that $\eta_t(x) = 1, \eta_t(y) = 0$ and in time

$[t, t + dt]$ the jump takes place with $p(x, y)dt$ being the jump probability. Thus, the average velocity associated with a jump from the site x is

$$v(t, x) = \sum_{y:y \neq x} \mathbb{E}(\eta_t(x)(1 - \eta_t(y)))p(x, y)(y - x)$$

If q denotes the charge placed on each particle in this process, then the average current density is proportional to $qv(t, x)$ and one can in principle calculate the EM field produced by the exclusion process field. More generally, conditioned on the state η_t at time t, the velocity process of a particle at the site x is

$$V(t, x) = \eta_t(x)(1 - \eta_t(y))p(x, y)(y - x)$$

and we can calculate the EM field produced by the associated random current $qV(t, x)$ and then compute the moments of this EM field, like the mean EM field and the space-time correlations of the resulting EM field.

18.5 MHD and quantum antenna theory

The basic MHD equations for a conducting fluid are the Maxwell equations along with the Navier-Stokes equation with EM force terms. Assuming incompressibility, these equations are:

$$div J + \rho_{q,t} = 0, div v = 0$$

$$J = \sigma(E + v \times B),$$

$$v_{,t} + (v, \nabla)v = -\nabla p/\rho + \nu \nabla^2 v + J \times B/\rho$$

$$div B = 0, curl E = -B_{,t}, curl B = \mu J + \mu \epsilon E_{,t}, div E = \rho_q/\epsilon$$

These equations cannot be derived from a field theoretic Lagrangian or a Hamiltonian because of the terms involving the conductivity σ which cause damping. They may however, be quantized using an appropriate Hamiltonian combined with Lindblad damping terms to take into account the dissipative effects of the conductivity.

Consider the special case when the EM field is given and the velocity and pressure field are to be determined using the incompressibility equation and the MHD-Navier-Stokes equation. For quantization purposes, we first analyze how a first order state variable system

$$X'(t) = F(t, X(t)), X(t) \in \mathbb{R}^n$$

may be derived from a Lagrangian and then from a Hamiltonian. The Lagrangian is elementary, simply introduce a Lagrange multiplier $\lambda(t) \in \mathbb{R}^n$ and the Lagrangian

$$L(X, X', \lambda, \lambda') = \lambda(t)^T(X'(t) - F(t, X(t)))$$

Then the Euler-Lagrange equations are

$$\lambda'(t) + F(t, X(t)) = 0, X'(t) - F(t, X(t)) = 0$$

These are respectively the co-state and state equations. However, if we try to introduce a Hamiltonian by defining the canonical momenta

$$p_\lambda = \frac{\partial L}{\partial \lambda'} = 0,$$

$$p_X = \frac{\partial L}{\partial X'} = \lambda$$

then we cannot construct the Legendre transormation

$$H = p_\lambda^T \lambda' - L$$

since there is no way to solve for λ' in terms of X, p_X, p_λ. However, we note that by introducing a small perturbation parameter δ and modifying the Lagrangian to

$$L(X, X', \lambda, \lambda') = \lambda^T(X' - F(t, X)) + \delta.\lambda'^T X'$$

we get

$$p_X = \lambda' + \delta X', p_\lambda = \delta X'$$

so that

$$X' = \delta^{-1} p_\lambda, \lambda' = (p_X - p_\lambda)$$

we can write the Hamiltonian as

$$H(X, \lambda, p_X, p_\lambda) = p_X^T X' + p_\lambda^T \lambda' - L =$$

$$p_X^T X' + p_\lambda^T \lambda' - \lambda^T(X - F) - \delta\lambda'^T X'$$

$$= \delta^{-1} p_X^T p_\lambda - \lambda^T X + \lambda^T F(t, X)$$

and this Hamiltonian can be quantized immediately and δ set to zero at the end of all the calculations.

By adopting the same approach, we shall write down the Lagrangian density of the velocity and pressure field as

$$L_v = \lambda^T(v_{,t} + (v, \nabla)v + \nabla p/\rho + \nu\nabla^2 v - \alpha(E + v \times B) \times B)$$

$$+\delta(\lambda_{,t}^T v_{,t} + \mu_{,t} p_{,t}) + \mu.divv$$

where $v = v(t, r), p = p(t, r), \lambda = \lambda(t, r), \mu = \mu(t, r)$ are fields. To this we add the standard Lagrangian density of the electromagnetic field interacting with the current density:

$$L_{EM} = (\epsilon/2)|E|^2 - (1/2\mu)|B|^2 - \rho_q\Phi + J.A$$

where we substitute

$$J = \sigma(E + v \times B), \rho_q = -\int_0^t div J dt$$

and

$$E = -\nabla\Phi - A_{,t}, B = \nabla \times A$$

The total Lagrangian density must be taken as the sum of both the Lagrangian densities and the position fields are $v, p, \lambda, \mu, A, \Phi$. It is instructive to carry out the variation in the resulting action and see how the resulting field equations differ from the basic MHD equations coupled to the Maxwell equations. Once we have quantized the fluid and electromagnetic fields, we can regard the electric, magnetic and fluid velocity fields as field operators and calculate the far field antenna Poynting vector pattern and its expectation, correlations and more generally higher order moments in any given initial state of the fluid and electromagnetic field. This would be another way of looking at a quantum MHD antenna.

18.6 Approximate Hamiltonian formulation of the diffusion equation with applications to quantum antenna theory

The Lagrangian density of our theory is

$$L(u, u_{,t}, u_{,x}, \lambda, \lambda_{,t}) = \lambda(u_{,t} - Du_{,xx}) + \delta\lambda_{,t}u_{,t}$$

This is equivalent to

$$L = \lambda u_{,t} + D\lambda_{,x}u_{,x} + \delta\lambda_{,t}u_{,t}$$

The canonical momenta are

$$p_u = \frac{\partial L}{\partial u_{,t}} = \lambda + \delta\lambda_{,t}$$

$$p_\lambda = \frac{\partial L}{\partial \lambda_{,t}} = \delta u_{,t}$$

Thus, the Hamiltonian density is given by

$$H(u, \lambda, p_u, p_\lambda) = p_u u_{,t} + p_\lambda \lambda_{,t} - L =$$

$$\delta^{-1}p_u p_\lambda + \delta u_{,t}\lambda_{,t} - \lambda u_{,t} - D\lambda_{,x}u_{,x} - \delta\lambda_{,t}u_{,t} =$$

$$\delta^{-1}p_\lambda(p_u - \lambda) - D\lambda_{,x}u_{,x}$$

Let us write down the canonical Hamiltonian equations:

$$u_{,t} = \delta H/\delta p_u = \delta^{-1}p_\lambda$$

$$\lambda_{,t} = \delta H/\delta p_\lambda = \delta^{-1}(p_u - \lambda)$$

$$p_{u,t} = -\delta H/\delta u = \partial_x(\partial H/\partial u_{,x}) = -D\lambda_{,xx},$$

$$p_{\lambda,t} = -\delta H/\delta\lambda = -\partial H/\partial\lambda + \partial_x(\partial H/\partial\lambda_{,x}) = \delta^{-1}p_\lambda - Du_{,xx}$$

From these equations, we can arrive at the modified heat equation:

$$u_{,t} = \delta^{-1}p_\lambda = p_{\lambda,t} + Du_{,xx} = \delta u_{,tt} + Du_{,xx}$$

which in the limit of $\delta \to 0$ becomes the heat equation

$$u_{,t} = Du_{,xx}$$

18.7 Derivation of the damped wave equation for the electromagnetic field in a conducting media in quantum mechanics using the Lindblad formalism

Consider first the damped wave equation in one dimension

$$u_{,tt} + \gamma u_{,t} - c^2 u_{,xx} = 0$$

This equation for $u(t,x)$ is valid for $x \in [0,L]$ and to quantize it, we expand the solution as a Fourier series in x:

$$u(t,x) = \sum_{n\in\mathbb{Z}} c_n(t)exp(2\pi inx/L)$$

Substituting this into the damped wave equation gives us a sequence of ode's

$$c_n''(t) + (2\pi in\gamma/L)c_n'(t) + (2n\pi c/L)^2 c_n(t) = 0, n \in \mathbb{Z}$$

We wish to derive these equations from a quantum mechanical formalism. To this end, we express

$$c_n(t) = a_n(t) + ib_n(t)$$

and express the above as two real equations:

$$a_n''(t) + \omega_n^2 a_n(t) - \gamma\omega_n b_n'(t) = 0,$$

$$b_n''(t) + \omega_n^2 b_n(t) + \gamma\omega_n a_n'(t) = 0$$

We take as our unperturbed Hamiltonian

$$H = (p_n^2 + \omega_n^2 q_n^2)/2$$

and Choose the Lindblad operators as

$$L_n = \alpha_n q_n + \beta_+ \beta_n p_n$$

where

$$[q_n, p_m] = i\delta_{nm}$$

Then apply the Heisenberg form the the Lindblad equations for any observable X:

$$dX/dt = i[H, X] - (1/2)\sum_n (L_n^* L_n X + X L_n^* L_n - 2L_n^* X L_n)$$

$$= i[H, X] - (1/2)\sum_n (L_n^*[L_n, X] + [X, L_n^*]L_n)$$

Taking $X = q_n$ gives

$$dq_n/dt = p_n - (1/2)(-i\beta_n L_n^* + i\bar{\beta}_n L_n)$$

$$= p_n - (i/2)(\bar{\beta}_n(\alpha_n q_n + \beta_n p_n) - \beta_n(\bar{\alpha}_n q_n + \bar{\beta}_n p_n))$$

$$= p_n + Im(\alpha_n \bar{\beta}_n)q_n$$

Taking $X = p_n$ gives

$$dp_n/dt = -\omega_n^2 q_n - (1/2)(i\alpha_n L_n^* - i\bar{\alpha}_n L_n)$$

$$= -\omega_n^2 q_n - (i/2)(\alpha_n(\bar{\alpha}_n q_n + \bar{\beta}_n p_n) - \bar{\alpha}_n(\alpha_n q_n + \beta_n p_n))$$

$$= -\omega_n^2 q_n - (i/2)(\alpha_n \bar{\beta}_n - \beta_n \bar{\alpha}_n)p_n$$

$$= -\omega_n^2 q_n + Im(\alpha_n \bar{\beta}_n)p_n$$

Writing $\gamma_n = Im(\bar{\alpha}_n \beta_n)$, we can express the above equations in the form

$$p_n' = -\omega_n^2 q_n - \gamma_n p_n, q_n' = p_n - \gamma_n q_n$$

so that

$$q_n'' = p_n' - \gamma_n q_n' = -\omega_n^2 q_n - \gamma_n p_n - \gamma_n q_n'$$

$$= -\omega_n^2 q_n - \gamma_n(q_n' + \gamma_n q_n) - \gamma_n q_n'$$

$$= -(\omega_n^2 + \gamma_n^2)q_n - 2\gamma_n q_n'$$

which is precisely the damped harmonic oscillator equation. We note that in the limit of large γ_n, this differential equation approximates to a first order differential equation in time like the heat equation:

$$q_n' \approx -(\gamma_n/2)q_n$$

More general kinds of damped oscillator equations involving vector operator valued functions of time as considered above by Fourier series expansion of the wave equation with damping can be obtained by considering

$$q(t) = (q_1(t), ..., q_N(t))^T, p(t) = (p_1(t), ..., p_n(t))^T$$

with

$$[q_i, p_j] = \delta_{ij}$$

and using the Hamiltonian

$$H = (p^T p + q^T K q)/2 = (1/2)(\sum_{k=1}^{n} p_k^2 + \sum_{r,s=1}^{n} K(r,s)q_r q_s)$$

with Lindblad operators

$$L_n = \alpha_n^T q + \beta_n^T p$$

We find that

$$dq_n/dt = i[H, q_n] - (1/2)\sum_m (L_m^* L_m q_n + q_n L_m^* L_m - 2L_m^* q_n L_m)$$

$$dp_n/dt = i[H, p_n] - (1/2)\sum_m (L_m^* L_m p_n + p_n L_m^* L_m - 2L_m^* p_n L_m)$$

We find that

$$[L_m, q_n] = [\beta_m^T p, q_n] = -i\beta_m[n],$$

$$[q_n, L_m^*] = [q_n, \bar\beta_m^T p] = i\bar\beta_m[n]$$

$$[L_m, p_n] = i\alpha_m[n],$$

$$[p_n, L_m^*] = -i\bar\alpha_m[n]$$

Further,

$$[H, q_n] = -ip_n, [H, p_n] = i\sum_r K(n,r)q_r$$

So we get for the noisy Heisenberg dynamics,

$$dq_n/dt = p_n - (1/2)\sum_m (-i\beta_m[n]L_m^* + i\bar\beta_m[n]L_m)$$

$$dp_n/dt = -(Kp)_n - (1/2)\sum_m (i\alpha_m[n]L_m^* - i\bar\alpha_m[n]L_m)$$

Equivalently,

$$dq_n/dt = p_n - (1/2)\sum_m [(-i\beta_m[n](\alpha_m^* q + \beta_m^* p) + i\bar\beta_m[n](\alpha_m^T q + \beta_m^T p)]$$

$$= p_n - \sum_m (Im(\beta_m[n]\alpha_m^*)q - Im(\bar\beta_m[n]\beta_m^T)p)$$

or equivalently in vector operator notation,

$$dq/dt = p + \sum_m (Im(\bar\beta_m \beta_m^T)p - Im(\beta_m \alpha_m^*)q)$$

and likewise,

$$dp/dt = -Kp + \sum_m (Im(\alpha_m \alpha_m^*)q - Im(\bar{\alpha}_m \beta_m^T)p)$$

Define the matrices

$$A_{11} = -\sum_m Im(\beta_m \alpha_m^*), A_{12} = \sum_m Im(\bar{\beta}_m \beta_m^T) = -\sum_m Im(\beta_m \beta_m^*)$$

$$A_{21} = \sum_m Im(\alpha_m \alpha_m^*), A_{22} = -\sum_m Im(\bar{\alpha}_m \beta_m^T) = \sum_m Im(\alpha_m \beta_m^*)$$

Then, the above noisy Heisenberg dynamics can be expressed as

$$dq/dt = A_{11}q + (A_{12} + I)p, dp/dt = A_{21}q + (A_{22} - K)p$$

Now consider a system of n charged quantum particles with $(q_1, ..., q_n)$ describing their positions and $(p_1, ..., p_n)$ their momenta. We assume that each q_k and each p_k is a 3-vector operator valued observable and that e_k is the charge placed on the kth particle. The equations of motion of these particles are described by the above noisy Heisenberg model of a system of damped harmonic oscillators. We wish to calculate the statistics of the radiation field produced by these particles when the initial wave function of the particles is $\psi_0(q_1, ..., q_n)$ or more generally, when the initial state of this system of particles is the mixed state with kernel $\rho_0(q|q')$ with $q = (q_1, ..., q_n)$ and $q' = (q'_1, ..., q'_n)$. The electromagnetic field produced by these particles is described by the four-potential

$$A(t, r) = \sum_k (\mu e_k / 4\pi) \int J_k(t - |r - r'|/c, r') d^3 r' / |r - r'|$$

$$\Phi(t, r) = \sum_k (e_k / 4\pi\epsilon) \int \rho_k(t - |r - r'|/c, r') d^3 r' / |r - r'|$$

where $J_k(t, r)$ and $\rho_k(t, r)$ are Heisenberg operator fields defined by

$$J_k(t, r) = (e_k / 2)(q'_k(t)\delta^3(r - q_k(t)) + \delta^3(r - q_k(t))q'_k(t))$$

$$\rho_k(t, r) = e_k \delta^3(r - q_k(t))$$

18.8 Boson-Fermion unification in quantum stochastic calculus

Let A_t, A_t^*, Λ_t denote the canonical noise processes in the Hudson-Parthasrathy quantum stochastic calculus. Let $W(u, U), u \in L^2(\mathbb{R}_+), U \in \mathcal{U}(L^2(\mathbb{R}_+))$ denote the Weyl operator in $\Gamma_s(L^2(\mathbb{R}_+))$. Define

$$dB_t = (-1)^{\Lambda_t} dA_t, \quad dB_t^* = (-1)^{\Lambda_t} dA_t^*$$

We have

$$(-1)^{\Lambda_t} = exp(i\pi\Lambda_t) = exp(i\pi\lambda(\chi_{[0,t]}I))$$
$$= W(0, exp(i\pi\chi_{[0,t]}I))$$

so that

$$dB_t|e(f) >= f(t)dt(-1)^{\Lambda_t}|e(f) >= f(t)dt|e(exp(i\pi\chi_{[0,t]})f >$$

$$= f(t)dt|e(-\chi_{[0,t]}f + \chi_{(t,\infty)}f) >$$

$$dB_s dB_t|e(f) >= f(t)dt.(-\chi_{[0,t]}(s)f(s) + \chi_{(t,\infty)}(s)f(s))ds.$$
$$\times|e((-\chi_{[0,s]} + \chi_{(s,\infty)})(-\chi_{[0,t]} + \chi_{(t,\infty)})f) >$$

Thus, for any s, t, we have

$$(dB_s dB_t + dB_t dB_s)|e(f) >=$$

$$(f(t)dt.(-\chi_{[0,t]}(s)f(s) + \chi_{(t,\infty)}(s)f(s))ds$$
$$+ f(s)ds(-\chi_{[0,s]}(t)f(t) + \chi_{(s,\infty)}(t)f(t))dt$$
$$\times|e((-\chi_{[0,s]} + \chi_{(s,\infty)})(-\chi_{[0,t]} + \chi_{(t,\infty)})f) >$$
$$= 0$$

since

$$\chi_{[0,t]}(s) + \chi_{[0,s]}(t) = 1 a.s.,$$
$$\chi_{(t,\infty)}(s) + \chi_{(s,\infty)}(t) = 1, a.s.$$

This proves that for all s, t,

$$B_s B_t + B_t B_s = 0$$

Taking the adjoint gives

$$B_s^* B_t^* + B_t^* B_s^* = 0$$

Now, for $s \leq t$,

$$< e(g)|dB_s^* dB_t|e(f) >=< dB_s e(g), dB_t|e(f) >$$

$$= \bar{g}(s)f(t)dsdt < e(-\chi_{[0,s]}g + \chi_{(s,\infty)}g), e(-\chi_{[0,t]}f + \chi_{(t,\infty)}f) >$$

and

$$< e(g)|dB_t dB_s^*|e(f) >=< e(g)|(-1)^{\Lambda_t}(-1)^{\Lambda_s} dA_t dA_s^*|e(f) >$$

When $s = t$, by quantum Ito's formula, this equals

$$< e(g), e(f) > dt$$

while if $s < t$, it equals

$$= f(t)dt < e(g)|(-1)^{\Lambda_t}(-1)^{\Lambda_s} dA_s^*|e(f) >$$

$$= f(t)dt < (-1)^{\Lambda_t} e(g), (-1)^{\Lambda_s} dA_s^*|e(f) >$$

$$= f(t)dt < dA_s(e(-\chi_{[0,t]}g + \chi_{(t,\infty)}g)), (-1)^{\Lambda_s}|e(f) >$$

$$= f(t)dt(-\bar{g}(s)ds) < e(-\chi_{[0,t]}g + \chi_{(t,\infty)}g)), |e(-\chi_{[0,s]}f + \chi_{(s,\infty)}f) >$$

It follows that for $s < t$,

$$< e(g)|dB_s^* dB_t + dB_t dB_s^*|e(f) >= 0$$

and

$$< e(g)|dB_t^* dB_t + dB_t dB_t^*|e(f) >= dt < e(g), e(f) >$$

From these relations, by quantum stochastic integration, we easily deduce the canonical anticommutation relations: For all $t, s \in \mathbb{R}_+$, we have

$$[B_t, B_s]_+ = 0, [B_t^*, B_s^*]_+ = 0, [B_t, B_s^*]_+ = min(t, s)$$

References

[1] Steven Weinberg, "The quantum theory of fields, vols. I,II,III. Cambridge University Press.

[2] Steven Weinberg, "Gravitation and Cosmology:Principles and applications of the general theory of relativity, Wiley.

[3] H.Parthasarathy, "General relativity and its engineering applications", Manakin press.

[4] K.R.Parthasarathy, "An introduction to quantum stochastic calculus, Birkhauser, 1992.

[5] Constantine Balanis, "Antenna theory", Wiley.

[6] P.A.M.Dirac, "Principles of quantum mechanics", Oxford University Press.

[7] V.S.Varadarajan "Harmonic analysis on semisimple Lie groups, Cambridge University Press.

[8] Naman Garg and H.Parthasarathy, "Belavkin filter applied to estimating the atomic observables from non-demolition quantum electromagnetic field measurements", Technical report, NSIT, 2017.

[9] Mark Wilde, "Quantum Information Theory".

[10] K.R.Parthasarathy, "Coding theorems of classical and quantum information theory", Hindustan Book Agency.

[11] M.Hayashi, "Quantum Information".

[12] D.Revuz and M.Yor, "Continuous Martingales and Brownian Motion", Springer.

[13] A.V.Skorohod, "Controlled Stochastic Processes", Springer.

[14] D.Stroock and S.R.S.Varadhan, "Multidimensional Diffusion Processes", Springer.

[15] Pushkar Kumar, Kumar Gautam, Navneet Sharma, Naman Garg and Harish Parthasarathy, "Design of quantum gates using the quantum stochastic calculus of Hudson and Parthasarathy", Technical report, NSIT, 2018.

[16] Vijay Mohan and Harish Parthasarathy, "Some versions of quantum stochastic optimal control", Technical report, NSIT, 2018.

[17] K.R.Parthasarathy, "An introduction to quantum stochastic calculus", Birkhauser, 1992.